Where Biology Meets Psychology

Where Biology Meets Psychology
Philosophical Essays

edited by Valerie Gray Hardcastle

A Bradford Book
The MIT Press
Cambridge, Massachusetts
London, England

This book was set in Times New Roman by Asco Typesetters, Hong Kong and was printed and bound in the United States of America.

Library of Congress Cataloging-in-Publication Data

Where biology meets psychology : philosophical essays / edited by
Valerie Gray Hardcastle.
    p.   cm.
"A Bradford Book."
Includes bibliographical references and index.
ISBN 0-262-08276-4 (hc. : alk. paper). — ISBN 0-262-58174-4
(pbk. : alk. paper)
1. Psychobiology—Philosophy. 2. Genetic psychology.
3. Teleology. 4. Philosophy of mind. I. Hardcastle, Valerie Gray.
QP360.W48   1999
570′.1—dc21   98-55769
                                                   CIP

Contents

Contributors

André Ariew
Department of Philosophy
University of Rhode Island

Mark A. Bedau
Department of Philosophy
Reed College

David J. Buller
Department of Philosophy
Northern Illinois University

Paul Sheldon Davies
Department of Philosophy
College of William and Mary

Stephen M. Downes
Department of Philosophy
University of Utah

Charbel Niño El-Hani
Research Group on History, Philosophy,
and Biological Sciences
Federal University of Bahia

Owen Flanagan
Department of Philosophy
Department of Psychology-Experimental
Department of Neurobiology
Duke University

Peter Godfrey-Smith
Department of Philosophy
Stanford University

Todd Grantham
Department of Philosophy
College of Charleston

Valerie Gray Hardcastle
Department of Philosophy
Virginia Polytechnic Institute and State
University and
Department of Philosophy
University of Cincinnati

Gary Hatfield
Department of Philosophy
University of Pennsylvania

Daniel W. McShea
Department of Zoology
Duke University

Karen Neander
Department of Philosophy
John Hopkins University

Shaun Nichols
Department of Philosophy
College of Charleston

Antonio Marcos Pereira
Research Group on History, Philosophy,
and Biological Sciences
Federal University of Bahia

Thomas Polger
Department of Philosophy
Duke University

Lawrence A. Shapiro
Department of Philosophy
University of Wisconsin-Madison

Kim Sterelny
Department of Philosophy
Victoria University of Wellington

Robert A. Wilson
Department of Philosophy
Beckman Institute for Advanced Science
and Technology
University of Illinois

William C. Wimsatt
Department of Philosophy
Committee on the Conceptual
Foundations of Science
Committee on Evolutionary Biology
Fishbein Center for History of Science
and Medicine
University of Chicago

Introduction

Valerie Gray Hardcastle

There is no question that the philosophy of biology has blossomed in the last twenty-five years. At the same time, philosophy of psychology and philosophy of mind have become more and more science-oriented, and more sensitive to biological issues in particular. Professionals working in both disciplines—as well as those in the actual sciences—have much to say to one another, for they are laboring in each others' backyards. Wheels need not be reinvented; instead, interdisciplinary connections should be forged.

This last claim was the premise behind a speaker series at the International Society for the History, Philosophy, and Social Studies of Biology conference held in Seattle in the summer of 1997. I had originally conceived of this session as for a small group of interested philosophical outliers and other misfits, working in the gray area between philosophy of psychology and philosophy of biology. I was not even sure that I would be able to generate enough interest for one panel discussion. Obviously, I was more outside the disciplines than I had thought. There was in fact great enthusiasm for what developed into almost a conference-within-a-conference. The project took on a life of its own, and this volume represents its ultimate product. Contained within are several papers, inspired by that conference, by most of the original participants.

Though the level of interest and excitement surrounding the interface between philosophy of biology and philosophy of psychology is high, the area itself is not well defined, nor is it well represented in mainstream philosophical publications (though perhaps the tide is turning there). As one of the first—if not the first—volumes that surveys the ways in which philosophy of biology and philosophy of psychology come together, this book should be considered a sampler of what one can get in this rather nebulous arena.

I have tried to cover the major intersections between philosophy of biology and philosophy of psychology through examples of what I think of as the major categories of study in this area: special topics of interest in both philosophy of biology and philosophy of psychology, ways in which considerations from biology and psychology can inform or constrain research in other philosophical fields, and parallels between arguments in philosophy of biology and those in philosophy of psychology. (Of course, there are lots of ways to slice this pie and I am not sure I am using the ideal criteria here. I redrew the boundaries between chapters several times and I remain dissatisfied, but there comes a time when one must simply move on to more productive activities.) In this very brief introduction, I summarize the connections and overlap between philosophy of biology and philosophy of psychology as I see them.

Philosophical studies in teleology and functions, in evolutionary psychology, and in innateness are three fairly well established areas in which biology and psychology currently cooperate. Many believe—including Karen Neander and Valerie Hardcastle—that the notion of function is fundamental to both biology and psychology, though whether that notion is fundamentally teleological divides the two authors. Neander holds that adaptive explanations underlie simple representational systems. In her piece, she defends an evolutionary account of function against charges of adaptationism in the pejorative sense. That is, she defends both that the function of a trait is what it was selected for and that the function of a trait is what it is supposed to do.

In contrast, Hardcastle provides a pragmatic account of function that could support uses in both biology and psychology. She argues that the selectionist account that Neander and others adopt might work for evolutionary biology, but it fails in other contexts. Her challenge is to provide a definition of function that would subserve all scientific uses.

Evolutionary psychology has received much bad philosophical press of late. In many respects, the present section is no different. However, though each of the chapters criticizes aspects of the current evolutionary psychology industry, each also remains supportive of the overall vision. Todd Grantham and Shawn Nichols divide evolutionary psychology into two projects, an explanatory one and a predictive one. They argue that the former project is theoretically viable, while the latter falls into adaptationism (in the pejorative Neanderian sense). They conclude that although it makes good sense to study the evolutionary origin and structure of our cognitive mechanisms, studying our ancestral problems will not shed much light on contemporary issues cognitive psychology.

Similarly, Paul Davies claims that evolutionary psychology really comprises two projects pushed together: evolutionary history and cognitive psychology. This fact explains why even though the methods of evolutionary psychology are barren, some hypotheses nonetheless appear correct. Davies agrees with Grantham and Nichols that evolutionary psychology will not help us understand our current cognitive architecture and goes on to claim that cognitive psychology can help us understand our evolutionary history.

Larry Shapiro challenges as too facile the inference from behavior to psychological mechanism, which many evolutionary psychologists rely on. He sketches when inferences to conclusions about the mind, as opposed to conclusions about physiology, are justified and when they are not.

In a related vein, David Buller holds that evolutionary psychology does not tell us about the hidden dynamics of the mind, despite what some evolutionary psychologists apparently think. In particular, it does not sanction inferences to unconscious or personal motivations, even though there may still be evolutionary explanations of some of our psychological traits.

The study of innateness has a venerable tradition in philosophy, especially in philosophy of mind and language. However, recent contributions to the debate have come from a different quarter, philosophy of biology. In this volume André Ariew gives an account of innateness based on Waddington's notion of canalization and contrasts his account with popular models, including Wimsatt's notion of generative entrenchment. Bill Wimsatt then explicates and defends his generative entrenchment account. Dan McShea uses a generative entrenchment model of innateness to explain how feelings constrain behavioral choices. He differentiates feelings or primitive motivators for behavior from action responses in animals and then argues that feelings are developmentally entrenched relative to behavioral consequences; that is, they come prior in the developmental sequence and hence are more robust.

Many of the themes developed in the three sections concerning the well-delineated areas of study in both philosophy of psychology and philosophy of biology—including the nature of representation, the evolution of psychological phenomena, the use of normative language, and the distinction between consciousness and unconsciousness—are explored in alternative ways in the remainder of the book, which focuses on other ways in which the studies of biology and psychology can overlap. Work in philosophy of biology and philosophy of psychology can inform both philosophy of mind and philosophy of science; and there are interesting, mutually beneficial parallels to be drawn between philosophy of biology and philosophy of psychology.

Kim Sterelny's piece on the evolution of our preferences bridges the two halves of the book. He examines the transition from environmental detection to full-blown representation. Like Neander and McShea, he looks at our primitive cognitive and representational capacities, and in the spirit of evolutionary psychology, he speculates about the environmental conditions that led to our current cognitive economies. Nevertheless, this is a piece in philosophy of mind, and so it focuses on questions concerning how it is that minds (come to) represent.

Tom Polger and Owen Flanagan illustrate another way that biology and psychology can inform and transform philosophy of mind. They are concerned with how a process of reflective equilibrium among the special sciences could answer philosophical questions about mind. To demonstrate the value of their approach, they taxonomize the philosophical issues concerning the nature and status of consciousness and point to where biology or psychology will be useful in answering the questions. Their chapter also echoes a motif from McShea, namely the function and place of consciousness in cognition.

Gary Hatfield probes another question at the loci of the traditional mind-body problem and the analysis of mental representation, though he is concerned to answer a question from philosophy of science. He wants to know what the direction of constraint is between psychology and neuroscience: What sort of theory or evidence trumps what, when does it, and why? His analysis helps to answer some of the considerations that Polger and Flanagan raise as well as to extend their analysis of visual perception.

A second way that considerations from biology and psychology can inform philosophy of science appears in Steve Downes's and Mark Bedau's chapters. Downes focuses on the debate concerning the putative parallel between scientific revolution and conceptual change within individuals. He uses the biological ideas of ontogeny and phylogeny to illuminate the psychological dimensions of the issue. Bedau also relies on biology in his explication of the notoriously problematic ceteris paribus laws in psychology. He uses recent models from the artificial-life community to help codify his notion of "supple" laws, which appear in both biology and psychology as well as in science in general.

Finally, though there are numerous parallels between debates in philosophy of biology and debates in philosophy of psychology, I have selected only three to illustrate this approach. Each enriches our current understanding of some important terms and trends in both biology and psychology.

Peter Godfrey-Smith uses discussions of mental representation to shed light on what it is that genes do. Do they transmit information or are they merely causal? How far into the environment does genetic information go? He speculates about how to approach answering these questions.

El-Hañia and Pereira look at biological causation through the lens of supervenience and emergence debates in philosophy of psychology. Like Godfrey-Smith, they try to bring advances in philosophy of mind to philosophy of biology. At the same time, they use considerations from biology to help solve the long-standing paradox of mental causation.

Rob Wilson explores the connections between biology and psychology in comparing Dawkins's extended phenotype with nonindividualistic or "wide" systems. He also explores the metaphors of causal powers and encoding and the notion of an individual as a locus of control.

Where biology meets psychology we find overlapping domains of study, larger philosophical implications, and even larger conceptual ties. Woven through these connections are shared concerns regarding the status of semantics, scientific law, evolution and adaptation, and cognition in general. One book can depict only a few examples of each of these connections and themes. But it is a beginning nonetheless.

I FUNCTIONS AND TELEOLOGY

1 Fitness and the Fate of Unicorns

Karen Neander

1 Is Teleosemantics Adaptationist?

Is teleosemantics adaptationist? People sometimes tell me so. But my own answer is that it all depends on what you count as adaptationist and on which version of teleosemantics you consider. Certainly, the general teleosemantic enterprise is not doomed to fail because of any commitments to dubious empirical assumptions of an adaptationst kind. However, I don't recommend that concerns about implausible adaptational assumptions be dismissed out of hand. There are particular versions of teleosemantics that have implausible adaptationist assumptions, and we need to be clearer about just what the adaptational commitments of teleosemantic theories are.[1]

The teleosemantic theories that I am interested in here are those in a class of theory in psychosemantics, as opposed to semantics proper. That is, they are an attempt to account for (what is often called) *original meaning*. A representation counts as having original meaning if it has meaning that does not depend on its (the representation's) being metarepresented by some further representation (e.g., in the intentions of someone using it to communicate with someone else). In this chapter I'll be assuming that original meaning is possessed by some brain states of evolved creatures such as ourselves. The open question is whether teleosemantic theories, which attempt to derive original meaning from a creature's evolutionary history, are bound to fail because they require implausible adaptationist assumptions.

In brief, teleosemantic theories attempt to move from historical facts about what some trait was selected for by natural selection to constitutive facts about what some (possibly other) trait (or state of a trait) represents. The attempted move is made via the teleological or teleonomic notion of a *function*, a notion that is involved in our, in some sense, normative talk about a biological system functioning properly or malfunctioning—a notion that entails a distinction between what a system is disposed to do and what it is supposed to do. Standardly, in teleosemantic theories, this notion of a proper function is understood by means of an etiological analysis, according to which the proper function of a trait is what it (i.e., that homologous type) was selected for by natural selection. And according to teleosemantic theories, the psychosemantic norms of original meaning are derived from these norms of proper functioning, although there is little consensus about how exactly this should be done. The extant theories are quite varied and there is scope for more variation still. For instance, extant theories tend to be versions of causal or informational theories, but teleonomy could also be used to add normativity to a conceptual role theory of mental content. All that teleosemantic theories need have in common is the idea that psychosemantic norms ultimately derive from the norms of teleonomic functions.

Given this, it is however true that all teleosemantic theories appeal to adaptational explanations, since these are explanations in terms of what something was selected for. So to this extent it is clearly appropriate that concerns about adaptationism arise. But not just any appeal to adaptational explanations counts as adaptationist, let alone as adaptationist in the pejorative sense of the term. There is a real problem for teleosemantics only if it too readily assumes that the adaptational explanations it requires are available. The open question is whether the most plausible teleosemantic theory or theories must do so.

Of course, teleosemantic theories are not in the business of providing such adaptational explanations themselves, for that's a task for the evolutionary biologist and not for the philosopher. Instead, the aim of teleosemantic theories is to explain how original meaning can in principle be derived from the natural facts, and in particular from those adaptational explanations that are true, whichever ones those are. If teleosemantics is doomed to fail because it has implausible adaptationist assumptions, then it must be the case that, whichever ones those are, they are insufficient as a basis for teleosemantics. I know of no good or even half good argument to this effect. At best, there are some good arguments with much more specific conclusions, to the effect that some particular teleosemantic theory (at least on the critic's construal) makes some unwarranted or even impossible adaptationist assumptions. At worst, there is a great deal of confusion and ignorance.

What follows is in three main sections. Section 2 reviews the adaptationism debate and explains that the biologists who are known as anti-adaptationists never seriously meant to question the central importance of adaptational explanations. We can accept what the anti-adaptationists say (about constraints on selection, and so on) and yet still believe that natural teleology is pervasive in modern biology. Section 3 then explains why the modern version of the argument from design, namely the argument for selection, holds good for complex organized systems like the brain. No alternative kind of explanation is sufficient to explain the accumulation of coadapted traits that is required for this sort of organized complexity. I claim that we know enough to know that there is a correct naturalistic yet normative neurophysiological description of the brain's representational functions. Whether this is enough for teleosemantics then depends on the details. Rather than trying to survey extant teleosemantic theories, section 4 addresses some more detailed concerns that have been raised in the literature. To do so, it looks at how at least one (fabricated pseudo-Dretskean) teleosemantic theory has commitments to implausible (and even impossible) adaptational explanations. The question is, however, whether it is the teleonomy or some other aspect of the theory (its atomism or nativism, for instance) that should be dropped.

2 The Adaptationism Debate

Before we can say whether teleosemantics is unduly adaptationist, we need some clarification of the latter term. Not just any appeal to adaptationist explanations deserves to be called "adaptationist." Adaptationism is an ill-defined thesis about adaptational explanations, but however it is defined it is clear that both the so-called adaptationists and the so-called anti-adaptationists agree that some adaptational explanations are perfectly legitimate.

An *adaptational explanation* is a kind of causal explanation, one that explains a trait by appeal to the fact that past traits of that type had adaptive effects which caused the phenotype (or its underlying genotype) to be selected by natural selection. So, for example, the gloss on the leaves of the eucalyptus tree can probably be explained as an adaptation for reducing dehydration in a hot, dry climate. If so, the gloss was responsible for less water loss, which in turn was responsible for the underlying genes being replicated more than they would otherwise have been, which in turn was responsible for the trait being preserved in the population. These adaptational explanations are entirely respectable stock in trade of modern biology.

What is the adaptationist debate about, then, aside from a lot of breast beating? It is often said that it is about the relative importance of natural selection as compared to other evolutionary forces or factors in determining and explaining evolutionary outcomes. For reasons that should become clear by the end of this section, it is more accurate to say that it is about the relative importance of adaptational explanations as compared to other explanations in explaining evolutionary outcomes. Someone is accused of being an adaptationist if she is seen as ignoring or giving too little attention to other kinds of explanations. The accusers are the so-called anti-adaptationists. At first, the adaptationist label was virtually by definition pejorative, but now, as the rhetorical tide turns, we find some people describing themselves as adaptationists, presumably because they think that adaptational explanations have begun to be underrated. So put, the debate must seem a very vague one, and to a large extent it is. The main problem with this debate is that everyone is really somewhere in the middle ground, but almost no-one is calmly acknowledging that fact and trying to clarify the less radical differences that really exist.

For instance, Dan Dennett (1995) has signed on as an adaptationist, stoutly declaring in doing so that adaptational explanations have a central and essential role in modern biology. I am with him there, one hundred percent. Except that if this is enough to make one an adaptationist, then even the so-called anti-adaptationists qualify. Beyond their surface rhetoric, they've never really denied that adaptational explanations belong on center stage; they've only insisted that they shouldn't hog the limelight. They take themselves to be addressing people who too frequently talk as if every trait has an adaptational explanation and as if adaptational explanations are the only kind of explanation to be considered in explaining evolutionary outcomes. Perhaps it is understandable that in such a context the anti-adaptationists do not stress the importance of adaptational explanations. But even if they are at fault here, the charitable interpretation is not that they think them unimportant, but rather that they think their appreciation of their importance will be taken for granted. At worst, the so-called anti-adaptationist have probably been guilty of being carried away at times by their rhetoric, of making their position sound more radical and original than it really is. In any case, to my knowledge, no so-called anti-adaptationist has ever seriously tried to argue that adaptational explanations do not have a central and essential role in modern biology.

Lately we have seen several explicit denials of the radical reading (or misreading) of their position. For instance, Niles Eldredge tells us in no uncertain terms that:

no rational evolutionary biologist feels that most change is not adaptive, or that adaptive change is not caused by natural selection.   (1995, p. 56)

And Stephen Jay Gould (1997), in a review of Daniel Dennett (1996), says, in the weary tone of one forced to state the all too obvious:

may I state for the record that I (along with all other Darwinian pluralists) do not deny the existence and central importance of adaptation, or the production of adaptation by natural selection. Yes, eyes are for seeing and feet are for moving. And yes, again, I know of no scientific mechanism other than natural selection with the proven power to build structures of such eminently workable design.   (p. 35)

So, what then is it that distinguishes the adaptationists from the anti-adaptationists? Gould chooses to call himself a Darwinian pluralist, rather than an anti-adaptationist. This might seem a better name for the so-called anti-adaptationists, since it does not suggest that they deny the importance of adaptational explanations. However it does have the unfortunate implication that their opponents, the adaptationists, deny the importance of other forms of explanation, and this is just as misleading in a different direction.

One target of anti-adaptationist criticism is Richard Dawkins (see especially his 1986). He has been an influential advocate of the power of natural selection to create the wondrous and intricate designs for which God previously took all the credit. But this in itself does not make him an adaptationist. As we saw in the quotes from Gould and Eldredge, they will wearily agree that natural selection and natural selection alone has this power. But nor does it make Dawkins a Darwinian monist. He will, I am sure (and probably just as wearily) allow that, as every second year biology student should know, there are constraints on the adaptive power of natural selection. I am quite sure he knows that natural selection must tinker with the available materials, that there are spandrels, that there is allopatric speciation with relatively rapid change due to migration, reproductive isolation, and the increased power of drift (see, for example, his 1986, chapter 9) and so forth. Dawkins is a Darwinian pluralist too, if that just means that he acknowledges other mechanisms besides natural selection and that he appreciates that there are constraints on the adaptive power of natural selection.

Dawkins, like Gould, addresses a popular audience, but he is more concerned to correct a different way in which his popular audience can and does fall into error. More than fifty percent of university entry-level biology students in America are creationists. Among other things, Dawkins is trying to persuade them and their neighbors of the power of natural selection, to convince them of the fact that natural selection can do the job. Again, perhaps it is understandable that in this context he does not emphasize the limitations on the adaptive power of natural selection (although it is true that evidence of such limitations—evidence in the form of an imperfect design product—is also good evidence against creationism).

Fodor complains of Dennett that he concedes too much to the anti-adaptationists to be an adaptationist in good standing, for Dennett concedes that the power of natural selection is seriously constrained in various ways. "Good adaptationist thinking is always on the lookout for hidden constraints, and in fact is the best method for uncovering them," he quotes Dennett (1995, p. 261) as saying. Fodor objects, "That makes it look as though there is practically nothing that an adaptationist in good standing is required to believe about how evolution works; he's only required to buy into a methodological claim about how best to find out how it does" (Fodor, 1996, p. 248). But Fodor's complaint is inappropriate. We may as well complain of Eldredge and Gould that they concede too much to the adaptationists to be anti-adaptationists in good standing. Either way, it's a silly exercise. As far as I can see, no one in the debate has seriously denied that adaptational explanations are of central importance *or* that the adaptive power of natural selection is seriously constrained, leaving aside any fictional caricatures created to lampoon one's enemies.[2] What real differences there are between adaptationists and the so-called anti-adaptationists concern *degrees* of importance. The debate is over how important certain influences on evolutionary outcomes are. These differences are hard to quantify precisely and the more visible side of the debate is far removed from any genuine attempt to do so.

I am not trying to suggest that there are no interesting and substantial issues in the vicinity of the adaptationism debate. But they are not, I think, what people have in mind when they accuse teleosemantics of dubious adaptationist assumptions. For instance, John Maynard-Smith (1978) believes that optimization theory is useful and he might be described as an adaptationist because of this. In optimization theory, an evolutionary trend with respect to some trait is modeled as if the adaptive power of natural selection were unconstrained. That is, it is modeled as if drift were absent, as if the relevant traits can be selected for (or against) independently, and so on. Maynard-Smith argues that this style of modeling is useful, not because nature is this way, but because deviations from the predictions provided by the model give us some measure of the effect of factors other than natural selection. So-called adaptationists and so-called anti-adaptationists can disagree about the usefulness of such models and the disagreement sometimes seems to depend on beliefs about the degree to which the model can be treated as an accurate picture of reality (even though both sides agree that it is far from exact). Whether or not such models are useful is an interesting issue on which sensible biologists can disagree, but it is hard to see exactly how it concerns teleosemantics. No one has yet tried to argue that it does, so I'll assume that this methodological issue is not what we are concerned with here.

The critics of teleosemantics seem to have a cruder notion of adaptationism in mind. Their adaptationist is the one targeted by Stephen Jay Gould and Richard Lewontin (1979) in their classic spandrels paper. This adaptationist is supposed to believe three things: (1) that virtually every trait that we can identify has an independent adaptational explanation; (2) that the form of these traits is perfectly fashioned for their function; and (3) that in modeling, explaining and predicting evolutionary outcomes, we can, in practice, ignore everything but natural selection, because natural selection is powerful enough, at least in the long run, to swamp the effects of other forces. For such an adaptationist, natural selection has more or less slipped right into the role vacated by God in guiding the design of living creatures. Although it lacks foresight and is therefore far from omniscient, it is pretty well omnipotent and omnipresent. This is the position Gould and Lewontin label 'Panglossian'. It is an extreme that no real biologist has avowed, although we might accuse some of too nearly approaching it in practice. The Panglossian sin is to err too far in this direction.

3   Teleonomy and Panglossian Sin

Does teleosemantics commit the Panglossian sin? This question will take a while to answer, but we can make a start on it in this section by noting that it does not do so merely by being committed to natural teleology. In this case, natural teleology, or teleonomy, is a teleological notion of proper functioning that can be cashed out in naturalistic terms, and more specifically, in terms of natural selection (as opposed to the purposeful intentions of a designer). Many readers will think it too obvious to need saying, but judging from what I have heard in discussions, it does need saying: natural teleology does require some adaptational explanations but it is not in itself Panglossian. To think otherwise is simply confused.

The standard means of introducing the idea of natural teleology is by the now familiar distinction between selection *of* and selection *for*. A trait with two properties *P* and *Q* counts as having been selected for *P* and not for *Q* if its being *P* contributed causally to its being selected and if its being *Q* did not contribute causally to its being selected, either because *Q* was neutral or because *Q* worked against selection of the trait.[3] Along the same lines, a gene sequence that's responsible for two traits, *R* and *S*, counts as having been selected for *R* and not for *S* if its being responsible for *R* contributed causally to its being selected whereas its being responsible for *S* did not contribute causally to its being selected, either because *S* was neutral or because *S* worked against selection of the gene sequence. If the gene sequence is selected, there is selection of *S*, even though there is no selection for *S*, only selection for *R*. In such circumstances, *S* is said to be a piggy-back trait because it rides on the back of some other trait that is adaptive and which does the work involved in getting them selected. Some piggy-back traits have been nicknamed "spandrels," for like the spandrels of cathedrals they are the inevitable architectural outcomes of selection for other traits (e.g., the human chin is the outcome of selection for the jaw). Other piggy-back traits are the result of pleitropy (where two potentially separate phenotypic traits are caused by the same gene sequence) and gene-linkage (where two potentially separate phenotypic traits are caused by gene sequences that are close together on the same chromosome). The possibility of piggy-back traits is thus standardly recognized in introductions to teleonomy for the purpose of explaining teleosemantic theories.

As it is normally understood in teleosemantic theories, a (or the) teleonomic function of a trait (type) is what it did that it was selected for.[4] This obviously does not imply that everything has an adaptational explanation, or even that most things do, or even, strictly speaking, that anything does. It entails that if something has such a teleonomic function then it has an adaptational explanation, an explanation in terms of what it was selected for.

Further, there is no implication from this definition that if something has a teleo-nomic function, and therefore has an adaptational explanation, then it is perfectly fitted for its function. Nor is it implied that its adaptational explanation is its complete evolutionary explanation. The Panda's "thumb" was selected for stripping leaves off bamboo, and that therefore is its teleonomic function. But it does not follow that the Panda has the best bamboo leaf-stripper it could possibly have, or that natural selection worked toward an adaptive outcome in this case in the absence of all constraints. Far from it (Gould 1980). While the Panda's thumb has a clear teleonomic function, it is a wonderful illustration of the fact that natural selection is a tinkerer and a satisficer, heavily constrained by the past and by the alternatives that are presently available. The Panda's thumb—in fact, an elongated wrist bone—is an imperfect thing from a design engineering point of view. To explain its imperfections we need to consider the limitations on the adaptive power of natural selection. An explanation in terms of these limitations is entirely compatible with and complimentary to the adaptational explanation. There is nothing remotely Panglossian about this combination of explanations and teleonomy is in no way committed to the Panglossian error.

As long as there is variation, there can be selection for one variation over another. Even if there is just one barely workable variation available, one workable variation among others that are worse than useless, that one workable variation can be selected for working as well as it does. Selection for a trait can occur as long as that trait increases fitness relative to alternative alleles, and that is consistent with the fitter trait falling well short of perfection. There is no incompatibility in the following two theses: that most traits have teleonomic functions and that the adaptive power of natural selection is severely constrained.

Actually, it is worth being clear at this point just in what sense the adaptive power of natural selection is constrained. People often speak of the various so-called constraints as constraints on natural selection, but they are really aspects of natural selection, and they are not sensibly viewed as opposing forces, or as other evolutionary mechanisms. I am speaking here of what are often referred to as developmental, phylogenetic and architectural constraints.

Consider the so-called developmental constraints first. These concern the fact that mutations that alter processes early on in embryonic development are less frequently beneficial than those that alter processes that occur later in development. This is because random changes to complex organized systems are more likely to be damaging if they are large and widespread than if they are small and localized, and the earlier the developmental process is, the larger and more widespread the ramifications of any changes to it are likely to be. But it can be misleading to see this as a constraint on natural selection, unless one also understands that this is itself an aspect of natural selection. When a mutation is selected against because it causes widespread changes that are on balance disadvantageous, this is just part of the selection process. Natural selection is just doing its thing, selecting against an unfit allele. We describe this aspect of selection as a constraint, I suppose, because it limits the kind of changes that natural selection can, given its nature, bring about. These facts about development explain why it is hard for natural selection to alter the fundamental design of a system, why it tends to be somewhat conservative in its designs, and why its designs will often be imperfect from an engineering point of view, because it cannot start from scratch when new design demands arise. But the conservative maintenance of early developmental pathways itself has an adaptational explanation. Or, to put it another way, to explain something as due to developmental constraints is to give a special form of adaptational explanation. Perhaps this is not what people generally have in mind when they think of adaptational explanations. But it is a matter of the maintenance of already established adaptation, especially the internal, delicate and massively complicated coadaptation of traits required for the functioning organism.

The so-called phylogenetic constraints concern macroevolutionary outcomes and explain why natural selection has ignored large areas in the space of apparent design possibilities. Natural selection cannot explore all possible kinds of workable designs because it has to get from where it is to where it is going via pathways consisting of actual functional creatures, choosing the most adaptive alternatives at each step along the way. As it happens there are lots of some kinds of creatures (mammals, for example) and none of some other possible kinds of creatures (I'll leave them to the reader's imagination) and so we get modifications on existing kinds and not modifications on kinds that don't exist. But once again, that's just the way natural selection works, and there is no separate force involved. The phylogenetic constraints are not like a dam holding back a river's natural flow. They are instead like deep river beds, eroded over vast time, created by as well as channeling the river's natural flow.

The so-called architectural constraints do not even limit design optimization in the way that the other two so-called constraints might be claimed to do, since architectural constraints just are elements of good design. Just as large office buildings need strong skeletal support and good circulation, so does something the size of an elephant, and that's one reason mosquitoes will never grow to an elephantine size. While a human architect can design unworkable buildings despite these architectural constraints, what is unworkable cannot be selected by natural selection. So I suppose we describe the architectural constraints as constraints because they limit possible designs, even though they do not limit optimization. But once again, of course, this is just the way natural selection works.

None of these so-called constraints on natural selection is an outside force, separate from or external to natural selection. It is senseless to ask whether they are more important than natural selection in determining evolutionary outcomes, since they are themselves aspects of natural selection. That there are such constraints is reason not to be a Panglossian, but it is no reason at all to think that adaptational explanations are not pervasive. Maybe one needs to be a little in the grip of the Panglossian perspective to even view these so-called constraints as constraints. They prevent natural selection from being an omnipotent designer (from being able to start from scratch, summon materials from thin air, explore all possible workable designs, and even unworkable ones). But to think that natural selection would otherwise be an omnipotent designer is definitely to err too far in the Panglossian direction. In the absence of these so-called constraints, natural selection would not be natural selection. It would be God.

There are mechanisms of change involved in evolution that can be counted as other than or external to selection itself. Mutation, migration, geographic isolation and drift are perhaps the most obvious ones. Of course, natural selection can only select from the available variation and the variation available is randomly generated by mutations and chromosomal cross-overs, so what occurs in this way both provides the raw materials for selection and limits its potential: it is both empowering and constraining. It limits optimization of design for the obvious reason that unless the optimal allele arises it can't be selected. But once again, if this were not so, we would not be talking about natural selection. And while the dependence of natural selection on the availability of mutations is a good reason not to be Panglossian it is no reason at all to think that adaptational explanations are not pervasive. If we want to explain why no mammal has green fur, it might have something to do with the fact that some mutation that could have occurred has not occurred (I don't know). But suppose that, in a given case, green fur would have been selected had the alternative been available. And suppose that, given that it was not available, brown fur was instead selected because it provided better camouflage than the alternatives that were available (orange, black and white stripes, etc.). It remains true that the brown fur was selected for camouflage. Teleonomy does not require perfection. It is enough that there was selection for a trait over its actual competition.

Migration, geographic isolation and drift are thought by many biologists to be of great importance in speciation. Drift is usually thought to include two components (i) deviation from the statistical norm in the random fertilization of gametes, and (ii) inequalities in parent sampling due to accident rather than variations in fitness.[5] Drift tends to tend in a different direction to natural selection because there are far more damaging mutations than beneficial ones, and while natural selection tends to favor the beneficial ones, drift is indiscriminate. Drift therefore competes with selection. It can also play a decisive role in determining the direction of evolution by eliminating certain variations when their representation is small, as is the case of new mutations, and in allopatric speciation where a small part of population separates from the rest. It can tend toward conservatism, as it probably does when it eliminates new mutations, but it can also tend toward change and phylogenetic diversity, as it probably does in allopatric speciation. When the overall numbers are small, relatively small fluctuations can eliminate a gene sequence or drive it to fixation. But just because drift (et al.) were decisive in determining an evolutionary trajectory does not mean that natural selection was not similarly decisive. As long as drift leaves more than one variation available for selection, then selection can still occur. And, to use an oft used metaphor, while drift (et al.) can force the train onto a different track it cannot drive the engine.

In brief, even if we assume that adaptational explanations are in principle available for most morphological change, there is nothing in the least Panglossian about such an assumption. Let us now turn to the somewhat more specialized question of whether the same holds true when we are considering the evolution of perceptual and cognitive capacities, for this is where we are concerned with specifically representational functions.

4 Teleosemantics

Yes, eyes are for seeing ... (yawn). The eye is almost the paradigm case of intricate, organized complexity in need of an adaptational explanation. Yet it is a relatively simple thing compared to the visual cortex that processes its input. The human brain is sometimes said to have the highest degree of organized complexity of anything in the known universe. Perhaps so. It is anyway, of all known things, one of the thing most in need of an adaptational explanation. And teleosemantics could hardly be based on a more solid empirical assumption insofar as it assumes that there is in principle available a rich adaptational explanation of the representational capacities of the brain. I'll say why in this section, starting with a few thoughts about teleosemantics by way of a little motivation and elaboration.

A central part of the puzzle about original meaning is how it is possible to move from ordinary descriptive facts about cognitive systems to psychosemantic norms (for correctness of representation, truth of beliefs, and so on). In general, if something is a representation, then it can, in principle, *mis*represent. So to know the content of a representation is to know something about which circumstances would count as correct applications and which would count as incorrect applications of the representation. Traditionally, this move from the natural to the normative has seemed daunting and one major attraction of teleonomy as a basis from which to develop a naturalized theory of meaning is that teleonomy is already both natural and normative. Moreover, there is a significant sense in which both talk of natural functions and talk of original meaning are normative in the same sense. To attribute a natural function or to attribute original meaning to something is to attribute a certain kind of normative property to the thing. That is, it is to attribute an evaluative standard to it that it could fail to meet, even chronically (i.e., systematically and persistently and even under ideal external circumstances). I call it an evaluative standard since intuitively it is a standard that the thing is in some sense supposed to meet. In the case of natural functions we are speaking of a standard of proper functioning and in the case of original meaning we are speaking of a standard of correct representation. According to an etiological theory of functions, the intuitive sense in which a trait is supposed to perform its function is cashed out naturalistically in terms of what the trait was selected for. While these biological norms of proper functioning cannot be simply equated with psychosemantic norms, the hope is that they can be deployed to determine what states of the brain are supposed to represent.

Philosophers have for far too long now spoken as if our puzzle is how to understand the relation between the intentional properties of the mind, on the one hand, and the nonintentional physical, chemical and biological properties of the brain, on the other hand. Some philosophers even recommend the elimination of intentional properties in favor of making do with neurophysiology. But neurophysiology is already thoroughly intentional, at least in the broader sense, in which something counts as intentional if it is representational. Once we are above the level of microbiology, neurophysiology is steeped in descriptions of the representational functions of neural states and processes. Physiology is the relating of form to function, and the function of the brain is largely representational, so it is hard to even make sense of the idea of a complete but nonrepresentational neurophysiology. The brain is a biological organ selected by natural selection for various representational functions; for representing variable states of a creature's environment, for representing the creature's own body and its place in this environment, for processing information about these things, and for mediating between them and the creature's behavior.[6]

My hunch is that this natural and normative description of the brain's representational functions will form the foundation for a future psychosemantic theory adequate for the purposes of a mature cognitive science. In fact, my hunch is that this will happen and is already beginning to happen whether we philosophers approve of it or not. But that is more than I an attempting to establish in this chapter. My goal here is simply to defend teleosemantics from those who argue that it is doomed because it is committed to dubious adaptationist assumptions.

We have already seen that a commitment to teleonomy, and even a commitment to pervasive teleonomy, are not at all unduly adaptationist (or Panglossian). However, most teleosemantic theories require something more specific; that there be a fairly rich teleonomic description of natural representational systems, as such.[7] That is, they require that there be in principle available a fairly rich adaptational explanation of cognition and perception. Do we have good reason to believe that such an adaptational explanation is in principle available, given that we know little about the evolution of the brain or even (relatively speaking) about how the brain represents?[8]

We do. We have very good reason to believe that adaptational explanations are important in explaining cognition and perception. One reason we understand so little about how brains represent is the sheer complexity of brains. From this, and from what we do know about the brain functioning (and malfunctioning), and from our general knowledge of the mechanisms of evolution we can infer with moral certainty that there is in principle available a rich adaptational explanation of perception and cognition (or rather, a rich complex of such explanations). I am appealing here to the modern version of the argument from design; an argument that I'll call the argument for selection. I am of course making no claims to originality in offering such an argument. This argument is as old as Charles Darwin's appreciation of William Paley.[9] But I think that some philosophers have forgotten just how powerful this argument is.

By far the best explanation available for the existence of organized complexity—that is, of a system with heterogeneous parts that are harmoniously coordinated toward the production of some complexly achieved overall activity—is an adaptational explanation. William Paley (1802) and others used what has been known as the argument from design to support the existence of a Divine Designer on the basis of an analogous appeal to the best explanation. (It would have been more appropriate to call it the argument *for* design since it was an argument for design from the existence of organized complexity.) There is disagreement about just how powerful Paley's argument was in the absence of the alternative Darwinian explanation. Was God the best explanation available at the time? Contemporary philosophers and historians disagree. It can be argued that the appeal to a Divine Designer only shifted the explanatory burden, or that even the mechanistic explanations that were available at the time were more plausible, at least in the light of the scientific knowledge of the time. But none of these worries about the power of Paley's argument in the least trouble its descendent, the argument for selection. It is an immensely powerful argument and no new developments in biology, anti-adaptationist or otherwise, have done anything to challenge this fact.

Let me make it clear that I am not maintaining that organized complexity of the kind that brains have is strictly nomologically impossible without selection for it, but I am saying that there is no other available explanation that is plausible and that makes organized complexity of this magnitude anything other than extremely improbable, improbable almost to the point of impossible. The probability that adaptational explanations are not important in explaining cognition and perception is so tiny it can be ignored for all practical purposes.

To properly understand this point, we should note that there are two notions of "complexity" in currency these days. One is a simple notion of complexity as mere heterogeneity (e.g., see Peter Godfrey-Smith 1996). On this notion, the more the parts of a system vary, and the more various the kinds of things they do, the more complex the system is. A high degree of complexity, in this sense of the term, is compatible with a high degree of *dis*organization. In contrast, organized complexity, requires organization as well as heterogeneity. The parts are described as communicating, coordinating, and cooperating with each other. In organized complexity, the order of the parts and their interactions as well as their variety matters. And it is this latter notion of organized complexity that is relevant to the argument for selection. Mere heterogeneity can easily enough result from nonselectional processes (a tornado, for example).

The adaptationism debate has served to highlight the importance of factors other than natural selection in accounting for evolutionary outcomes: mutation, drift, migration, pleitropy, genetic linkage, heterozygous advantage, and so on. But these other phenomena are none of them candidates for explaining organized complexity. The first three tend away from adaptation, and hence do not bias change toward the accumulation of coadapted traits. The next three just prevent independent selection of traits, and only act to preserve the present or hinder the future coadaptation of traits.

Mutations are arbitrary, not in the sense that a mutation is as likely to occur as not, or that all mutations are equally likely to occur, but in the sense that whether a mutation occurs is insensitive to whether its occurrence is adaptive. Since most variation produced at random in a complex organized system will tend to disrupt its organization rather than improve it, mutation without selection tends away from, not toward, the accumulation of coadapted traits. On its own, it is a force for change, but it moves things in the wrong direction. Complex systems that cease to be positively selected for, such as the eyes of fish that take up residence in dark caves, tend to lose their organized complexity. Some people, not well educated in biology, can be overly impressed by the fact that a mutation or two sometimes results in profound phenotypic changes. They are tempted to think that saltation, where new complex organization arises suddenly without cumulative selection, is a common enough phenomenon. They might be impressed by the fact, for example, that two point mutations (*bithorax* and *postbithorax*) in *drosophila* can produce two whole extra wings, turning a two-winged fly into a four-winged fly. But this is not an example of saltation; the "design-work" has already been done by natural selection in cases like this. The genes responsible for the new wings are the ones that are responsible for the old wings, and they are the result of extensive cumulative selection for flight. What has been altered by the two point mutations is just some positional information controlling the expression of the relevant gene sequence.[10]

Drift is also random and for the same reason it also tends to favor disorganization, not organization. That is why it can be decisive in determining the evolutionary trajectory and yet cannot power the accumulation of coadapted traits. Drift increases the probability of complexity only in the sense of mere heterogeneity. It is, for example, heavily implicated in the variation in junk DNA (DNA that is either not transcribed or not translated into amino acids). But this is so precisely because junk DNA has no phenotypic outcome and hence no effect on the physiology and complex organization of the organism.

Anyone who denies that we can be confident that adaptational explanations will be important in explaining cognition and perception owes us a response to the argument for selection. These phenomena are the products of a system of immense organized complexity. Cumulative selection is the only thing that can in practice explain such complexity. And while this fact seems to have become blurred behind all the rhetoric over recent radical, or anyway supposedly radical, revisions to (neo-) Darwinian biology, this fact remains.

There is the in-principle possibility that a complex system could be selected for one function and then fortuitously be used to do something else instead or in addition. Such things happen. To borrow an example from Fred Dretske, a scale for measuring weights can, using fixed weights, be used for measuring altitude instead. So it doesn't *follow* from the fact that cognition is performed by a complex system that the complex system was selected *for cognition*, and therefore it doesn't follow that the adaptational explanation of the brain is one that will bestow representational functions upon it. However, while the in-principle point is certainly correct, its application is absurdly far-fetched in this case. For example, what else might the visual cortex have been selected for that just happened to have the right complex structure for visual perception? The ancients believed that brains were an organic radiator with the function of dissipating heat. But even if this were a function of the brain, its selection for the dissipation of heat could only explain at most a few of its features: for example, the large surface to volume ratio and the high concentration of capillaries near the surface. It would not begin to explain, for example, why the input to the retina was mapped on to V1 and from there on to V2 and V3, and so on and so forth. The suggestion that brains were not selected for representing variable features of the environment and for processing information about them is about as plausible as the suggestion that a car's radiator was really designed to be a portable navigational computer. In the second case, the radiator does not have the kind of structural complexity required for it to be the plausible outcome of such a process, and in the first case, the proposed process could not plausibly explain the outcome in need of explanation, namely the specific structural complexity of the brain.

There's no room for reasonable doubt that there is a rich and detailed adaptational story to be told about the organized complexity responsible for the representational capacities of our brains. Of course, not every mental trait will have an adaptational explanation. But the presence of spandrels, piggy-back traits, vestigial traits, and so on, does not prevent teleonomy from setting standards of proper functioning in ordinary somatic physiology. So we need a special argument to show that the presence of such things in psychological systems prevents teleonomy from setting standards of correct representation for cognitive science, specifically with respect to the representational capacities that are our natural endowment. No good argument so far has been forthcoming.

The general teleosemantic enterprise does have some empirical assumptions, but insofar as it just requires that there be a rich and detailed adaptational explanation of our innate representational capacities, it is on about as firm a footing as it can possibly be.

5 Specific Teleosemantic Theories

It is, however, easy to think up particular versions of teleosemantics that require utterly implausible adaptationist assumptions. I'll describe one in a moment.

The general idea of teleosemantics is very abstract. As I've already mentioned, all teleosemantic theories attempt to derive content from the notion of a (teleonomic) function, but this can be attempted in many different ways, and there is at present little consensus concerning how it should be done. I will not attempt to survey the variety of possible teleosemantic theories here. What I want to do is address some objections that have been made to specific versions of teleosemantics. However, I have no wish to defend these particular versions of teleosemantics. I want to avoid getting bogged down in exegetical details altogether by fabricating a simple pseudo-Dretskean teleosemantic theory. The point of doing so is to acknowledge that a version of teleosemantics can have unduly adaptationist assumptions while at the same time illustrating the fact that it is not the teleonomy alone that is responsible for the implausible adaptationist assumptions. In this case, at least, it is the combination of teleonomy with radical atomism and radical nativism. And in this case, in my view, these are the more obvious candidates for elimination.

My sacrificial theory is a single-factor theory of content for perceptual representations. On this theory, representations of a type, R, refer to some type of feature, F, iff instances of R have the function of indicating instances of F. An R indicates an F iff (reliably) if there is an instance of an R then there is an instance of an F. And Rs have the function of indicating Fs iff Rs were selected for indicating Fs (by natural selection) because past instances of Rs caused a characteristic movement, M, and doing M in the presence of Fs was (often enough) fitness enhancing. Furthermore, according to the sacrificial theory, radical atomism and radical nativism are true.

According to radical atomism, every (or virtually every) simple lexeme of a natural language has a corresponding Mentalese lexeme (i.e., a representation in the language of thought, or in whatever system of representations the brain employs) that is semantically simple. One way to express this is by saying that the meanings of such mental representations are not molecular, which is to say that they are not constructed out of the meanings of other simpler or more basic representations. For example, the Mentalese equivalent of "bachelor" does not have its meaning constituted out of the meanings of the Mentalese semantic equivalents of "male" and "married" and "not," and so on.

Atomism is not the same as nativism. The latter concerns whether or not we learn new concepts. Perhaps there is no uncontroversial way to spell out what a concept is or what it would be to learn one, but it may be enough here to note that there should be nothing in the definition of the relevant terms to rule out the possibility that a Mentalese lexeme can be simple and yet learned. For example, it could be learned by classical conditioning and have its meaning constitutively determined by that classical conditioning (e.g., by what it was recruited to indicate during conditioning, as in Dretske's actual psychosemantic theory). According to radical nativism, virtually all Mentalese concepts are somehow acquired or possessed without their having to be learned. The idea is that they come about as the result of maturation or are triggered by particular experiences, but that none of this counts as learning a new concept. Given the addition of radical atomism and radical nativism to the pseudo-Dretskean formula, the version of teleosemantics we are considering entails that every, or virtually every, Mentalese term-type has its content determined independently by its own individual evolutionary history. More specifically, it entails that its content is determined by past occasions in which instances of it occurred in ancestral creatures in the presence of instances of the things that belong in its extension.

This pseudo-Dretskean theory has many problems. Just one of them is that there will be many missing concepts on this theory. This is because it implies that virtually everything we can think about has been detected by and has had an impact on the fitness of our ancestors, which is of course nonsense. As Peacocke and Fodor have argued, content reaches beyond fitness. Nonexistent things, fictional things, impossible things, things that exist outside our light cone, minutely small and extremely distant things, things that only exist in the present or in the future; we can think about all of these kinds of things and our ability to do so is inexplicable on the pseudo-Dretskean theory we are considering. There are no true adaptational explanations in which such things appropriately figure in the ways specified by the theory.

Of course, thoughts about unicorns could certainly affect our fitness. A hunter who only hunts unicorns will go hungry. And someone might misperceive a small deer as a unicorn and succeed in spearing it anyway and so be happily well fed. But the pseudo-Dretskean theory requires members of the relevant represented kind themselves to impact upon our fitness. We cannot employ the pseudo-Dretskean theory to provide appropriate content for nonreferring but purportedly referring terms. No type of Mentalese lexeme was ever selected because it indicated unicorns and because it caused some characteristic movement that was fitness enhancing in the presence of unicorns, for the simple and obvious reason that there never were any unicorns. Equally obvious problems also arise for the representations of many actually instantiated kinds. We can be quite sure that no innate Mentalese lexeme was ever selected because in our ancestors it indicated quasars and caused some characteristic movement that was fitness enhancing in the presence of quasars. For one thing because quasars are not detectable with our naked senses, nor are they even resolvable with a telescope. So our ancestors did not possess a quasar detector, and so our ancestors could not have had a Mentalese indicator of quasars. For another thing, although quasars are real, and the most energetic objects known in the universe, and an encounter with one would definitely be fatal to all of life on eath, they are luckily very distant from us and must have always been so.

So here is a teleosemantic theory that is definitely committed to implausible (and even impossible) adaptational explanations. But the interesting question is, what should we learn from this? One has to be very eager to see the end of teleosemantics to see its doom forecast in such flimsy tea leaves as these. Does anyone in their right mind seriously think that we have innate and simple concepts of quasars and unicorns? (No, seriously. Not even Fodor does.) It should be obvious that there are other elements of the pseudo-Dretskean theory that can be dropped instead of the teleonomy.

Fodor (1996) argues that teleosemantics won't work because it attempts to base intentionality on natural selection's selection *for*, and this won't work, he argues, because adaptational contexts, unlike intentional contexts, guarantee existential quantification. Now it is, of course, true that adaptational explanations guarantee existential quantification. That's just to say that if, for example, the frog's optic fibers were selected for helping the frog to feed on flies (or small moving black things), then flies (or small moving black things) must obviously have existed and must have fed the frog. Natural selection can only select on the basis of *actual* past causal contributions to fitness, and no kind of thing can causally contribute if it does not exist. But it doesn't even begin to follow from this truism that a teleological theory of content cannot account for the failure to guarantee existential quantification in intentional contexts.

Actually, failure of existential quantification comes fairly cheaply (let Brentano turn in his grave, as he may). It is as cheap, at least, as a minimal capacity for *misrepresentation*. Even our pseudo-Dretskean theory provides for some modest failure of existential generalization, assuming that it provides for the possibility of misrepresentation. (Fodor has vigorously denounced that latter assumption, or assumptions sufficiently like it, in earlier papers. But he intends this anti-adaptationist style of attack on teleosemantics to be a new and independent argument against it, so let's assume for the sake of the present discussion that the theory does permit the possibility of misrepresentation.) If R was selected for indicating Fs, according to the formula specified, there must have been Fs around during the selection process. But it doesn't follow from this that an F must be around every time an R is tokened. On the contrary, if misrepresentation is possible, then it is possible that an R be tokened in the absence of an F. And on such occasions, were they to occur, existential generalization would fail. Kermit's seeing something (or nothing) as a fly (or a small dark moving dot) before him does not guarantee that there really is a fly before him. He might be hallucinating.

However, the general critical point can be put this way. We humans can think about uninstantiated kinds of things, and about instantiated kinds of things that can have had no significant impact on our fitness or on the fitness of our ancestors. This is a valid objection to the pseudo-Dretskean theory (and perhaps to some actual versions of teleosemantics that have been seriously proposed).

Does this make the pseudo-Dretskean theory unduly adaptationist? Well, yes and no. Yes, it is certainly the case that the pseudo-Dretskean theory we are considering is committed to some extremely implausible adaptational assumptions (or alternatively to our having far fewer concepts than we thought we had). But to describe it as adaptationist is to do a severe disservice to any biologist who might be fairly thought of as an adaptationist (I doubt any of them were suggesting that we were really cavorting with unicorns or running away from quasars, or alternatively that natural selection could choose representations of such things for us, just in case a unicorn or a quasar happened along).

If we abandon radical atomism and radical nativism, however, we lose the implausible adaptational assumptions. A more modest teleosemantic theory (see, e.g., Sterelny 1990) is one that uses teleonomic functions to determine the content of a set of semantic simples, which are in turn used in different combinations to construct more complex concepts. This is the approach that I find most plausible. On such an approach, many of our concepts are learned or are aquired by a process involving (in some sense) construction out of simpler or more basic concepts (where simplicity is relative to a cognitive system). The idea is that ultimately all of our concepts are constructed out of semantic primitives. On a more modest theory of this sort, only the semantic primitives that purport to refer must refer to instantiated kinds that have had a significant causal impact on the fitness of our ancestors. The task of discovering what semantic simples humans and other creatures possess is a task for scientists, not for philosophers. To learn what the most plausible candidates are we should consult percepetual and other psychological theories. And I haven't heard of any that suggest that our concepts of unicorns or quasars might be primitive. The primitives of David Marr's (1982) theory of vision, for example, are such things as representations of edges, surface discontinuities, brightness, size, orientation, spatial arrangements, and so on. Here a more modest, revised pseudo-Dretskean theory will strike further problems, but the problem is no longer that the represented kinds are not instantiated or have not been detected by our ancestors or have not had a causal impact on our ancestors. (Now the problem is that there is no characteristic movement in response to these kinds of things, but this another aspect of the pseudo-Dretskean theory that I would drop.)

There are difficulties that the best of such modest proposals will have to face. But no extant theory of content is compatible with radical atomism and radical nativism, or not without costing us some concepts we thought we had.[11] The issues involved here are too large to be treated in the closing paragraphs of this chapter, but I will finish with some cursory comments on the difficulties to be faced.

One obstacle that is supposed to stand in the way of such proposals for molecular meaning is the traditional "problem of analysis": the problem of providing traditional philosophical analyses of concepts by means of providing a set of necessary and sufficient conditions that specify the conditions of application for the term under analysis. Good analyses of this kind are notoriously difficult to find; that's the problem. But these traditional philosophical analyses are not what is called for. The concepts employed in these analyses are a far cry from, for example, the primitives of the perceptual system that Marr proposed, and from which, he suggested, particular and canonical representations of objects are inferred and constructed. There is no reason to assume that the semantic primitives of Mentalese will have corresponding concepts in ordinary English (or in professional philosophical English either). Nor is there any reason to assume that there need be a set of necessary and sufficient conditions, expressed in terms of these primitives, that captures the meaning of complex concepts. Or at any rate, this is only a safe assumption if we are very lenient about what would qualify as such (e.g., if obtaining a certain outcome from the implementation of a complex program could count as satisfying a necessary and sufficient condition). Once again, we need to think outside of the philosophical tradition here. Consider, for example, Minskyean frames. Perhaps the concept of a "restaurant" is constituted by a complex data structure more nearly along these lines than along the lines of a philosophical analysis.

Another obstacle that is supposed to stand in the way of molecular meaning is the absence of an analytic/synthetic distinction. The problem is that without an analytic/synthetic distinction there is thought to be no distinction between meaning constitutive beliefs about things and other (non-meaning constitutive) beliefs about things. No distinction between the belief that bachelors are unmarried men and the belief that all bachelors have ears or that some bachelors are smelly. So, it is argued, there can be no molecular meaning, and we have to choose between meaning atomism and its opposite, meaning holism.

Of course, this is a hard problem, about which I can make only a few sketchy comments here. But I want to suggest that the problem is not as intractable as people sometimes suppose. For one thing, although traits that we can loosely refer to as innate traits need not be universal, they often are universal and will anyway generally be shared by a significant proportion of the population. This means that at the level of innate representations, there is no fast track from meaning holism (or rather, from a one-way meaning holism) to an absence of shared content. The idea needs more careful elaboration than I can give it here, but the idea is that concepts that are themselves innate, or that are formed fairly directly from concepts that are themselves innate, can be widely shared even if their meaning is determined by their place in the entire intricate conceptual framework at that level and below (where "below" means simpler or more basic). This limited one-way holism is compatible with a great deal of shared meaning. In any case, there are other mitigating factors, other major structural demarcations in the web. For example, if it is true that much early learning is modularized, then we can also exploit the vertical boundaries of these modules in elaborating a molecular theory of meaning at this next level.

This isn't the place to develop these admittedly very sketchy ideas. I am only trying to open up our sense of the possibilities, not trying to put these worries to rest. My main point in this last section is a much more modest one, to show that teleosemantics is not the obvious choice for elimination when we consider particular teleosemantic theories that have implausible adaptational assumptions. In my view, in the case of the pseudo-Dretskean theory we began by considering, it is its radical atomism and radical nativism that are the more obvious choice. I have tried to suggest some directions in which the less radical alternatives can be coherently developed. In the case of other teleonomic theories, the problems may well be otherwise. Each theory needs to be considered on a case by case basis.

I have explained why neither teleonomy per se nor teleosemantics per se have commitments to implausible adaptationist assumptions. I also very briefly indicated why I believe teleosemantic norms are a promising source of psychosemantic norms. In closing, let me just say that, whether or not the reader agrees with my brief positive claims regarding the promise of teleosemantics, there is little else actively being developed in psychosemantics these days. That's all the more reason why we should be very careful not to throw this baby out with the dirty bath water.

Notes

1. I'm assuming that the reader has some familiarity with teleosemantic theories, but I give the basic gist of things and I hope the chapter is intelligible for those new to the topic. A paper that is critical of teleosemantics along antiadaptationist lines is Fodor's (1996). His more specific target is Dan Dennett (1995) but he expresses his conclusion in more general terms. See also Dennett's (1996) response. For other versions of teleosemantics, see Dretske (1988), Millikan (1993), and Papineau (1993).

2. Actually, even Pangloss acknowledges the divine analog of a piggy-back trait. Why did Columbus bring syphilis to the New World? So that he could bring chocolate back to Europe. This is noted in an amusing spoof of Gould and Lewontin's (1979), but I am sorry to say that I cannot locate the reference.

3. The now classic statement of the distinction is in Sober (1984) pp. 97–102. Sober says that selection *for* pertains to properties and selection *of* pertains to objects, but I do not understand his motivation for drawing the distinction in this way. We can, example, have selection for and mere selection of properties, since we can distinguish between properties that were and were not causally efficaceous in getting their bearers selected.

4. See Wright (1976) and Neander (1991a,b) for elaboration of etiological theories of function.

5. I am grateful to Adam Goldstein for discussion on this point.

6. Fodor (1996) appeals to the alleged gulf between intentional psychology and nonintentional physiology to argue that we have special reason to doubt the importance of adaptational explanations in explaining cognition. He notes that small genetic changes can result in large physiological effects and he comments that small physiological changes might also result in large psychological effects. The idea seems to be that, in the case of perception and cognition, there is a further level of potential magnification of modification. It is unclear what is supposed to follow (saltation?). It is also unclear whether it is true. Why do the representational functions of the brain count (in the relevant sense) as being at a higher level than the highest level of ordinary physiological functions. The argument needs to be further spelled out. Such claims are usually about levels of description. Crudely, the entities referred to at the higher levels are composed of or are reductively explained by (and are perhaps multiply physically realized by) the entities at the lower levels of description. But ordinary somatic physiology has many levels of description, in just the same way. Why are representational functions higher than, say, reproductive functions?

7. Not all teleosemantic theories, perhaps. Millikan (1993) for example, claims that the content of a represention is not determined by its function, or by the function of its producer, but rather by the function of its consumer. Just what the consumer is remains vague. It seems plausible that it would often be some further component of the cognitive system, but Millikan has suggested (in conversation) that the consuming system, in the case of a frog detecting a fly, could be the frog's digestive system, or even the whole frog after its catching of the fly.

8. Fodor would have us believe that we know next to nothing about the evolution of the mind. In his (1996), he says, "Pace Dennett, even a relaxed adaptationism about our psychological traits and capacities isn't an article of scientific faith or dogma; we'll just have to wait and see how, and whether, our minds evolved. At the time of writing, the data aren't in." It isn't an article of faith or dogma, but I will assume that the evolution of our minds is taken for granted. That is to say that I'll assume that our minds are in some sense the product of our brains and that our brains evolved. It is hard to know what Fodor could possibly have in mind (but see note 6 above). Evolution should of course be distinguished from evolution by natural selection. Evolution occurs just in case a change in the proportion of genes occurs, or more weakly, perhaps, if a change in the proportion of genotypes occurs.

9. Some of these points are also made in section III of Neander (1997). For far more extended presentation of the general argument, I recommend Dawkins (1986, especially chapter 9) and for a more specific treatment in relation to the evolution of language see Pinker and Bloom (1990).

10. Dawkins makes these same points. See his discussion of the stretched DC8 macromutation (1986, p. 234).

11. Fodor's (1990) asymmetric dependency theory also requires the resort to construction sometimes, for otherwise, on his theory, necessarily uninstantiated concepts cannot be primitives and nor can necessarily coinstantiated concepts be distinguished. Quarks and electrons also present obvious problems that, as far as I can see, he can only deal with if he resorts to molecular meaning.

References

Barrash, D. (1979) *Sociobiology: The Whisperings Within*. New York: Harper and Row.

Dawkins, R. (1986) *The Blind Watchmaker*. New York: W. W. Norton.

Dennett, D. (1995) *Darwin's Dangerous Idea*. New York: Simon and Schuster.

Dennett, D. (1996) "Granny versus Mother Nature—No Contest," *Mind & Language* 11:263–269.

Dretske, F. (1988) *Explaining Behavior*. Cambridge, MA: The MIT Press.

Eldredge, N. (1995) *Reinventing Darwin*. New York: John Wiley and Sons.

Fodor, J. (1990) "A Theory of Content, Part 1," in Fodor, J., *A Theory of Content and Other Essays*. Cambridge, MA: The MIT Press.

Fodor, J. (1996) "Deconstructing Dennett's Darwin," *Mind & Language* 11:246–262.

Godfrey-Smith, P. (1996) *Complexity and the Function of Mind in Nature*. New York: Cambridge University Press.

Gould, S. and Lewontin, R. (1979) "The Spandrels of San Marco and the Panglossian Paradigm," *Proceedings of the Royal Society of London* 205:281–288.

Gould, S. J. (1980) "The Panda's Thumb," in Gould, S. J., *The Panda's Thumb: More Reflections in Natural History*. New York: Norton.

Gould, S. J. (1997) "The Darwinian Fundamentalists," *The New York Review of Books*, vol. 44, no. 10, June 12, pp. 34–37.

Marr, D. (1982) *Vision*. New York: Freeman and Company.

Maynard-Smith, J. (1978) "Optimization Theory in Evolution," *Annual Review of Ecology and Systematics* 9:31–56. Reprinted in Sober (ed.), 1994.

Millikan, R. (1993) *White Queen Psychology and Other Essays for Alice*. Cambridge, MA: The MIT Press.

Neander, K. (1991a) Functions as Selected Effects, *Philosophy of Science* 58:168–184.

Neander, K. (1991b) "The Teleological Notion of 'Function,'" *Australasian Journal of Philosophy* 69:454–468.

Neander, K. (1995) "Misrepresenting and Malfunctioning," *Philosophical Studies* 79:109–141.

Neander, K. (1997) "The Function of Cognition: Godfrey-Smith's Environmental Complexity Thesis," *Biology and Philosophy* 12:567–580.

Papineau, D. (1993) *Philosophical Naturalism*. Oxford: Blackwells.

Pinker, S. and Bloom, P. (1990) "Natural Language and Natural Selection," *Behavioral and Brain Sciences* 13:707–784.

Sober, E. (1984) *The Nature of Selection*. Cambridge, MA: The MIT Press.

Sterelny, K. (1990) *The Representational Theory of Mind*. Oxford: Blackwells.

Wright, C. (1976) *Teleological Explanation*. Berkeley, CA: University of California Press.

2 Understanding Functions: A Pragmatic Approach

Valerie Gray Hardcastle

In an article celebrating the twentieth anniversary of Larry Wright's seminal paper, "Functions" (1973), Peter Godfrey-Smith asserts that "much of the literature [on functions] has ... engaged in the refinement of Wright's original idea" (1993, p. 196). Colin Allen, Marc Bekoff, Karen Neander, Ruth Millikan, and others concur with that assessment, calling Wright's characterization "the Standard View" (Allen and Bekoff 1995a, 1995b; Millikan 1989; Neander 1991a; see also Ayala 1977; Griffiths 1993; Mitchell 1993; Walsh forthcoming; and Walsh and Ariew 1996). I believe that the optimism that all we have to do is tweak Wright's conception of function to finish a project in philosophy is unwarranted. Only by focusing on a very narrow use of the term is the apparent unanimity among philosophers of biology possible. Really, to capture the notion of function as it is used in science generally, we have to go back to the drawing board.

After outlining the philosophical project of defining functions, I shall then discuss the sort of questions a functional explanation in the biological and social sciences is supposed to answer. I believe that how we understand these questions is crucial for any philosophical analysis of functions and that many philosophers of biology construe them much too narrowly. Next, I shall sketch the three sorts of analyses currently in vogue in philosophy, noting that they all are at risk of death from a thousand failures. Last, I shall point toward a different approach to understanding functions, one that is more faithful to science as it is actually practiced and to how functions are actually assigned.

1   The Philosophical Project

Here is the project as I see it. Many disciplines within the biological and social sciences use the notion of function with a perfectly straight face in their explanations of phenomena. This may be a distinguishing mark of the "softer" sciences. Philosophers, at any rate, are quick to point out that physics and chemistry don't use functional language in their explanations, although "this reagent functions to remove potassium from the gel" seems like a perfectly good explanation to me. But regardless of deep philosophical insight into the nature of chemistry's explanations, anthropology, biology, linguistics, medicine, psychology, psychiatry, neuroscience, and sociology use functional explanations with impunity and could only renounce them with great difficulty, if at all. These disciplines ascribe functions to objects, traits, or properties and assume that doing so aids in answering some explanatory challenge.

On the face of it, talking about something's function is a perfectly legitimate scientific move in many fairly unrelated disciplines.

On the other hand, on the face of it, the concept of function is an odd one, for it seems to refer to possible future effects as relevant to current properties. The function of my house-key is to let me into my house. That is, the purpose of this key now is to let me into my house later, when I go home. How are we supposed to talk about possible future effects as a defining property, without being mysterious? Suppose this key is broken and I can't get in my house. Suppose that the key has always been and will always be broken. Nevertheless, this key would still have the same function. How are we supposed to know this, if there aren't any effects that we can point to as the keys fulfilling its function?

Moreover, when picking out something's function, we highlight one effect among many. This key also takes up space in my pocket, will be left on the dresser when I sleep, belongs to me, and so on. How is it that we can distinguish the accidental features or effects of an object from its function? What is the extra ingredient that makes functions special? If we want to use functions in scientific explanations, we need to answer these puzzles. In gross terms, this is philosophy's project: We are charged with giving a naturalistic account of function that is applicable to any discipline in science that might want to use it.

But once we move beyond a coarse outline of what philosophers are supposed to be doing here, disagreements about what exactly our task is emerge. Is this some form of conceptual analysis (see, e.g., Amundson and Lauder 1993; Enç and Adams 1992; Neander 1991b; Wright 1973)? Are we interested in remaining faithful to our intuitions about what has a function and what that function is? Or are we devising a theoretical definition (as Millikan 1984, 1989 believes), so that it would restrict and guide our usage, intuitions aside? Or are we regimenting an explanatory structure (Mitchell 1995; Nordmann 1990)? Are we making clear what scientists already do, regardless of popular sentiment or necessary and sufficient conditions?

I actually see us as doing a hybrid of all three. Functional explanations are already established in the scientific community—we can point to many paradigm examples of them in the various disciplines—so it makes sense to analyze how functions are already being used as tools. In doing so, we should rely on the set scientific standards for ascribing functions. But since scientists haven't operationalized their definition of function, this is going to be a bit like sexing chickens or distinguishing rocks from mines by sonar. We are going to be relying on scientists' educated opinions, as it were, as our touchstone in analysis. I won't pump your intuitions with cases of the function of a schmanda's thumb on Twin Earth, but I will push on what the experts claim are perfectly good descriptions of functions. And our ultimate goal is to devise some sort of definition or reach some understanding of function that clarifies what it is we are doing when we say that a panda's wrist-bone functions as a thumb. This regimentation could then feed back into scientific discourse and regulate or constrain use.

I should note at the outset, though, that I diverge considerably from some in taking our project to be devising a general account of functions. Peter Godfrey-Smith (1993, 1994), Ken Schaffner (1993), Ruth Millikan (1989), and Sandra Mitchell (1995), for example, all believe that evolutionary biology can use one notion of function and the rest of science maybe another. In all honesty, I see little reason for adopting this strategy, except that it saves their particular analyses and it allows them to be graciously ecumenical. But when a biologist claims that the function of crossover is to increase diversity, on the one hand, and an anthropologist claims that the function of pig feasts in Papua New Guinea is to distribute protein, or a linguist claims that the function of noun suffixes is to decrease sentence ambiguity, or a medical doctor claims that the function of the thymus is to control cellular immunity, or a neuro-scientist claims that the function of myelin is to increase the rate of information transmission among neurons, or a psychologist claims that the function of dreams is to consolidate memory, or a psychiatrist claims that the function of repression is to minimize psychical damage, on the other, they do appear to be using "function" in more or less the same way. Certainly it is not transparent that a biologist's use differs substantially from everyone else's. Indeed, if we take the tool-kit analogy seriously, then we should be searching for a general conception of function, just as we have general notions of cause or force or standard deviation that remain constant across the sciences. I will have more to say about why some philosophers of biology are led astray in a moment, but I am not going to defend my take on this project further. This is just where we will begin.

2 Functional Explanations in Science: A First Pass

Consider the query: Why does the thalamocortical system in the brain have an oscillatory firing pattern? There are at least two sorts of questions we could be asking here, depending on our contrast classes (cf. van Fraassen 1980). On the one hand, we might want to know why the thalamocortical system, as opposed to some other system, is doing the oscillating. On the other, we might want to know why the system oscillates, instead of firing randomly or doing nothing. The second sort of question generates a functional response. We want to know why some previously fixed type of system or object or process, o, possesses some trait or property, T. One answer is that having that trait does something for the system or object or process. That is a functional answer.

But notice the query of why To (as opposed to $[Uo \lor Vo \lor Wo]$) is still ambiguous regarding the sorts of answers one might be looking for. On the one hand, we could be asking about how it is that o came to have the property T. In this case, we would be looking for some historical story about o's coming to acquire and maintain T. To use Fred Dretske's phrase, we would be looking for the "triggering cause," the events in the world that led o to have T (Dretske 1989). Or to use Ernst Mayr's (1982) phrase, we would be looking for a "genetic cause." So, in the case of the oscillatory firing patterns of the thalamocortical system, we might explain how a coordination among action potentials allowed organisms to distinguish several different objects in their environment simultaneously. These organisms now had a selective advantage over other creatures who could only perceive objects serially or not at all, for they could recognize mates and avoid predators more easily than before. Consequently, the genes underlying this trait spread throughout populations, and here we are today with oscillating brains.

On the other hand, we could be asking about how the structure works. We want to know what reasons there could be for o having T. This is Dretske's "structuring cause" and Mayr's "ecological cause." If this is the case, then a causal etiology would not be an appropriate response to the query; here we would be seeking a list of effects of T that are somehow important to o. So, for example, we might note that particular thalamocortical oscillations co-occur with particular environment stimuli (the presence of discrete objects, say) and are absent under different conditions (viewing grid patterns, say). We might also the note that organisms with oscillating brains can distinguish objects from backgrounds, but organisms whose brains either don't oscillate or are prevented from firing synchronously cannot, and that the oscillating organisms fare much better. Hence we would conclude that uniting features into perceptions of objects is the important effect of thalamocortical oscillations.

Both sorts of answers, I submit, could be functional in character. We could explain o's having T in virtue of some advantage that T had historically given o and which caused o to maintain T. This effect of T for o would be the function of T. The function of the thalamocortical oscillations is to bind together disparate sensory inputs into meaningful bundles, for this particular effect is responsible for the traits being maintained in the brain. Or we could explain o's having T in virtue of some advantage that T will give (or currently gives) to o. This effect of T for o too would be a function of T. The function of the thalamocortical oscillations is to bind together disparate sensory inputs into meaningful bundles, for this particular effect is important for what the system is doing. In both cases, T does something for o and that something is explanatorily important. The difficulty before us is how to cash out "doing something for o" naturalistically and without begging any questions.

Many philosophers of biology have focused exclusively on the etiological approach as the proper sort of functional explanation in biology. But they are just missing the mark here. Clearly other sorts of functional explanations are involved. Hubel and Weisel's edge-detectors help calculate the shape of an object from shading. The peacock's tail attracts peahens. The basal ganglia fine-tunes motor impulses. Assuming light sources come from above shortens the processing time of complex visual inputs. All these sentences list effects that are valuable for some biological system, and we can talk about these effects' contribution as their functions without ever worrying how or if or when they came into being. They each give one answer for why o has some T.

Indeed, the analysis of a trait's role is much more prevalent in the biological and social sciences, if for no other reason than one can perform the requisite laboratory experiments to test causal-role functional hypotheses much more easily. In sum, I just disagree with Robert Brandon when he writes that "ahistorical functional ascriptions only invite confusion, and that biologists ought to restrict the concept [to] its evolutionary meaning" (1990, p. 187 n. 24). Ahistorical functional ascriptions may be confused (though I doubt that too), but given the sorts of explanations that biology as a whole engages in, we can't restrict its meaning to an evolutionary perspective and still remain faithful to how science is actually practiced. We do want to know why some trait, rather than some other trait, exists, but we don't always want to know how that thing came to exist, which is what evolutionary explanations cover.

3 Philosophical Analyses

There have been three major strands in the philosophical discussion of how to understand functions: champions of a backward-looking etiological approach, which is what is now regarded by some as the Standard View and which relies on selective histories to isolate the important effects as functions; advocates of a forward-looking propensity approach, which looks to future advantages to define functions; and proponents of a causal-role analysis, which examines what the trait or property currently is doing for the system. I'll examine each only briefly, for none of them is up to the task before us.

Let us take the etiological view of functions first, since this is the most popular in philosophy of biology today (cf. Ayala 1977; Brandon 1981, 1990; Canfield 1964;

Godfrey-Smith 1994, 1995; Griffiths 1993; Kitcher 1993; Millikan 1989; Neander 1991a,b; Ruse 1971; Sober 1993; Williams 1966; Wimsatt 1972; Wright 1973, 1976). This perspective assumes that the function of a trait explains the continued existence of that trait. Wright's original formulation is canonical:

The function of X is Z means

(a) X is there because it does Z,

(b) Z is a consequence (or result) of X's being there.   (1976, p. 81)

The typical example given in the Standard View is that the function of the heart is to pump blood. Pumping blood is one thing that the heart does, and the fact that it pumps blood (as opposed to making thumping noises) explains why we have hearts. The fact that organisms evolved to have hearts means that blood circulates more efficiently. More efficient circulation means better distribution of chemicals and nutrients to the system and quicker waste disposal. This in turns leads to better survival rates, and hence, more progeny. These "backward-looking" functional explanations account for why something is there in natural systems by picking out the property of that thing most valuable to previous generations plus some selection mechanism. In biological or social systems, this usually translates into an appeal to evolution by natural selection.

   Already we know that this cannot be the entire story regarding function, because functional explanations can do more than explain why something continues. We still should be able to talk about the function of some trait, even if the individual who has it is sterile and got it through some series of random mutations. But we can see that even on its own terms, this etiological approach relies on too simple a notion of evolution. It is easy to spin "just-so stories" about why some T is maintained in o in virtue of the advantage T offers o, but it is nearly impossible to fit these stories to our very complex biological world, even ignoring the difficulties of connecting phenotypes to genotypes. In the world, what we actually observe—as a trend—is differential reproduction toward one of several possible local maxima of fitness—and even that is questionable in cases where fitness of a genotype is a function of its prevalence (see Hull 1988; Lewontin 1965)—but that is a far cry from nature carefully selecting the organisms whose set of traits is most adaptive. As Stebbins notes, "Looking at the whole sweep of evolution, one sees not progressive trends, but rather successive cycles of adaptive radiation followed by wholesale extinction, or by widespread evolutionary stability" (1968, p. 34). For example, "organisms like bacteria and blue-green algae have perpetuated their evolutionary lines for at least three billion years without evolving into anything more complex. Moreover, even in those phyla which we regard as progressive, the proportion of species which have evolved into more complex adaptational types is minuscule compared to those which either became extinct or evolved into forms not very different from themselves" (1968, p. 33). Toss in pleiotropy, exaptations, random drift, and other forms of genetic hitchhiking, and we just can't assume in this world a natural selection of the fittest. That is but a toy model of evolution.

At best, then, this analysis could give us a view of function as a heuristic or placeholder explanation until we know the details of the real story. But it gets worse. Bigelow and Pargetter (1987) are right: this account is vacuous under certain conditions. According to them, on the etiological view, we retrospectively judge that T has some function when its having that effect contributed to the past survival or proliferation of o. But then, on pain of circularity, it would no longer be possible to explain why T has persisted by saying that it serves some function—for we would be explaining the persistence of T in terms of persistence of T.

This point has been roundly criticized in the literature for assuming that "persistence" has to be translated as "evolved by natural selection" (cf. Enç and Adams 1992; Godfrey-Smith 1994; Mitchell 1993). So, as long as there are other evolutionary forces that might account for the persistence of T, claiming that something did evolve by selection as opposed to these other things is explanatory. If we eschew strict adaptationism, we can avoid any explanatory vacuity.

But this criticism does not answer the charge. Our immediate project is to define or otherwise characterize functions, not to explain the presence of traits tout court. Certainly there are many ways to account for the presence of some property, many of which are not functional explanations. But these do not concern us here. Bigelow and Pargetter's point, as I understand it, is that if we restrict our scope to functional explanations of T, and if we define functions in terms of the past selection of T, then we cannot use functions to explain the past selection of T for that would be using the past selection of T to explain the past selection of T. In other words, if we use a backward-looking approach to understanding functions, then we simply cannot give functional explanations of things we know are adaptations. This circularity remains.

Bigelow and Pargetter's solution is to use a forward-looking account of functions conjoined with the additional claim that the functions were active in the past to explain evolution by natural selection. They claim that functions are dispositions or propensities to succeed under future selection pressures in a normal environment (see also Bechtel 1989; Horan 1989; Staddon 1987; Tinbergen 1963). "Something has a (biological) function just when it confers a survival-enhancing propensity on a creature that possesses it" (1987, p. 192). So the function of the heart is to pump blood and not to make lub-dub noises because creatures whose blood is pumped have a greater chance of surviving and reproducing than creatures whose hearts merely thump. If we define function in terms of its propensity to enhance fitness in the future, then we should be safe in using this notion to explain events from the past. That difficulty is avoided.

But does the propensity view of functions actually work? I think not. Even if we once again ignore the fact that this view, as with the etiological view, cannot account for the all the uses of the term "function" in science—it cannot explain how a system works either—it too fails on its own terms. Indeed, its difficulties are analogous to the problems with the etiological view. The backward-looking approach cannot assign functions to new things, to things that have not undergone selection, or to things have been selected for but are now used differently (Williams 1966; Ruse 1971; Wright 1976). Similarly, a forward-looking interpretation of function can't assign functions to things that cannot be selected for (because the individuals in question don't reproduce, say). Further, such an interpretation has no mechanism to prevent attributing functions to structures that might aid in selection, but whose benefits are fortuitous (Borse 1976; Achinstein 1983; Adams 1979; Adams and Enç 1988; Enç and Adams 1992). Young children play peekaboo with rags and other soft items. Doing this can shield a child from scalding water being dropped on its head. Hence, covering one's head with a rag has the propensity to help one survive being scalded, but we wouldn't want to say that this is function of peekaboo. Finally, this view can't determine what the function is of something whose use changes over time, because it has no way of determining the appropriate chunk of the future to use in assessing functions. To take Enç and Adams's (1992) example, we use our teeth now to pulp food, but if the Ice Age cometh and all our food freezes over, then the food-pulping propensity of teeth will no longer contribute to our survival. Does this mean that our teeth no longer have that function now?

Moreover, Bigelow and Pargetter's "solution" is subject to the same charge of circularity they brought against the standard view. Sometimes we want to explain how some T is maintained in a system functionally. Of course, there are other ways to account for maintaining certain properties, but let us restrict our scope for the moment to a functional sort of explanation. So, for example, we might say that ritual taboos against incest have the function of maximizing variation in the gene pool of that population and that that fact explains why Western cultures forbid marriage between siblings. If we understand functions as certain dispositions to enhance survival, then we would say that maximizing the variation in the gene pool of that population enhances the survival of offspring. This premise, conjoined with the fact that this function operated in the past, would explain nonvacuously why these ritual taboos existed and proliferated in the past. However, it cannot explain why the taboos continue to exist today, for that turns on the implications of having the trait now. We want to say that the taboos are adaptive today because they have the function of maximizing the variation in the gene pool. But if function means propensity to enhance survival, then we would be saying that proscriptions against marriage between siblings will continue to be selected because it will continue to be selected. In other words, if we define functions in terms of future selection of T, then we cannot use functions to account for the future selection of T for that would be using the future selection of T to explain the future selection of T. A propensity interpretation does not avoid the problem of the vacuity of some functional explanations.

My diagnosis is that it is a mistake to tie functions to selection in the first place, since often selection is exactly what we want to explain in functional terms. And if we explicate function in terms of selection, there will always be some case of selection we want to explain functionally but can't, because doing so would be empty in that it would "explain" the selection in terms of exactly that selection. We need an entirely different sort of approach to understanding functions.

Robert Cummins (1975) offers one (see also Bechtel 1989; Rosenberg 1985). He suggests that we should replace the selection views with inference to the best explanation. We would infer the presence of some trait T from our best explanation of the system o which contains it. So, ascribing a function to something means that we are ascribing a capacity to it that is singled out by its role in our analysis of the system of which it is a part. Loosely speaking, then, the causal role T plays in o is its function. So, the function of the heart is to pump blood because our understanding of complex biological organisms dictates that there be a blood-pumping mechanism and the heart fulfills that role.

This approach has not been picked up by philosophers of biology, largely because it is not restrictive enough for their tastes. They believe that it allows just about anything in as a function, as long as you can describe it in terms of its role. In other words, a conceptual-role approach doesn't distinguish functions from mere dispositions or accidents. My hips retained their maternal store and thus contribute to the capacity of my body cast a pear-shaped shadow. But we don't want to say that the function of the fat in my thighs is to cut a certain silhouette. But nothing about a causal-role view seems to rule this out.

However, this reaction is not quite fair. Cummins is careful to restrict functional descriptions to the subject matter of science, which takes care of the more extreme counterexamples (see also Amundson and Lauder 1994). Still, this approach does allow the same trait to have different functions, depending upon one's perspective. But maybe that is okay. We can think of birth, for example, as a flooding of hormones, the second stage of labor, a period of confinement, the fittest surviving, the beginning of life, the emergence of an identity, the creation of a family, the transfer of power, an alteration in social structure, the saving of an inheritance, the healing of wounds, or the end of an era.

On the other hand, identifying function with causal role—even if we restrict its domain to scientific discourse—confuses higher-level abstractions for talk about functions (see also Schaffner 1993). And this is too broad. But Cummins is not alone here; many philosophers of mind make this mistake. Sam's lust for Beck's Dark contributed to his rummaging in my cooler, and we would infer his desire from his begging me for a Beck's Dark and his looking behind the cabbage after I told him the beer was there, but we wouldn't want to identify any of these effects as the function of Sam's want. It is not clear to me that this is a functional description at all, at least in the sense most scientists use the term.

At the same time, a causal analysis is too narrow. Restricting analysis to how a system actually behaves means that causal-role descriptions will misdescribe the functions of malformed or otherwise broken things (Millikan 1989; Neander 1991b). The function of lungs is to pass oxygen to and remove waste from the bloodstream. But John's lungs are covered in scar tissue, so they are not doing that anymore. Still, we should be able to say that that is what they are supposed to be doing, even if they can't. A pure causal analysis prevents us from making this claim, for John's lungs are not making that contribution to his system.

For similar reasons, a causal role approach overlooks interspecies variation (Neander 1991a). Cells in the lateral geniculate nucleus in primates and in birds serve the same function—they both enhance the antagonism between the center and the surround in the receptive fields of visual neurons—but because their respective causal roles are different—the LGN in primates is connected to MT, while it feeds into the tectum in birds—a causal-role analysis would consider them as having different functions. This approach forces us to miss many interesting and important generalizations.

And these generalizations are the point of functional explanations. They permit us to capture abstract patterns and relations between otherwise diverse phenomena. They allow us to see biological or social convergences. If we just explain o or o's interactions locally, then we will miss important similarities between behaviors that are formally very different. What we care about here are the patterns of effects found in groups of items, not individual causal roles, for the general effect doing something across the group explains the particular tokens of the trait in particular organisms (Enç and Adams 1992). We need these higher-level generalizations for a complete understanding of complex systems.

This is one reason it would be nice to clear a path for functions that is not merely heuristic. But to do this, we need some expression of the extra ingredient that separates dispositions from functions, and we need to appeal to things beyond the structure of the system in delineating its functional components in order to capture important regularities (cf. Enç and Adams 1992). Cummins is right to start with causal roles instead of selection, for that avoids the problem with vacuous explanations. But the causal role approach comes up short by neither culling the functional effects from the rest, nor providing us a way to generalize across like individuals. Consequently, we have to go back to the drawing board.

4 A Different Approach

To understand functions, I propose to start from a different direction. Let us begin in the laboratory instead of the armchair and look at how scientists build functional descriptions in the first place. I want to challenge the starting assumptions of some philosophers of biology anyway, that "it is the 'proper function' of a thing that puts it in a biological category" (Millikan 1984, p. 17), and that "most biological categories are only definable in functional terms" (Neander 1991b, p. 180; see also Davies 1994). If this were true, it would be no wonder that some functional explanations end up being circular. Fortunately, though, as Amundson and Lauder (1994) spend many pages arguing, it isn't.

Science, in fact, creates many, if not most, biological and social categories out of structural components. Homologies, features or traits that characterize the monophylectic clade of a set of species, for example, do not depend on functional characteristics. We detect them through low-level anatomical similarities, similarities in their position in the larger patterns of physiological interactions, and through common embryological elements (Amundson and Lauder 1994; see also Patterson 1982). We do not need to know at all what effects these items have in an organism or across organisms in order to distinguish them. Sparrows' wings are homologous to owls' wings because they exhibit the same structural configuration of bones, muscles, and feathers and because they both belong to birds, but they are not homologous to bees' wings because birds and insects form different natural evolutionary clades.

Homologous organs are actually too wide to distinguish functional units for our purposes, so I don't want to push this example too far. The forelimbs of humans, dogs, bats, moles, and whales are morphologically homologous, but they function in very different ways. (I take this example from Amundson and Lauder 1994.) But my claim is that scientists use these sorts of structural categories to bootstrap their way into functional descriptions. Certainly, they can pick out the forelimbs of a single species or group or population and go on to figure out what the normal distribution of the trait is, how it might be malformed or diseased or truncated, and so on.

So, we identify some traits and their places in the causal stream purely morphologically. We can pick out hearts and cells and glottal stops, and tattoos and warfare and implicit memory, just in terms of their structural characteristics. If we were going to be anatomists, then we might just stop here, for they identify all the effects of a trait with its function (cf. Amundson and Lauder 1994). However, the rest of the biological and social sciences add some constraints so that only one or a few effects of each trait are catalogued as its function(s). How do scientists winnow out the choices?

Bill Wimsatt (1972) and Ken Schaffner (1993) both argue that we need to include an intentional "goal state" explicitly in our analysis of function (Kitcher 1993 makes a similar move with his notion of design; see also Dennett 1987). The function of some trait is to help attain the goal for o. Wimsatt does not believe that this goal or purpose is nonnatural, since evolution can serve as the primary purpose for biological systems (see also Walsh forthcoming; Walsh and Ariew 1996). "Increasing the fitness (or long-term survival) of an evolutionary unit" is the basic goal of natural systems, from which we can derive other subsidiary goals. In response, Schaffner argues that natural evolution is very different from natural selection and cannot sustain teleological assumptions. He concludes that all functional ascriptions are heuristic and turn on our own goals and other intentional states.

I wish to skate a middle line between these two views. On the one hand, as I have argued above, evolution is too complicated and messy and we know too little about how it has actually unfolded to use it to ground our analysis of functions. On the other hand, I do not want to acquiesce that we get to set the goal for some trait or some organism more or less arbitrarily, which would render all functional explanations merely place-holders for when we finally figure out the real story behind the interaction. On my view, which effects to privilege depends on the discipline generating the inquiry.

The nature of the query itself defines the sorts of effects scientists are supposed to be tracking. In evolutionary biology and anthropology, scientists do focus on selection and satisficing selection constraints. But in medicine, doctors focus on health; in physiology, neuroscientists emphasize information processing; and in cognitive psychology, psychologists look at cognition. In each case, they highlight the effects that contribute to, enhance, or sustain that aspect of the system.

Which parameters are chosen as important in assigning functions depends on the domain doing the abstracting. Our philosophically classic (and perhaps intuitively plausible) accounts of theorizing do not touch this issue, but the pragmatic aspects of explanation account for how this is done. Local social and historical facts determine which disciplinary orientation biological and social scientists designate as the correct one. In brief, previously accepted theories, the community in which research occurs, the background context of investigations, and the particular biases of the individuals doing the explaining define the factors germane to functional hypotheses.

Consequently, different disciplines can analyze the same property or disposition quite differently. There is not exactly one function appropriate for each trait. But this seems right to me. Most of us have an enzyme in our gut called lactase, and we can ask why our stomachs contain it (as opposed to no enzymes or a different one). An evolutionary biologist might answer that lactase helps breaks down lactose, the sugar in milk. This enzyme became prevalent in dairy-based populations because it helped the organisms digest one of their main foods. As dairy products proliferated and spread, so did people with this enzyme. The function of lactase is to decompose lactose because that is the effect or the role that accounts for its selection. But a medical doctor might answer that it helps the body digest milk and milk products. Persons with this enzyme living in Western cultures fare much better than those without because they are able to continue to consume milk after weaning. Because dairy products are so prevalent in the West and are able to provide so many of the nutrients that the body requires, those with lactase can take advantage of a natural source of vitamins, minerals, and proteins. On the medical story, the function of lactase is to help digest milk and milk products because that is its effect or role that promotes health. And an anthropologist might answer that this enzyme helped the Scandinavian cultures to take over other European societies and eventually the societies in America because it gave them a food source that other cultures could not tap into. The function of lactase was to help one culture dominate and proliferate at the expense of others, for that is its effect or role that explains cultural transformation. And an economist might answer that lactase is key in the balance among Western consumers, dairy farmers, and the cows themselves. We have bred heifers so that they have to be milked twice daily. This provides a livelihood for the dairy farmers, who sell their product to the consumers able to use milk. The function of lactase is to sustain a mutual dependency among producers, a product, and consumers, for that is its effect or role that accounts for the microeconomic cycle.

Letting the discipline set the boundaries on relevant effects is pragmatic, but it is neither mysterious nor arbitrary. Within their fields, scientists define exactly what they mean by fitness or health or exploitation or an economy or whatever. Ideally, these are operationalized to sets of recognizable, rigorous, and robust criteria. And once the criteria are fashioned and accepted by the relevant agents, they then set the agenda for the research of the discipline. Of course, scientists can (and do) refine or completely replace their criteria for the boundaries and interests of particular fields, but that does not occur swiftly, at the behest of one or even a few, or in response to particular experimental outcomes.

Moreover, relying on pragmatic considerations to cull functions from other effects does not distinguish functional explanations from other sorts of higher-level accounts. It is important to keep in mind that scientists in general have no intention of accounting for all the details of their subject matter when they construct their theories or hypotheses. They regularly disregard complexities of the actual world and define artificial domains isomorphic to the world in only a few aspects. To take a clear example from "basic science," classical particle mechanics uses point-masses, velocities in frictionless environments and distances traveled over time, to characterize falling bodies. Theoretical physicists want to explain only the general pattern of behavior of moving objects; hence they rightly ignore the color of the body, the date of its falling, interference from gravitational attraction, and so on. Similarly, the law of diminishing returns in economics, Chomskian theories of competence in linguistics, the Nernst equation in neurophysiology, theories of attention in psychology, and the notion of psychogenic pain in psychiatry all describe the behavior of a few "fundamental" mechanisms under ideal conditions that only approximate the behavior of real phenomena. In sum, "softer" sciences work no differently than any of the more "basic" sciences. They all identify only a few effects and properties as crucial, focus on those exclusively when theorizing about their phenomena, and ignore the rest.

What does separate a functional account from, for example, a morphological one? Cummins (1975) argues that the most useful functional explanations occur when the explanans is simpler and different in type than the explanandum. But many sorts of explanations are given in terms of things less complex and of a different type than what is to be explained. That is just what a good explanation does: It finds the few parameters that account for some phenomenon and traces the relationships between those variables. And whittling down the number of parameters we need to pay attention to means that it is trivially true that whatever is doing the explaining is going to be simpler and different than what is explained. That isn't going to distinguish (good) functional explanations from the rest.

But level of abstraction and the scope of the explanation do seem to matter. The more general descriptions, at the more abstract levels of analysis and with a relatively small scope, are of functions and reasons; descriptions at a lower level of analysis with a larger scope are of causes and structures; and the most fundamental descriptions, at the lowest level of analysis and with universal scope, are of pure correlations (see also Schaffner 1993). HCC stimulates ovulation—that is one of its functions—and that explains why some women who experience difficulty getting pregnant take the hormone artificially. This explanation is fairly well removed from the structures instantiating these events. That is, we can think of this token instance of ovulation as occurring at the level of the physical shape of the three-dimensional protein, its meshing with other chemicals and biological structures, and so forth. Our functional description of this type of event—stimulating ovulation—abstracts over these details. Moreover, what we are explaining is fairly narrow; we want to account for the activities of only a few woman. And we do so by taking our functional ascription as giving us a reason for the womens behaving similarly.

Most functional explanations gloss over the details, cover relatively few items, and are simply correlations discovered between the parameters defined in theoretical models of phenomena picked out by the historical interests of the discipline. To understand functions requires a pragmatics of explanation, but that move distinguishes neither the biological and social sciences from physics and chemistry, nor functional explanations from any other. Functions are simply what T is doing in o, relative to a domain of inquiry.

Acknowledgments

Earlier versions of this paper were presented to the Philosophy Department at Duke University and at the International Society for the History, Philosophy, and Social Studies of Biology Conference in Seattle. My thanks to both audiences for their lively discussions. Special thanks go to André Ariew, Majorie Grene, Karen Neander, and Geoffrey Sayr-MacCord for their detailed comments and nagging worries.

References

Achinstein, P. (1983) *The Nature of Explanation.* Oxford: Oxford University Press.

Adams, F. (1979) "A Goal-State Theory of Function Attribution," *Canadian Journal of Philosophy* 9:493–518.

Adams, F. and Enç, B. (1988) "Not Quite by Accident," *Dialogue* 27:287–297.

Allen, C. and Bekoff, M. (1995a) "Function, Natural Design, and Animal Behavior: Philosophical and Ethological Considerations," in N. S. Thompson (ed.), *Perspectives in Ethology* 11, pp. 1–46. New York: Plenum Press.

Allen, C. and Bekoff, M. (1995b) "Biological Functions, Adaptation, and Natural Design," *Philosophy of Science* 62:609–622.

Amundson, R. and Lauder, G. V. (1993) "Function Without Purpose: The Uses of Causal Role Function in Evolutionary Biology," *Biology and Philosophy* 9:443–469.

Ayala, F. J. (1977) "Teleological Explanations," in T. Dobzhansky (ed.) *Evolution*, pp. 496–504. San Francisco: W. H. Freeman and Co.

Bechtel, W. (1989) "Functional Analyses and Their Justification," *Biology and Philosophy* 4:159–162.

Bigelow, J. and Pargetter, R. (1987) "Functions," *Journal of Philosophy* 84:181–196.

Borse, C. (1976) "Wright on Functions," *The Philosophical Review* 85:70–86.

Brandon, R. N. (1981) "Biological Teology: Questions and Explanations," in P. Asquith and T. Nicles (eds.), *PSA 1980*, volume 2, pp. 315–323. East Langsing, MI: Philosophy of Science Association.

Brandon, R. (1990) *Adaptation and Environment*. Princeton, NJ: Princeton University Press.

Cummins, R. (1975) "Functional Analysis," *Journal of Philosophy* 72:741–765.

Davies, P. S. (1994) "Troubles for Direct Proper Functions," *Nous* 28:363–381.

Dennett, D. C. (1987) "Intentional Systems in Cognitive Ethology: The 'Panglossian Paradigm' Defended," in D. C. Dennett *The Intentional Stance*, pp. 237–286. Cambridge, MA: The MIT Press.

Dretske, F. (1989) *Explaining Behavior: Reasons in a World of Causes*. Cambridge, MA: The MIT Press.

Enç B. and Adams, F. (1992) "Function and Goal Directions," *Philosophy of Science* 59:635–654.

Godfrey-Smith, P. (1993) "Functions: Consensus without Unity," *Pacific Philosophical Quarterly* 74:196–208.

Godfrey-Smith, P. (1994) "A Modern History View of Functions," *Nous* 28:344–362.

Griffiths, P. E. (1993) "Functional Analysis and Proper Functions," *The British Journal for the Philosophy of Science* 44:409–422.

Hall, B. K. (ed.) (1984) *Homology*. San Diego: Academic Press.

Horan, B. (1989) "Functional Explanations in Sociobiology," *Biology and Philosophy* 4:131–158.

Hull, D. (1988) "Progress in Ideas of Progress," in M. H. Nitecki (ed.), *Evolutionary Progress*, pp. 27–48. Chicago: University of Chicago Press.

Kitcher, P. (1993) "Function and Design," *Midwest Studies in Philosophy XVIII*: 379–397.

Lewontin, R. C. (1965) "Selection in and out of Populations," in J. Moore (ed.), *Ideas in Modern Biology*, pp. 299–311. New York: Natural History Press.

Mayr, E. (1982) *The Growth of Biological Thought*. Cambridge, MA: Harvard University Press.

Millikan, R. (1984) *Language, Thought, and Other Biological Categories*. Cambridge, MA: The MIT Press.

Millikan, R. (1989) "An Ambiguity in the Notion of Function," *Biology and Philosophy* 4:172–176.

Millikan, R. (1993) *White Queen Psychology and Other Essays for Alice*. Cambridge, MA: The MIT Press.

Mitchell, S. D. (1993) "Dispositions or Etiologies? A Comment on Bigelow and Pargetter," *Journal of Philosophy* 90:249–259.

Mitchell, S. D. (1995) "Function, Fitness, and Disposition," *Biology and Philosophy* 10:39–54.

Neander, K. (1991a) "The Teleological Notion of Functions," *Australasian Journal of Philosophy* 69:454–468.

Neander, K. (1991b) "Functions as Selected Effects: The Conceptual Analyst's Defense," *Philosophy of Science* 58:168–184.

Nitecki, M. H. (ed.) (1988) *Evolutionary Progress*. Chicago: University of Chicago Press.

Nordmann, A. (1990) "Persistent Propensities: Portrait of a Familiar Controversy," *Biology and Philosophy* 5:379–399.

Patterson, C. (1982) "Morphological Characters and Homology," in K. A. Joysey and A. E. Friday (eds.), *Problems of Phylogenetic Reconstruction*. London: Academic Press.

Rosenberg, A. (1985) *The Structure of Biological Science*. New York: Cambridge University Press.

Ruse, M. (1971) "Function Statements in Biology," *Philosophy of Science* 38:87–95.

Sober, E. (1993) *Philosophy of Biology*. Boulder, CO: Westview Press.

Schaffner, K. F. (1993) *Discovery and Explanation in Biology and Medicine*. Chicago: University of Chicago Press.

Staddon, J. E. R. (1987) "Optimality Theory and Behavior," in J. Dupre (ed.), *The Latest on the Best: Essays on Evolution and Optimality*. Cambridge, MA: The MIT Press.

Stebbins. G. L. (1968) "Integration of Development and Evolutionary Progress," in R. Lewontin (ed.), *Population Biology and Evolution*, pp. 17–36. Syracuse: Syracuse University Press.

Suppe, F. (1989) *The Semantic Conception of Theories and Scientific Realism*. Chicago: University of Illinois Press.

Tinbergen, N. (1963) "On the Aims and Methods of Ethology," *Zeitschrift für Tierpsychologie* 20:410–433.

van Fraassen, B. (1980) *The Scientific Image*. Princeton, NJ: Princeton University Press.

Walsh, D. M. (forthcoming) "Fitness and Function," *The British Journal for Philosophy of Science*.

Walsh, D. M. and Ariew, A. (1996) "A Taxonomy of Functions," *The Canadian Journal of Philosophy*. 25:493–515.

Williams, G. C. (1966) *Adaptation and Natural Selection*. Princeton: Princeton University Press.

Wimsatt, W. (1972) "Teleology and the Logical Structure of Function Statements," *Studies in the History and Philosophy of Science* 3:1–80.

Wright, L. (1973) "Functions," *Philosophical Review* 82:139–168.

Wright, L. (1976) *Teleological Explanation*. Berkeley: University of California Press.

II EVOLUTIONARY PSYCHOLOGY

3 Evolutionary Psychology: Ultimate Explanations and Panglossian Predictions

Todd Grantham and Shaun Nichols[1]

In the wake of the strident controversy over sociobiology (see e.g., Caplan 1978), a band of social scientists have recently developed a new approach to understanding human behavior in evolutionary terms. Led by Leda Cosmides and John Tooby, these evolutionary psychologists have put forward a framework to unify our understanding of human behavior and culture. This approach has already attracted significant scholarly attention—much of it sharply critical (e.g., Barkow et al. 1992; Sterelny 1995; Davies et al. 1995; Griffiths 1997; Davies 1996). While we too will criticize Cosmides and Tooby's ambitious conception of evolutionary psychology, we maintain that part of their approach offers a promising research agenda.

Cosmides, Tooby, and other evolutionary psychologists have proposed two rather different projects, one focused on explanation and the other on prediction. After explaining and delineating these projects in section 1, we devote the majority of the paper to a careful analysis of the arguments offered in support of each project. We maintain that while evolutionary psychologists have compelling arguments to support the explanatory project (section 2), the arguments for the predictive project collapse into Panglossian adaptationism. In particular, we argue that Cosmides and Tooby's defense of the predictive project overestimates the precision of evolutionary predictions and underestimates the precision of description already available to us (section 3). In the final section, we briefly consider the relation between explanation and prediction in evolutionary biology.

1 The Two Projects of Evolutionary Psychology

The main claim of evolutionary psychology is that the human mind is a set of cognitive mechanisms that are adaptations to the environment of the Pleistocene. This claim emerges from two convictions. First, evolutionary psychologists maintain that the mind is largely a set of cognitive mechanisms or "organs" that are devoted to solving specific problems. The second conviction is a commitment to adaptationism, the view that natural selection is the only important cause of most phenotypic traits (Sober 1993, p. 122). The combined force of these claims leads evolutionary psychologists to maintain that natural selection is the only important cause of most mental organs. According to Cosmides and Tooby, mental organs have been shaped by natural selection in just the way bodily organs have. They claim that "cognitive psychologists, like physiologists, are usually studying adaptations and their effects, and they can find a productive new analytic tool in a carefully reasoned adaptationist approach" (Cosmides and Tooby 1994, p. 86). If we grant that the mind is a collection of organs, then, just as we suppose that kidneys are well adapted, so should we assume that our mental organs are adaptations.[2]

Barkow, Cosmides, and Tooby distinguish two quite different projects that emerge from the view that the mind is a collection of adaptations. The first project aims to provide "ultimate" evolutionary explanations (*sensu* Mayr) of cognitive mechanisms that are independently known to exist. In this project, they claim that researchers

start with a known phenotypic phenomenon, such as pregnancy sickness, language, or color vision, and try to understand what its adaptive function was—why that design was selected for rather than alternative ones. To do this, one must show that it is well designed for solving a specific adaptive problem, and that it is not more parsimoniously explained as a by-product of a design that evolved to solve some other problem.   (Barkow et al. 1992, pp. 9–10; see also Tooby and Cosmides 1992, p. 76)

The second project is to predict the existence and structure of currently unrecognized cognitive mechanisms. Barkow, Cosmides, and Tooby write, "One can use theories of adaptive function to help one discover psychological mechanisms that were previously unknown" (p. 10). Cosmides (1989) contrasts the predictive project with the "speculative" or explanatory project:

In a speculative approach, one first discovers a psychological mechanism, and then one speculates about what adaptive problem it evolved to solve. The approach advocated here is the reverse: first, one uses existing and validated theories from evolutionary biology to define an adaptive problem that the human mind must be able to solve, and to deduce what properties a psychological mechanism capable of solving that problem must have. Then one tests to see whether there is evidence for a psychological mechanism having the hypothesized properties. It is a constrained and predictive approach, rather than a compilation of post hoc explanations for known phenomena   (p. 190n2)

Although this simple contrast between the explanatory and predictive projects does not constitute an exhaustive classification of all forms of evolutionary psychology, it is a useful distinction. In our view, if we look to the supporting arguments offered by evolutionary psychologists, these two projects have markedly different prospects.

2   The Explanatory Project

Perhaps the easiest way to explicate the explanatory project is to consider an example. Cognitive psychologists are currently exploring the hypothesis that the human brain contains a distinct "theory of mind" module (Leslie and Thiass 1992; Baron-

Cohen 1995). Imagine that further research confirms this hypothesis: cognitive psychologists provide evidence that "theory of mind" is a distinct module with a complex internal structure of domain-specific inference mechanisms and representations. Once such a structure has been documented, evolutionary biology might be able to explain the origin of this "mental organ," as well as some of its properties. The attempt to provide such evolutionary explanations of the origin of cognitive mechanisms is the core of the explanatory project.

Although this project is fundamentally explanatory, the attempt to give an ultimate explanation for a cognitive mechanism can serve to generate important experiments. Barkow, Cosmides, and Tooby note that

asking functional questions and placing the phenomenon in a functional context often prompts important new insights about its organization, opening up new lines of investigation and bringing to light previously unobserved aspects and dimensions of the phenomenon.   (1992, p. 10)

Consider, for example, Wilson and Daly's (1992) claim that male sexual jealousy is actually an adaptive solution to the problem of cuckoldry avoidance. Because human males provide significant parental investment but cannot determine paternity, they run the risk of being "cuckolded" into devoting significant energy toward the care of other males' offspring. Wilson and Daly postulate that sexual jealousy helps the male to avoid cuckoldry by making him pay careful attention to any potential infidelity on the part of his mate. Tooby and Cosmides treat this as an example of the explanatory project because it provides an evolutionary explanation of a trait (jealousy) which is already known to exist (Tooby and Cosmides 1992, p. 76). This explanation can, however, issue in new predictions about jealousy that can confirm the evolutionary explanation. For example, David Buss hypothesized that since males and females face different adaptive problems (males face paternity mistakes; females face loss of paternal resources), they should be sensitive to different aspects of infidelity: men should be concerned about sexual infidelity, whereas women should focus on emotional attachment. As it happens, Buss found that males showed greater physiological response when imagining their partner having sex with someone else than when imagining their partner developing an emotional relationship with someone else. Women showed the opposite pattern (Buss 1995, Buunk et al. 1996). Testing the proposed adaptation explanations in this way is a crucial aspect of the explanatory project.

By presupposing that mental organs are adaptations, the explanatory project opens itself to the objection that it is Panglossian. In the remainder of this section, we explain the problem of Panglossianism and argue that the explanatory project need not depend on a Panglossian form of adaptationism.

2.1 Panglossianism

Biologists have often assumed that all (or at least most) traits are adaptations. This assumption—which Gould and Lewontin (1978) call "adaptationism"—has become the focus of considerable debate within evolutionary biology and philosophy of biology (see, e.g., Dupre 1987; Orzack and Sober 1994; Brandon and Rausher 1996). Gould and Lewontin argue that it is naively optimistic or "Panglossian" to assume that each trait is an optimal solution to an adaptive problem. The problem is that many nonselectionist mechanisms have important bearing on evolution. Consider the following examples. (1) The way a trait is genetically encoded can prevent the evolution of an "optimal" solution (as in the case of heterozygote superiority). (2) When two traits are developmentally linked, some combinations of traits—including optimal phenotypes—may not be possible. (3) As Gould and Lewontin (1979) point out, some traits are not themselves the objects of selection. Because the chin was not directly selected to perform any function, it is not an adaptation. Instead, it is the effect of selection on two independent growth fields. Applying a similar argument in the realm of human behavior, Kitcher argues that "it is highly unlikely that a propensity to incest-avoidance is a trait that has been shaped or maintained by natural selection" (1990, p. 109) because incest-avoidance probably arises from a complex interaction of more fundamental dispositions which are directly shaped by selection. (4) If natural selection is thought of as a force driving a ball (a population) up fitness peaks in an adaptive landscape, it is clear that a population can become stranded on a "local optimum" and be unable to evolve toward higher peaks of greater adaptedness.

Aside from these perfectly general worries, evolutionary accounts of human traits face an additional problem: the prevalence of cultural factors in human evolution. Several theorists have argued that cultural evolution can override biological evolution (Boyd and Richerson 1985; Cavalli-Sforza and Feldman 1981). For example, the power of the cultural ideal of a small family seems to have overcome biological pressure to produce more offspring (see Sober 1991). Furthermore, cultural practices can alter the biological-fitness reward for certain kinds of behavior. If men who display sexual jealousy were to be so severely ostracized that they would not receive the group support necessary to sustain life, then the "cultural environment" would significantly alter the *biological* fitness of jealous males. While evolutionists continue to disagree about both the definition and viability of the adaptationist research program, most would agree that an adaptive explanation of a trait must avoid Panglossianism to be convincing. Thus, the explanatory project of evolutionary psychology—like any area of evolutionary biology that aims to provide adaptive explanations for traits—must show that it does not depend on Panglossian assumptions. Can the explanatory project meet this burden?

2.2 The Complexity Defense

Adaptationism in evolutionary biology has been defended on the grounds that natural selection is the only process that can explain the evolution of complex physiological traits. A familiar example comes from Maynard-Smith (1978). Sharks have several globular structures (the ampullae of Lorenzini) that are located throughout the head region and are associated with the nervous system. For years after the discovery of these structures, biologists didn't know what they were for. But the complexity of the structures was so apparent that it was assumed that the ampullae must have a biological function. Biologists eventually found that the ampullae functioned to aid sharks in finding prey in the mud. (See also Dawkins 1986.)

Pinker and Bloom exploit the analogy between mental organs and bodily organs to extend this argument to evolutionary psychology. Just as selection is the only explanation of complexity in physiological organs, the claim is, it is the only explanation of complexity in mental organs: "The key point that blunts the Gould and Lewontin critique of adaptationism is that natural selection is the only scientific explanation of adaptive complexity" (Pinker and Bloom 1990, p. 709). This solution is also endorsed by Tooby and Cosmides, who appeal to the compelling example of the eye as a paradigm of complex functional design. They write:

It would be a coincidence of miraculous degree if a series of ... function-blind events, brought about by drift, by-products, hitchhiking and so on, just happened to throw together a structure as complexly and interdependently functional as an eye (Dawkins 1986; Pinker and Bloom 1990). For this reason, nonselectionist mechanisms of evolutionary change cannot be seen as providing any reasonable alternative explanation for the eye or for any other complex adaptation. Complex functional organization is the signature of selection.   (1992, p. 57)

Ultimately, we believe that this "complexity defense" deflects adaptationist worries in the context of the explanatory project. If we have confirmed the existence of a module with a "complex functional organization," we can be confident that natural selection was an important force creating that structure. Nonetheless, we have a couple of concerns about the argument.

First, some biologists have challenged the central premise of this argument by suggesting that complexity can arise without natural selection. Kauffmann (1995), for example, has presented mathematical models intended to show that some kinds of complexity can increase in biological systems even in the absence of selection. Similarly, Page and Mitchell (1990) argue that some of the complexity of honey-bee colonies may be the result of the dynamics of complex systems rather than natural selection. At present, we have mere suggestions of how such evolution might proceed. If further research demonstrates that nonselective forces can create functional complexity, this would strongly undermine the complexity defense.

In addition, even if one assumes that (at least some) cognitive mechanisms have a degree of complexity that requires a selective explanation, that only shows that natural selection must have been *an* important causal factor in the creation of this trait. It does not show that natural selection can (by itself) explain all aspects of the trait. Even if natural selection is a necessary part of the evolutionary explanation, nonselective factors may also play a crucial role in explaining some features of the trait (Sober 1993).

It's important to note that these problems are not limited to evolutionary psychology. From a theoretical standpoint, these problems with the complexity argument affect evolutionary explanations of both morphological and psychological traits. And if evolutionary psychology shares its fate with evolutionary physiology, one can hardly expect more. There is, however, a further concern about how to characterize the relevant notion of complexity and this concern seems especially troubling for evolutionary psychology. Pinker and Bloom assert that "natural selection is the only scientific explanation of adaptive complexity" (1990, p. 709), but surely no one would deny that "adaptive complexity" is adaptive. For the complexity principle to be substantive, we must be able to identify complex organization independently of any knowledge of selective history. That is, there must be some way of identifying complexity without identifying it as adaptive.

Pinker and Bloom clarify their notion of adaptive complexity by stipulating that " '[a]daptive complexity' describes any system composed of many interacting parts where the details of the parts' structure and arrangement suggest design to fulfill some function" (1990, p. 709).[3] One crucial feature of adaptive complexity, then, is the presence of "interacting parts." Indeed, what makes the eye such a forceful example of adaptive complexity is largely the fact that it is made of a number of functionally discrete parts (lens, muscles, iris, rods, cones, etc.) that interact smoothly as parts of the visual system. Similarly, then, to demonstrate the relevant complexity of a cognitive capacity, we need to demonstrate that the capacity involves the interaction of functionally discrete submechanisms. But of course, this just raises a difficult question: How do we discern functionally discrete cognitive submechanisms? This question hasn't received adequate attention from evolutionary psychologists, as is evidenced by some of their arguments.[4] Nonetheless, cognitive science might provide a way to demonstrate the relevant kind of cognitive complexity. One important source of evidence that a cognitive mechanism is functionally discrete comes from evidence that the mechanism is "informationally encapsulated" (Fodor 1983; Pylyshyn 1984). A mechanism is informationally encapsulated if information "outside" the mechanism doesn't affect the internal processing. To take the standard example, the knowledge that the Muller-Lyer lines are of the same length doesn't change the visual effect that one line appears longer than the other. This indicates that (at least some) visual processing is informationally encapsulated. Another source of evidence for functionally discrete cognitive submechanisms might come from psychopathology—if we find that a cognitive capacity can be selectively damaged, this might indicate that there is a corresponding discrete submechanism.

If we have accurately characterized the operative notion of complexity for the complexity argument, providing evidence that a mental organ has this kind of complexity is extremely demanding. Nevertheless, we *do* have reason to think that the capacity for understanding language has the relevant kind of complexity. Psycholinguists maintain that language understanding involves phoneme recognition, phonological analysis, lexical analysis, and parsing, among other subprocessors (see, e.g., Caplan 1992). Further, we have reason to think that at least some of these subprocesses are functionally discrete. For instance, Swinney found that semantic lexical analysis is informationally encapsulated (Swinney 1979). And syntactic comprehension is selectively damaged in Broca's aphasics (Schwartz et al. 1980). These hypothesized submechanisms must interact smoothly and quickly to explain the speed of language understanding. These sorts of considerations make the language faculty a plausible place to launch a complexity argument. For in the language faculty, we do seem to have a set of smoothly interacting, functionally discrete subprocessors.

Thus, despite the above concerns, we have no serious doubts that some cognitive mechanisms (e.g., the language and perceptual faculties) are complex in the relevant sense, and we believe that the existence of complex mechanisms demands an adaptation explanation. This extension of the complexity defense into evolutionary psychology seems perfectly reasonable. Still, if the eye provides the paradigm, then there are as yet precious few cognitive mechanisms that have been shown to have the requisite kind of complexity. Given our current knowledge of how the mind works, the complexity argument has a rather limited range for evolutionary psychology. Thus, insofar as the explanatory project hangs on the complexity defense, evolutionary psychologists will need to wait until cognitive scientists provide us with a much more detailed picture of the functionally discrete cognitive submechanisms that make up the mind.

In light of the foregoing, if Pinker and Bloom are right that the complexity argument is the "key point that blunts" the Panglossian threat, then it seems that the current reach of the explanatory project is extremely limited. However, focusing solely on the complexity defense needlessly restricts the explanatory potential of evolutionary psychology. The revival of comparative methods (see, e.g., Harvey and Pagel 1991) provides a second means of rebutting charges of Panglossianism. Of course, the comparative method is unavailable to evolutionary psychologists interested in cognitive capacities that are presumed to be distinctively human, like theory of mind (e.g., Cosmides and Tooby 1994) and metarepresentation (Sperber 1994). For such putatively species-specific capacities, complexity really is the key defense. As a result, for distinctively human capacities, evolutionary psychologists need to wait for cognitive science to provide the details necessary to mount an argument from complexity. However, there are interesting and important evolutionary questions to ask about psychological capacities that we might share with other species. These questions can be approached by evolutionary psychologists using the comparative methodology of looking at closely related species in different niches and looking at quite distantly related species in similar niches. Indeed, one of the most celebrated cases in evolutionary psychology, male sexual jealousy (Wilson and Daly 1992), provides a good example of this approach. There is virtually no evidence that human males have a sexual jealousy module that is composed of several functionally discrete interacting parts. Rather, the best evidence for Wilson and Daly's claim that male sexual jealously is an adaptation comes from (1) comparative data about males in relevantly similar species, and (2) data on the differences between males and females in our own species (Buunk et al. 1996).

3   The Predictive Project

In addition to their promotion of the explanatory project, Cosmides and Tooby provide a spirited defense of the predictive project, which they describe as follows:

> ... evolutionary theory ... allows one the pinpoint the kinds of problems the human mind was "designed" to solve, and consequently should be very good at solving. And although it cannot tell one the exact structure of the cognitive programs that solve these problems, it can suggest what design features they are likely to have. It allows one to develop a "computational theory" for that problem domain: a theory specifying what functional characteristics a mechanism capable of solving that problem must have.   (1987, p. 285)

The idea is that by looking at the adaptive problems that our ancestors faced, we can "predict" what sorts of cognitive mechanisms have evolved.

Tooby and Cosmides present Hamilton's rule as a paradigm of the predictive project. Hamilton's rule (Hamilton 1964) states that an individual can maximize his inclusive fitness by helping his relatives when and only when the cost to himself is less than the benefit to the relative, multiplied by the "coefficient of relatedness." In mammals and birds, the coefficient of relatedness is 0.5 for offspring and siblings, 0.25 for cousins and nieces. So, according to Hamilton's rule, an individual bird should help its offspring when the cost to the parent is less than half the benefit to the offspring. Tooby and Cosmides claim that the example of Hamilton's rule

shows how knowledge drawn from evolutionary biology can be used to discover functional organization in our psychological architecture that was previously unknown. Hamilton's rule is not intuitively obvious; no one would look for psychological mechanisms that are well designed for producing behavior that conforms to this rule unless they had already heard of it. (1992, p. 68)

As Tooby and Cosmides note, behavioral ecologists guided by Hamilton's rule found that the rule captured behavior in some nonhuman animals (Krebs and Davies 1984).

It's quite possible that reflections on evolutionary biology will lead to the discovery of unexpected cognitive mechanisms in humans. Cosmides and Tooby want to make a more radical claim—that evolutionary biology will predict a wealth of cognitive mechanisms and that this research program will revolutionize cognitive psychology (Cosmides and Tooby 1994, p. 85). It's a fool's errand to guess the etiology of future discoveries in science, but we maintain that Cosmides and Tooby don't have any good *arguments* to support their claim that reflecting on adaptive problems will lead to the discovery of an abundance of unexpected cognitive mechanisms. Indeed, we think that their arguments depend crucially on Panglossian assumptions.

Cosmides and Tooby's principal argument for the predictive project seems, at first glance, to be a causal one: "Because the enduring structure of ancestral environments *caused* the design of psychological adaptations, the careful empirical investigation of the structure of environments from a perspective that focuses on adaptive problems and outcomes can provide powerful guidance in the exploration of our cognitive mechanisms" (Cosmides and Tooby 1994, p. 103, our emphasis). That is, since our cognitive mechanisms are adaptations to an ancestral environment, understanding the selective pressures of that environment will help us to predict the structure of these mechanisms.

The causal argument might be stated more fully as an argument from evolutionary function. The argument seems to depend on two claims. First, Cosmides and Tooby maintain that it is extremely difficult to identify (or discover the structure of) cognitive mechanisms without some understanding of their function. Second, Cosmides and Tooby argue that because natural selection "designed" our cognitive mechanisms, analyzing the environment of the Pleistocene will provide insight into the adaptive problems our ancestors faced. This analysis of function will, they claim, provide a clearer and more detailed account of function than anything we currently possess. Cosmides and Tooby conclude that analysis of evolutionary function provides a crucial resource for the project of discovering and predicting the structure of un- expected cognitive mechanisms. In brief, since evolutionary biology can (according to Cosmides and Tooby) provide a clearer and more detailed account of function than anything else we currently possess, and since an analysis of function is very helpful to cognitive psychology, evolutionary biology will provide powerful guidance.

We maintain that Cosmides and Tooby's defense of the predictive project depends on understating the promise of *nonevolutionary* cognitive psychology and exaggerat- ing the specificity of evolutionary predictions. In the rest of this section, we'll argue that Cosmides and Tooby rely on an untenably strong adaptationism, and that when the adaptationism is appropriately tempered, it's unlikely that the analyses of func- tion provided by evolutionary biology will be significantly better than those we can glean without the help of evolutionary biology, for example, from common sense and cognitive science.

3.1   Cognitive Science and Evolutionary Function

Cosmides and Tooby's first claim is that cognitive psychology is really at sea without evolutionary biology. They write:

It is nearly impossible to discover how a psychological mechanism processes information unless one knows what its function is, what it was "designed" or selected to do. Trying to map out a cognitive program without knowing what its function is, is like attempting to understand a computer program by examining it in machine language, without knowing whether it is for editing text, accounting, or launching the Space Shuttle.... It is far easier to understand the architecture of a "black box" if one knows what it was designed to do....
   This is exactly the question that evolutionary theory allows one to address.   (1987, p. 285; see also Barkow et al. 1992, p. 11.)

More recently, Cosmides and Tooby have written:

Suppose you have to figure out how an appliance works by studying some of the things it *can* do. I tell you that it can be used as a paperweight, that you can use it to warm your hands on a cold day, and that you can kill someone who is taking a bath by throwing it into the tub with him. By studying each of these uses of the appliance, you will learn a little bit about its structure ... but you won't get a very coherent idea of what it is or how it works.
   Suppose on the other hand, that I tell you that the appliance is a mechanism that was designed to toast slices of bread—it is a "toaster." Your research strategy for discovering how it works would be completely different.   (1994, p. 95)

These analogies tend to ridicule nonevolutionary characterizations of cognitive mechanisms. But if the implication of this is supposed to be that cognitive psychology is virtually impossible without evolutionary biology, then this claim must be regarded as simple hyperbole even by the lights of Cosmides and Tooby. For Cosmides and Tooby (1994) approvingly cite Chomsky, Marr, and Leslie as having discovered innate cognitive mechanisms for language, vision, and theory of mind. But the accomplishments of Chomsky, Marr, and Leslie did *not* develop through reflection on the precise adaptive problems our ancestors faced. Indeed, Chomsky has been positively hostile to evolutionary considerations (e.g., 1988). We concede that some understanding of the function of language, vision, and theory of mind is crucial to characterizing the cognitive mechanism, but we deny that cognitive scientists require a specifically *evolutionary* account of the function of language, vision, or theory of mind. Our mundane, intuitive understanding of the function of these capacities was sufficient to guide cognitive science in these early successes. This highlights a subtle equivocation in Cosmides and Tooby's characterization of cognitive science and evolutionary function. While it is true that cognitive psychologists must have some idea of the function of a mechanism in order to develop a computational theory of its structure, there's no reason to think that cognitive scientists must have a specifically evolutionary account of the function of a mechanism. That is, there's no reason to think that cognitive scientists must rely on evolutionary considerations to determine the function of a mechanism.[5]

3.2 Ultra-adaptationism

Cosmides and Tooby's second claim is that evolutionary analyses provide novel predictions about which cognitive mechanisms can be expected to exist. Cosmides and Tooby's defense of this claim depends not only on traditional adaptationism—which might be a reasonable research strategy (Sober 1993)—but on a more ambitious doctrine we'll call "ultra-adaptationism." To fully understand this new departure, it is helpful to review traditional adaptationism.

In the traditional adaptationist research program, one begins with a limited understanding of a trait and its function. The researcher seeks to refine her understanding of the trait (and its evolution) by performing an optimality analysis. Models of optimal foraging, for example, try to determine which search strategy will maximize nutritional gain based on assumptions about the nutritional value of various food items, the spatial distribution of food items, and the energetic costs of the various search strategies. The fact that an organism follows a (nearly) optimal search strategy provides evidence that the search strategy is an adaptation.

In traditional optimality models, one knows (1) that a phenotypic trait (e.g., foraging behavior) exists and (2) that it has likely been shaped by selection. But in the ultra-adaptationist predictive project, we do not know that a phenotypic trait (cognitive mechanism) exists. Consider the following from Tooby and Cosmides (1995): *"Detailed theories of adaptive function can tell cognitive scientists what modules are likely to exist*, what adaptive information-processing problems they must be capable of solving, and—since form follows function—what kind of design features they can therefore be expected to have" (pp. xv–xvi; emphasis ours). The idea seems to be that evolutionary biology will guide psychologists' search for modules by telling the psychologist which adaptive problems must be addressed. Tooby and Cosmides provide a remarkable list of such adaptive problems:

Over the course of their evolution, humans regularly needed to recognize objects, avoid predators, avoid incest, avoid teratogens when pregnant, repair nutritional deficiencies by dietary modification, judge distance, identify plant foods, capture animals, acquire grammar, attend to alarm cries, detect when their children needed assistance, be motivated to make that assistance, avoid contagious disease, acquire a lexicon, be motivated to nurse, select conspecifics as mates, select mates of the opposite sex, select mates of high reproductive value, induce potential mates to choose them, choose productive activities, balance when walking, avoid being bitten by venomous snakes, understand and make tools, avoid needlessly enraging others, interpret social situations correctly, help relatives, decide which foraging efforts have repaid the energy expenditure, perform anticipatory motion computation, inhibit one's mates from conceiving children by another, deter aggression, maintain friendships, navigate, recognize faces, recognize emotions, cooperate, and make effective trade-offs among many of these activities, along with a host of other tasks.   (1992, p. 110)

Cosmides and Tooby are claiming that if we simply look to the adaptive problems that our ancestors faced, we will be able to generate predictions about the mechanisms that must exist to solve those problems. In our view, Cosmides and Tooby's ultra-adaptationism is Panglossian in several respects.

Not All Adaptive Problems Are Solved.   It's unlikely that every adaptive problem has been (or even could be) solved. Once a solution begins emerging, it will be shaped by natural selection (assuming it's heritable and variable); but the mere presence of an adaptive problem does not guarantee a solution. Many of the adaptive problems that Tooby and Cosmides claim *must* be solved have this characteristic: solving the problem would enhance fitness, but the creature might nevertheless survive and reproduce without solving the problem. For example, at least some groups of humans might have survived and reproduced without having been able to capture animals, acquire grammar, understand and make tools, or help relatives.

It is true that *some* adaptive problems must be solved or else the organism will not be able to survive or reproduce. For instance, an organism must be able to find food and find a mate. However, the range of possible solutions for these problems is so great that merely saying that our species had to solve these problems will be of little use to cognitive psychologists.

Impediments to Optimality.   As we noted in section 2.1, the principal problem for adaptationism is that many forces other than natural selection are known to affect the course of evolution, thereby preventing the achievement of an "optimal" pheno-type. Genetic, developmental, and cultural constraints are some of the important obstacles to achieving an optimal solution to an adaptive problem. As we noted, evolutionary biologists think that this problem can be met by the complexity de-fense, and Pinker and Bloom (1990) make a plausible extension of this defense into the domain of evolutionary psychology. But it's crucial to note that this defense has merit for the explanatory project only when we know that the trait under consider-ation is sufficiently complex to require an adaptation explanation. In the context of the predictive project, the complexity defense is unavailable since we have not (yet) identified a complex cognitive mechanism.

If we were justified in expecting optimal solutions to all adaptive problems, then this might carve out a small range of possible mechanisms which approximate an optimal solution. But if we accept that many factors influence the course of evolution—often blocking the evolution of optimal solutions to adaptive problems—then the range of possible cognitive mechanisms increases significantly. That is, as one softens the optimality assumption to a reasonable level, recognizing the variety of nonselective forces that may have affected the evolution of cognitive mechanisms, the range of near optimal solutions expands dramatically. As a result, evolutionary considera-tions become much less powerful in predicting the structure of previously unknown cognitive mechanisms.

Multiple Realizability Problems.   The above problems are compounded by the fact that each of these near-optimal behavioral solutions can be cognitively realized in a multitude of ways. As Rosenberg (1989) has pointed out, selection for function is blind to structure. That is, if two underlying structures (e.g., genes, proteins, cogni-tive mechanisms) produce the same outward structure or produce the same net effect on fitness, then selection cannot distinguish between them. This is essentially a prob-lem of multiple realization—the same function can be carried out by multiple lower-level mechanisms. And to the extent that this is true, evolutionary biology will not be a powerful guide in determining which of these structures is realized in human brains. Cosmides and Tooby concede that evolutionary biology doesn't predict the precise structure of cognitive mechanisms (Cosmides and Tooby 1987, p. 285), but the problem may be worse than they expect.

Consider Kitcher's (1990) discussion of incest-avoidance. Kitcher suggests that the adaptive problem of incest-avoidance might be solved partly in virture of a general-purpose tolerance mechanism that would make those who are very familiar un-attractive. Presumably this proposed mechanism might really be part of a solution to the adaptive problem of incest-avoidance, yet it's unlikely that the discovery of such a mechanism would be expedited by a search for a solution to the adaptive problem of incest-avoidance. The point is simply that there are many cognitive structures that would constitute solutions to the adaptive problem of incest-avoidance.

Indeed, some adaptive problems don't even require a *cognitive* solution. For some animals, the adaptive problems of predator-avoidance and aggression-deterrence are solved by physiological features. And in humans, it's possible that the adaptive problem of attracting potential was solved partly by physiological features (see, e.g., Buss 1992).

Finally, the solutions to adaptive problems can be so utterly surprising that no one could have expected the mechanism that nature settled on. Ironically, this is a feature of natural selection that Tooby and Cosmides explicitly note. They write: "evolution can contrive subtle solutions that only a superhuman, omniscient engineer could have intentionally designed" (Tooby and Cosmides 1992, p. 61).

Ancestral Environment.   The problem of multiple mechanisms is exacerbated by the limitations of our understanding of the ancestral environment. As Sterelny has pointed out,

The Pleistocene is not a single environment ... There must have been very considerable variations in physical environment: climate, terrain and food resource. Even more important were the changes in social environment. If Corballis (1991) is right, first there was the invention of language, then somewhat later a technocultural Great Leap Forward. There may have been important changes in mating systems.... (1995, p. 377)

Furthermore, evolution often operates by "escalation." For example, as a clam develops a harder shell (to avoid predation), predators face increased selection pressure. Similarly, if selection were to improve a male's ability to determine time of ovulation, then females would face a new environment—one in which new techniques for concealing ovulation time might prove adaptive. Thus the selective environment can change significantly even when the physical environment remains constant (Brandon 1990).

The ultra-adaptationist predictive project—the attempt to predict the structure of cognitive mechanisms by studying the adaptive problems faced by our ancestors—faces several problems. Not all adaptive problems are solved. To determine which solutions are optimal, we need to know a great deal about the then-current environment—information which is often very hard to come by. Even if we can clarify our understanding of the Pleistocene environment(s) in which we evolved, our cognitive mechanisms may not be optimal solutions to those problems. Finally, the problem of multiple realizability further increases the set of cognitive mechanisms that are compatible with an analysis of the adaptive problem. Despite all these criticisms, though, we don't mean to suggest that evolutionary biology provides *no* predictions about human cognitive mechanisms. Rather, the point is that in light of the foregoing, the best we can hope for are extensively qualified, rather vague predictions. Evolutionary biology cannot provide precise and well-justified predictions about the structure of cognitive mechanisms.

3.3 Summary of Our Critique

Cosmides and Tooby defend the predictive project with the following argument: Since it is very difficult to discover the structure of a cognitive mechanism without understanding the function of the mechanism, and since evolutionary biology can provide a much more precise analysis of the function than we can glean using traditional methods in psychology, evolutionary biology is poised to make significant contributions to cognitive psychology. We are skeptical of this argument for two reasons. First, cognitive scientists have succeeded in identifying and characterizing cognitive mechanisms using nonevolutionary analyses of function (section 3.1). Second, several considerations suggest that the predictions of evolutionary psychology will be vague (section 3.2). Since there is no reason to think that the functional analyses provided by evolutionary biology are significantly clearer or more detailed than the functional understanding already available to us, Cosmides and Tooby's argument is unconvincing.

This general problem is well illustrated by considering the predictive potential of Hamilton's rule, the central example used in Tooby and Cosmides (1992).[6] Only the most roseate Panglossian would predict that humans have cognitive mechanisms that in the Pleistocene would have led us to help our kids ($c < b/2$) and our nieces ($c < b/4$). On the other hand, it's not implausible to use Hamilton's rule to predict that people should be able to distinguish kin from non-kin and that people will be motivated to help kin more than non-kin. We might even go so far as to predict that people will be motivated to help their children significantly more than their nieces.

But, of course, those are hardly useful predictions—they don't predict anything we didn't already know. Note that there is a crucial difference here between cognitive science and behavioral ecology. Evolutionary considerations might be quite predictively useful for behavioral ecologists and comparative psychologists, since they often know relatively little about the species under consideration. So even the vague predictions of a temperate adaptationism might be informative in behavioral ecology. However, since cognitive psychologists already know a considerable amount about the species under consideration, it is much more difficult for evolutionary considerations to generate useful predictions about us.

4 Concluding Reflections on Explanation and Prediction

Cosmides and Tooby's program for evolutionary psychology contains two distinct projects. Whereas we find the attempt to explain the evolutionary origin and structure of cognitive mechanisms (the explanatory project) promising, we have criticized Cosmides and Tooby's claim that studying the adaptive problems faced by our ancestors will provide powerful guidance for cognitive psychology (the predictive project). Together, these two claims seem to imply that evolutionary biology can explain but cannot predict.[7] Although this conclusion may be surprising to some philosophers of science, Cosmides and Tooby's prized analogy between cognitive psychology and physiology suggests that this is exactly what we should expect.

   Cosmides and Tooby often write as though cognitive psychology and physiology have roughly the same relation to evolutionary biology. They say, for example, that "cognitive psychologists, like physiologists, are usually studying adaptations and their effects, and they can find a productive new analytic tool in a carefully reasoned adaptationist approach" (1994, p. 86; see also Tooby and Cosmides 1992, pp. 68–69). On their view, physiology and cognitive psychology have similar goals. Just as anatomy and physiology aim to determine the structure and function of organs like the heart and pancreas, cognitive psychologists want to identify the structure of mental organs and determine how they function. In section 2 we argued that the analogy between bodily organs and mental organs has been extremely useful in deflecting objections to the explanatory project of evolutionary psychology. And evolutionary biology, has, of course, played an essential role in explaining the origin and properties of bodily organs. However, the history of human physiology provides little support for the idea that evolutionary biology guides our discovery of the structure and function of human organs. Physiologists were, as a matter of fact, able to uncover the structure and function of the major organs without appealing to evolutionary predictions. The structures and functions of the major organs (e.g., the heart, kidneys, and lungs) were understood well before Darwin published the *Origin* (see Knight 1980). Furthermore, contemporary work in comparative physiology does not seem to be powerfully guided by analyses of adaptive problems. We see no reason to think that the situation will be different in psychology. In the case of physiology as in the case of cognitive psychology, the predictions on offer from evolutionary biology are relatively vague, especially when compared to the amount of information about structure and function that can be gleaned from other traditional methods.

As is typical in our discipline, the bulk of our paper has been critical. We've argued that Cosmides and Tooby have no persuasive defense of the predictive project. The predictions provided by evolutionary biology are notoriously vague, and the traditional methods of cognitive science offer at least as much precision in characterizing the functions of mental organs. Further, the predictive successes of behavioral ecology provide no support for the predictive project; behavioral ecology isn't a relevantly analogous precedent since we typically know very little about the species under consideration in behavioral ecology. We believe that physiology provides a more telling analogy, and there, the lesson seems clear: Evolutionary biology played no significant role in guiding physiology, and we see no reason to expect that evolutionary biology will radically transform our attempt to discover the structure of psychological organs. Nonetheless, we don't mean to suggest that evolutionary biology will play no role in cognitive science. Quite the contrary: We have maintained that the explanatory project of evolutionary psychology shows considerable promise. Once we have uncovered important features of a mental organ, evolutionary psychologists can appeal to the complexity of cognitive structures and to the comparative method to develop and test adaptation explanations of cognitive capacities. That is the truly exciting prospect that evolutionary psychology offers.

Acknowledgments

We wish to acknowledge the advice of several colleagues. The philosophy department of the College of Charleston and an audience at the 1997 International Society for the History, Philosophy, and Social Studies of Biology provided helpful comments on earlier drafts of this paper. Paul Davies, Paul Griffiths, David Hull, Harold Kincaid, Greg Morgan, Eddy Nahmias, and Hugh Wilder all provided useful comments on the penultimate draft.

Notes

1. Both authors contributed equally to this paper.

2. Cosmides and Tooby (1987, 1994) have argued that important cognitive mechanisms will be universally shared by all humans. In reply, Sterelny (1995) and Griffiths (1997) have argued that this "monomorphic mind" thesis ignores the prevalence of polymorphism. For present purposes, we won't address this aspect of evolutionary psychology.

3. For this proposed definition of adaptive complexity to be substantive, "function" would have to refer to accomplishing a task (e.g., vision, sound recognition) rather than an etiological notion of function. More on this in section 3.1.

4. For instance, Pinker and Bloom catalog the complexity of the parts of *language*, and they claim that "The fact that language is a complex system of many parts … is so obvious in linguistic practice that it is usually not seen as worth mentioning" (pp. 461–462). But the fact that language is a complex system doesn't prove that the underlying cognitive mechanisms have the relevant kind of complexity. After all, we can perfectly well note the complexity of set theory without thinking that the parts of set theory will reflect divisions in cognitive mechanisms. In other words, showing the complexity of linguistic output doesn't directly show the kind of cognitive complexity that is relevantly analogous to the complexity of the eye.

5. This is not to say that the function determined by cognitive scientists won't partly converge with the functional analysis provided by evolutionary considerations.

6. Tooby and Cosmides (1992) also allude to Cosmides (1989) as a paradigm of the predictive project. Cosmides (1989) has been critically treated at length by Davies et al. (1995). But in any case, the claim that this is a paradigm of the predictive project doesn't quite live up to its billing. For Cosmides's case is embedded in a rich body of psychological knowledge about inferences over material conditionals. Indeed, Cosmides's experiments are based on previous results (Griggs and Cox 1982).

7. Our view is actually compatible with the symmetry thesis. We have argued that evolutionary biology can (in principle) provide well-grounded explanations of the origin and structure of cognitive mechanisms. If one accepts the symmetry thesis, then the explanans of any complete explanation of a cognitive mechanism must be sufficient to predict the existence of that mechanism. In our view, evolutionary explanations do make predictions—and these predictions are important for our efforts to test them. But even if the symmetry thesis is true, it is an open question whether predictions based on evolutionary considerations will be more useful in guiding cognitive psychology than other social science disciplines (e.g., anthropological observation, cognitive science, neuropsychology, etc.). Our primary claim has simply been that given the vagueness of evolutionary predictions and our rich body of background knowledge about humans, there is no compelling reason to think that evolutionary biology will provide the strong guidance that Cosmides and Tooby suggest.

References

Barkow, J., Cosmides, L., and Tooby, J., eds. (1992) *The Adapted Mind: Evolutionary Psychology and the Generation of Culture*. New York: Oxford University Press.

Baron-Cohen, S. (1995) *Mindblindness*. Cambridge, MA: The MIT Press.

Boyd, R. and Richerson, P. 1985. *Culture and the Evolutionary Process*. Chicago: University of Chicago Press.

Brandon, R. N. (1981) "Biological Teology: Questions and Explanations." In P. Asquith and T. Nicles (eds.), *PSA 1980*, volume 2, pp. 315–323. East Langsing, MI: Philosophy of Science Association.

Brandon, R. (1990) *Adaptation and Evironment*. Princeton, NJ: Princeton University Press.

Brandon, R. and Rausher, M. (1996) "Testing and Adaptationism: A Comment on Orzack and Sober," *American Naturalist* 148:18–201.

Buss, D. (1992) "Mate Preference Mechanisms: Consequences for Partner Choice and Intrasexual Competition," in Barkow et al. (eds.), op. cit. pp. 249–266.

Buss. D. (1995) "Psychological Sex Differences: Origins Through Sexual Selection," *The American Psychologist* 50: 164–169.

Buunk, B., Angleitner, A., Oubaid, V., and Buss, D. (1996) "Sex Differences in Jealousy in Evolutionary and Cultural Perspective," *Psychological Science* 7:359–363.

Caplan, A. (1978) *The Sociobiology Debate*. New York, NY: Harper and Row.

Caplan, D. (1992) *Language*. Cambridge, MA: The MIT Press.

Cavalli-Sforza, L. and Feldman, M. (1981) *Cultural Transmission and Evolution: A Quantitative Approach*. Princeton: Princeton University Press.

Chomsky, N. (1988) *Language and Problems of Knowledge: The Managua Lectures*. Cambridge, MA: The MIT Press.

Cosmides, L. (1989) "The Logic of Social Exchange: Has Natural Selection Shaped How Humans Reason? Studies with the Wason Selection Task," *Cognition* 31:187–276.

Cosmides, L. and Tooby, J. (1987) "From Evolution to Behavior: Evolutionary Psychology as the Missing Link," in J. Dupre, op. cit.

Cosmides, L. and Tooby, J. (1994) "Origins of Domain Specificity: The Evolution of Functional Organization," in L. Hirschfeld and S. Gelman (eds.), *Mapping the Mind*. Cambridge: Cambridge University Press.

Daly, M., Wilson, M. and Weghorst, S. (1982) "Male Sexual Jealousy," *Ethology and Sociobiology* 3:11–27.

Davies, P. (1996) "Discovering the Functional Mesh: On the Methods of Evolutionary Psychology," *Minds and Machines* 6:559–1585.

Davies, P., Fetzer, J., and Foster, T. (1995) "Logical Reasoning and Domain Specificity: A Critique of the Social Exchange Theory of Reasoning," *Biology and Philosophy* 10:1–38.

Dawkins, R. 1986. *The Blind Watchmaker*. New York: Norton.

Dupre, J. (ed.) 1987. *The Latest on the Best: Essays on Evolution and Optimality*. Cambridge, MA: The MIT Press.

Fodor, J. (1983) *The Modularity of Mind*. Cambridge, MA: The MIT Press.

Gould, S. and Lewontin, R. (1978) "The Spandrels of San Marco and the Panglossian Paradigm: A Critique of the Adaptationist Programme," *Proc. Royal Soc. London* 205:581–598.

Griffiths, P. (1997) *What Emotions Really Are: The Problem of Psychological Categories*. Chicago, IL: Chicago University Press.

Griggs, R. and Cox, J. (1982) "The Elusive Thematic-Materials Effect in Wason's Selection Task," *British Journal of Psychology* 73:407–420.

Hamilton, W. (1964) "The Genetical Theory of Social Behavior," *Journal of Theoretical Biology* 7:1–52.

Harvey, P. and Pagel, M. (1991) *The Comparative Method in Evolutionary Biology*. Oxford: Oxford University Press.

Kauffman, S. (1995) *At Home in The Universe: The Search for Laws of Self-Organization and Complexity*. New York: Oxford University Press.

Kitcher, P. (1990) "Developmental Decomposition and the Future of Human Behavioral Ecology," *Philosophy of Science* 57:96–117.

Knight, B. (1980) *Discovering the Human Body*. New York, NY: Lippincott and Crowell.

Krebs, J. and Davies, N. (1984) *Behavioral Ecology*, 2nd ed. Oxford: Blackwell.

Leslie, A. and Thaiss, L. (1992) "Domain Specificity in Conceptual Development: Neuropsychological Evidence from Autism," *Cognition* 43:225–251.

Maynard Smith, J. (1978) "Optimization Theory in Evolution," *Ann. Rev. Ecol. Syst.* 9:31–56.

Orzack, S. and Sober, E. (1994) "Optimality Models and the Test of Adaptationism," *American Naturalist* 143:361–380.

Page, R. and Mitchell, S. (1990) "Self-Organization and Adaptation in Insect Societies," in A. Fine et al. (eds.), *PSA 1990*, vol. 2, pp. 289–298. East Lansing, MI: Philosophy of Science Assn.

Pinker, S. and Bloom, A. (1990) "Natural Language and Natural Selection," *Behavioral and Brain Sciences* 13:707–784. Reprinted in Barkow et al., op. cit. Page references to Barkow et al.

Pylyshyn, Z. (1984) *Computation and Cognition*. Cambridge, MA: The MIT Press.

Rosenberg, A. (1985) *The Structure of Biological Science*. New York: Cambridge University Press.

Rosenberg, A. (1989) "From Reductionism to Instrumentalism," in M. Ruse (ed.), *What the Philosophy of Biology Is*, pp. 245–262. Dordrecht: Kluwer.

Schwartz, M., Saffran, E., and Marin, O. (1980) "The Word Order Problem in Agrammatism," *Brain and Language* 10:249–262.

Sober, E. 1991. "Models of Cultural Evolution." Reprinted in E. Sober (ed.), *Conceptual Issues in Evolutionary Biology*, pp. 477–492. Cambridge, MA: The MIT Press.

Sober, E. (1993) *Philosophy of Biology*. Boulder, CO: Westview.

Sperber, D. (1994) "The Modularity of Thought and the Epidemiology of Representations," in L. Hirschfeld and S. Gelman (eds.), *Mapping the Mind*. Cambridge: Cambridge University Press.

Sterelny, K. (1995) "The Adapted Mind," *Biology and Philosophy* 10:365–380.

Swinney, D. (1979) "Lexical Access during Sentence Comprehension: Reconsideration of Context Effects," *Journal of Verbal Learning and Verbal Behavior* 18:645–660.

Tooby, J. and Cosmides, L. (1992) "The Psychological Foundations of Culture," in Barkow et al., op. cit., pp. 19–136.

Tooby, J. and Cosmides, L. (1995) Foreword to S. Baron-Cohen (1995).

Wilson, M. and Daly, M. (1992) "The Man Who Mistook His Wife for a Chattel," in Barkow et al., op. cit., pp. 289–322.

4 The Conflict of Evolutionary Psychology

Paul Sheldon Davies

Recent work in so-called evolutionary psychology is conflicted. The conflict, which is neither on the surface nor all that deep, arises from the truth of the following claims. On the one hand, some research in evolutionary psychology appears promising; some hypotheses concerning the evolutionary origins of our psychological capacities appear coherent and subject to testing. On the other hand, the methods of inquiry distinct to evolutionary psychology are unpromising. As I have argued elsewhere (Davies 1996), it is practically impossible to discover the evolutionary origins of our cognitive capacities without first knowing the correct architecture of those capacities, and thus the aims of our psychological inquires must be accomplished before the evolutionary inquiries commence. How is it, then, that a form of inquiry, the methods of which are demonstrably barren, nevertheless gives the appearance of bearing theoretical fruit? This is the conflict within evolutionary psychology.

The aim of this chapter is to explicate this conflict and trace its source. I first defend the claims that generate the conflict—that some hypotheses in evolutionary psychology appear promising and that the methods distinctive to evolutionary psychology are not promising. I then defend my central thesis. I shall argue that the source of the above conflict results from a conflation of two quite distinct research projects, accompanied by the false belief that there is a single, unified project under-way. One research project is evolutionary history; the other is cognitive psychology. Once these projects and their respective domains are disentangled it is clear, I claim, that the apparent plausibility of some recent work derives not from applying the methods of evolutionary psychology but rather from engaging in a bit of evolutionary history. Its apparent plausibility derives from the fact that this work falls within the scope of evolutionary history, not evolutionary psychology. The conflict, then, arises from the fact that evolutionary psychology is not what it takes itself to be.

I conclude more generally that appeals to our evolutionary history are of little value to the aims of cognitive psychology. No doubt considerations of our psychology are important in pursuing the evolutionary history of our species. Indeed, the recent work discussed below illustrates—even if unwittingly—the importance of psychology to the study of our history. But the importance of psychological considerations in the study of our history should not tempt us to the view that evolutionary considerations are important in the pursuit of the architecture of our cognitive capacities.

1   The Methods of Evolutionary Psychology[1]

Evolutionary psychology is a species of cognitive psychology generally and, as such, shares its basic aim. What distinguishes evolutionary psychology are its methods. I begin with the aims and methods of cognitive psychology and then turn to the methods specific to evolutionary psychology.

The fundamental aim of cognitive psychology is to discover the information-processing mechanisms that underwrite our psychological capacities. The goal is to uncover the architecture within the black box. Background to this aim is the assumption that the modern digital computer serves as a useful model for understanding the mind. Computers consist of hardware components that enable them to run software programs that endow them with the capacities to execute certain tasks. Software programs consist of algorithms and representations, and these underwrite capacities such as graphics, word processing, and the like. The computational model requires that we think of the human animal as a complex information-processing machine: Our central nervous system contains hardware components that enable the organism to run software programs that endow the organism with various psychological capacities. The aim of cognitive psychology is to discover the software programs that underwrite our various psychological capacities, including memory, vision, language acquisition, and the like. The aim is to discover the algorithms and representations of our minds.

The fundamental method of cognitive psychology, at least since Marr (1982), is to generate hypotheses about mechanisms inside the box on the basis of the tasks or functions fulfilled by our psychological capacities. An account of these functions, it is assumed, provides valuable design constraints on the sorts of mechanisms within. Background to this method is the assumption that each of these three elements— our neurological hardware, information-processing software, and psychological capacities—can be investigated more or less independently of the rest. The ultimate goal, of course, is an account that integrates all three, but for the purposes of our present inquiry it is enough to focus upon any one of the following:

(1)   A computational theory: A systematic characterization of the relevant psychological capacities in terms of their functions or goals within the organism's interactions with its environment;

(2)   An information-processing theory: A systematic characterization of the information-processing mechanisms (algorithms and representations) that underwrite the functions of the psychological capacities characterized in (1);

(3)   A neurological theory: A systematic characterization of the neurological mechanisms that underwrite the information-processing mechanisms described in (2).   (Marr 1982)

While the central aim of cognitive psychology is to accomplish (2)—to develop a theory of the information-processing mechanisms that underwrite our psychological capacities—the methodological strategy is to approach (2) by way of (1). Marr insists that the best way to get at the mechanisms hidden within the black box is to develop a taxonomy of the functions of our various psychological capacities. An account of the functions of our capacities should provide valuable clues to the design of the underlying mechanisms.

In his approach to (1), Marr construes functions nonhistorically. He does not explicitly eschew considerations of evolutionary history; rather, he restricts his attention to the tasks accomplished within the organism's current environment:

A pigeon uses vision to help it navigate, fly, and seek out food. Many types of jumping spider use vision to tell the difference between a potential meal and a potential mate....

The frog ... detects bugs with its retina; and the rabbit retina is full of special gadgets, including what is apparently a hawk detector.... Human vision, on the other hand, seems to be very much more general, although it clearly contains a variety of special-purpose mechanisms.   (Marr 1982, p. 32)

For the purpose of accomplishing (1), it is enough to observe the organism in its typical environment performing those behaviors that tend toward the satisfaction of its needs and toward the avoidance of danger and pain. Such observations serve as evidence for the attribution of quite specific functions. They do so, moreover, whether or not we concern ourselves with or possess any knowledge of the organism's natural selective or evolutionary history.

It is precisely here, however, on the issue of function-attributions in level (1), that evolutionary psychologists diverge from Marr. They agree that the best way to accomplish (2) is by way of (1), but they disagree with Marr's nonhistorical approach to functions. The best way to accomplish (1), they claim, is to take a historical and, in particular, evolutionary approach to the discovery of functions. What distinguishes the methods of evolutionary psychology, then, from those of cognitive psychology generally, is their assumption that the best way to discover the psychological functions of our psychological capacities is to discover first the *adaptive functions* of those capacities. Adaptive functions, evolutionary psychologists claim, are our best guide to psychological functions; inquiry into our selective and evolutionary history is our best strategy for discovering how our cognitive systems are built.[2]

Adaptive functions are functional properties acquired as a consequence of evolution caused by natural selection. The mammalian heart, for example, plausibly has the adaptive function of pumping blood, on the assumption that ancestral hearts pumped blood and such pumpings were selectively and evolutionarily efficacious.

The methods distinctive to evolutionary psychology thus incur the burden of uncovering evidence concerning our natural selective and evolutionary history. These
methods and their evidential commitments may be represented by the following
stepwise strategy. Begin with

(i)   speculations concerning the adaptive problems facing our ancestors and, on the
basis of these speculations,

(ii)   attribute adaptive functions to our psychological capacities and, on the basis of
these attributions,

(iii)   generate hypotheses concerning the information-processing mechanisms that
underwrite our psychological capacities.

Step (ii) is the crucial step, for it is in the course of attributing adaptive functions
that we accomplish level (1) in Marr's taxonomy of levels; it is here that we construct
a computational theory of our psychological capacities. It is also here that evidence
from our natural selective and evolutionary history must be brought to bear. Finally,
it is assumed that the adaptive functions attributed to our psychological capacities
are identical with, or intimately related to, the psychological functions of our psychological capacities.[3] I conclude, then, that steps (i)–(iii) represent the core methods
of so-called evolutionary psychology.

It may be objected that, in attempting to characterize the methods of evolutionary
psychology, I have succeeded only halfway. For Cosmides, Tooby, and Barkow
(1992, pp. 9–11) assert that the methods of evolutionary psychology consist in two
complementary approaches. The first approach begins with speculations concerning
certain adaptive problems and, on the basis of these speculations, attributes adaptive
functions and generates hypotheses concerning the mechanisms of our minds. This is
the approach represented above in steps (i)–(iii). The second approach, by contrast,
begins with a working understanding of certain psychological mechanisms and
inquires into their selective and evolutionary history, with an eye toward identifying
their adaptive functions and with a further eye toward uncovering unexpected clues
concerning hitherto unknown psychological mechanisms. These approaches are
thought to complement and feed upon one another: Historical considerations in the
first approach help us discover hitherto unknown cognitive mechanisms, and inquiry
into the adaptive functions of known mechanisms enriches the historical considerations of the first approach, which in turn fuels further discoveries of hitherto
unknown mechanisms. And so on and so on.

But despite its utility, this second approach does not help distinguish the methods
of evolutionary psychology from those of cognitive psychology. As we have seen, the

central aim of cognitive psychology is the discovery of the information-processing mechanisms of the mind. But Cosmides et al. concede that their second approach *presupposes* knowledge of the relevant cognitive mechanisms. They concede that, on their approach, we inquire into the adaptive functions of certain mechanisms already understanding how they work. But this means that the aim of our cognitive psychological inquiry is already satisfied and hence the historical inquiry is quite beside the point. This is not to say that such historical inquiry is unimportant or uninteresting; it is to say only that it is unimportant to fulfilling the aims of cognitive psychology.

In response, Cosmides et al. may repeat their claim that inquiry into the history of known psychological mechanisms may lead to further information about our ancestral psychology and that this, in turn, may lead to discoveries of hitherto unknown psychological mechanisms. I do not deny this; in fact I agree that this is possible. What I deny is that this cuts any methodological ice. I agree that inquiry into our history may generate clues about hitherto unknown psychological mechanisms. But of course the same is true of all sorts of inquiry, including inquiry that ignores considerations of evolutionary history, and yet none of these underwrites a new or distinct methodology. Inquiry into how humans behave under conditions of severe boredom, for example, may generate clues concerning the sorts of mechanisms that comprise our cognitive capacities. Similarly for studies of humans under extreme sensory deprivation, under conditions of various types of brain damage, under prolonged political oppression, and so on. All such studies may surprise us and prompt us to hypothesize hitherto unknown mechanisms. But the potential utility of these studies hardly underwrites a method of inquiry that is distinct from the general methods of cognitive psychology. Such potential, on the contrary, is already contained in the methods of cognitive psychology. And there is no reason to think that the second approach, as described by Cosmides et al., is any different in this regard.

Moreover, this second approach, when compared with available nonhistorical forms of inquiry, is inferior in two respects. The first is that some inquiries uninformed by evolutionary history offer insights that inquiries formed by such history are prone to miss. Neurologists and speech-language pathologists, for example, study language production and comprehension in brain-damaged patients. This is to study language under conditions such as strokes and significant head injury; this is to study language under conditions that were selectively inefficacious and thus conditions that did not lead to any adaptive functions. And such studies offer insights that cannot be garnered from the study of intact patients. We are likely to miss such insights when studying language under conditions that were selectively advantageous; we thus are likely to miss insights when studying language under conditions that led to adaptive functions. So the study of our cognitive mechanisms under conditions that were selectively disadvantageous is, at least in some instances, more fruitful than the study of such mechanisms under conditions that were selectively advantageous. Nonevolutionary inquiry into our psychology can bear fruit that evolutionary inquiry cannot.

The second way in which the historical inquiries favored by Cosmides et al. are inferior is in terms of our epistemic access. Nonhistorical inquiries, such as the study of language in brain-damaged patients, allow us to manipulate and repeat experiments in ways historical inquiries often do not. Neurologists and speech-language pathologists of course are limited in the sorts of manipulations they can impose, but nevertheless they study language under a wide range of neurological deficits and they intervene surgically, chemically, or therapeutically. Moreover, many neurological and linguistic deficits can be type-classified, thus providing substantial experimental repeatability. By contrast, the evolutionary speculations offered by Cosmides et al. are impoverished. As I describe in section 3 below, even the most compelling evolutionary explanations involve nontrivial assumptions for which our evidence is thin and difficult to reconstruct. The nonhistorical inquiries, in consequence, bear fruit where the historical inquiries cannot.

I conclude, therefore, that the second approach of Cosmides et al. fails to mark a method distinct from the general methods of cognitive psychology. The core methods of evolutionary psychology, in consequence, are those described in steps (i)–(iii). And if these steps do not express the methods distinctive to evolutionary psychology, then I have no idea what evolutionary psychology is and neither, I submit, does anyone else. I turn now to the first claim that leads to the conflict in evolutionary psychology.

2   Recent Work

Some recent work in evolutionary psychology is implausible on its face.[4] Other work, although perhaps initially plausible, nevertheless suffers significant epistemological challenges.[5] But among the more plausible attempts is Cummins's (1996) discussion of reasoning and the evolution of dominance hierarchies in primates. Her thesis is that humans today possess cognitive mechanisms, descended from those of distant primate ancestors, that govern the ways we reason in cooperative and in competitive social contexts. Cummins is explicit that this hypothesis emerges from speculations concerning the evolutionary pressures faced by such ancestors. Her line of reasoning fits cleanly into the steps described in (i)–(iii) above, steps that characterize the methods of evolutionary psychology. Cummins begins by (i*)   speculating about the evolution of dominance hierarchies in primates and, on the basis of such speculations,

(ii*)   attributes specific adaptive functions to specific psychological capacities and, on the basis of these attributions,

(iii*)   generates hypotheses concerning the existence and nature of information-processing mechanisms that underwrite such capacities.

I will explicate each step in turn.

Cummins speculates that evolutionary pressures on ancestral primates probably conduced to the formation of social groups. One way an organism may enjoy survival and reproductive advantage is by wielding superior physiological capacities, such as size or speed. Another is through the formation of social relations. Group living offers obvious advantages with respect to safety, hunting, mating, and the like; hence the pressure toward cooperative relations. But group living has its costs, especially competition between group members for food, mates, and so on. This competitive cost exerts pressure toward the formation of various social alliances and the formation of various distinctions of rank. As a result of forming shrewd, powerful alliances, some organisms come to dominate others, enjoying relatively greater access to resources such as food and mates. This occurs even when the dominating organisms enjoy less brute physical prowess than those they dominate. Hence we have Cummins's speculative story concerning the evolution of dominance hierarchies in primates.[6]

This pressure toward the evolution of dominance hierarchies, according to Cummins, imposes simultaneous pressure toward the evolution of requisite cognitive capacities. The requisite cognitive capacities acquire rather specific adaptive functions, including

(a)   the capacity to recognize the social rank of oneself and of one's fellows,

(b)   the capacity to recognize what sorts of actions are forbidden and what sorts are permitted, given one's rank, and

(c)   the capacity to decide whether to engage in or refrain from activities that may enable one to move up in rank.

Cummins's claim is that, given the truth of her evolutionary speculations, it is reasonable to view such functions as adaptive functions and thus as psychological functions. It further is reasonable to claim that humans today inherited the mechanisms that underwrite such functions, on the assumption that the same or functionally equivalent evolutionary pressures have persisted throughout, or that these mechanisms are sufficiently entrenched to resist elimination.

In the final step, Cummins appeals to the functions postulated in (a)–(c) to generate hypotheses concerning the mechanisms that underwrite such capacities. At a minimum, (a)–(c) suggest that ancestral primates possessed algorithms and representations with which they reasoned in terms of

(H1)   relations that are transitive and

(H2)   relations governed by deontic operators.

The capacity to recognize one's own rank and that of one's fellows plausibly involves the capacity to reason transitively. The capacity to recognize which actions are forbidden and which are permitted plausibly involves the capacity to reason deontically.[7] It is thus not unreasonable to construct experiments intended to test for the current existence and efficacy of the mechanisms described in (H1) and (H2).

This, then, is one recent attempt to employ the methods of evolutionary psychology as represented in (i)–(iii) above. It is an attempt that appears plausible on its face. I believe, however, that such appearances deceive. Cummins's account seems plausible only to the extent that she is *not* applying the methods of evolutionary psychology but rather engaging in a bit of evolutionary history. Or so I shall argue. Before pursuing that argument, however, I turn to the second claim that generates the conflict within evolutionary psychology.

3   Idle Methods

The methods distinctive to evolutionary psychology are, in my view, idle in our attempts to discover the architecture of the mind. My argument contains two uncontroversial premises, which I explicate in turn.

My first premise is that step (ii) in the methods of evolutionary psychology—the attribution of adaptive functions to psychological capacities—requires substantial historical evidence, in particular, evidence concerning our natural selective and evolutionary history. This should be obvious. If, for example, I claim that the adaptive function of the mammalian heart is to pump blood, the burden is on me to show that past hearts pumped and that such pumpings were selectively efficacious. After all, my claim is that the heart's function of pumping blood is a specific kind of function, one that emerges out of the history of the heart, rather than a function that derives from nonhistorical considerations.[8] This places on me a twofold burden. The first is to show that the historical processes specified within the theory of adaptive functions actually occurred for the specified trait. The second is to show that this historical approach to understanding functions, as opposed to competing approaches, is the correct one. For present purposes I waive the second burden and focus on the first, for the first is burden enough.[9]

To substantiate my claim that the heart's adaptive function is to pump blood, I must adduce evidence that the heart is a product of evolution due to natural selection. At least two specific forms of evidence are required. First, I must show that the heart is a product of selection as a consequence of *its own selective efficacy*; I must show that the heart was *selectively evolved for* and not merely *selected of*.[10] This involves showing that at least some causal effects of the heart contributed to the organism's being relatively better adapted than certain conspecifics. After all, the heart may have pumped blood and nevertheless failed to contribute to the organism's selective success. This may occur if the selective efficacy of some other trait within the organism overrode the potential selective efficacy of the heart's pumping blood. It also may occur if some other selective pressure within the organism's environment neutralized the selective efficacy of the heart's pumping blood. This also involves showing that certain effects of the heart, by virtue of being selected for, contributed to evolutionary change within the population. After all, a trait can be favored by selection and nevertheless not evolve; evolution by natural selection requires heritability, but selection alone does not. And even if the trait is heritable and selected for, it nevertheless may fail to evolve if its selective efficacy is neutralized by some other evolutionary process such as drift. Everything depends upon the efficacy of other traits within the organism and upon the full range of selective and evolutionary pressures within the organism's selective environment. Second, I must show that the heart was selected *specifically for* its pumping of blood. After all, the heart may have pumped and produced some other effect whose efficacy overrode the potential selective efficacy of the heart's pumping of blood. It all depends upon the relations between the heart's pumping and its other effects. My evidential commitments, then, in the attribution of an adaptive function to some capacity, are substantial indeed.

These commitments are all the more pressing given that our epistemic access to such evidence is woefully lacking. Much of the evidence involved has literally long since died and rotted away. Philosophers of biology, notably Brandon (1990), have responded by suggesting that we settle for explanations of our selective history that are evidentially less demanding. This may involve *how-possibly* explanations with relatively strict evidential standards concerning relevant possibilities or *how-actually* explanations with relatively relaxed evidential standards.[11] The important point is that there are better and worse approximations to these weaker forms of explanation.

Brandon advocates how-possibly explanations and illustrates with the following example. Kingsolver and Koehl (1985) studied the evolution of insect wings. Fossils from the Paleozoic show that ancestors of today's insects had small, proto-wings. Kingsolver and Koehl constructed physical models of those ancestral bugs and placed them in a wind tunnel, only to discover that the proto-wings were too small to produce any lift. Early insect wings, they discovered, *could not* have been selected for flight or for gliding; the physical and biological conditions that actually obtained made this impossible. But we also know from Paleozoic fossils that early insect wings were thicker than insect wings today and thus potentially more effective conductors of heat. So once again Kingsolver and Koehl constructed physical models, this time testing for thermoregulatory efficiency, and they discovered that these early proto-wings were indeed effective in maintaining relatively high body temperature, thus enabling the bugs to move about for relatively longer periods of time. Early insect wings, they discovered, *could* have been selected for thermoregulatory benefits. Kingsolver and Koehl conclude that early insect wings could have been selected for thermoregulation, though not for gliding or flight, and that subsequent mutations for increased size could have led to selection for gliding and flight. This is a compelling explanation of the evolution of insect wings that may suffice to justify the attribution of associated adaptive functions to a particular trait.

It is significant, however, that Kingsolver and Koehl could not have constructed and substantiated their how-possibly explanation without *prior knowledge* provided by the fossil record. Without prior knowledge of the size of the proto-wings, they could not have ruled out the possibility that these early wings were selected for flight. Without prior knowledge of the thickness of the proto-wings and the tissue out of which they were made, they could not have ruled in the possibility that these wings were selected for thermoregulation. The strength of their how-possibly explanation depends upon the data garnered from the manipulation of physical models of ancestral insects, where such experiments depend upon the extent of our prior knowledge of relevant organismic structures.

My second premise, then, is that the requisite inquiry into our selective and evolutionary history—the construction of a viable how-possibly explanation—requires that we *know in advance* the structure and the salient outputs of the traits under investigation. In the evolution of insect wings, the power of the explanation offered derives directly from physiological information provided by the fossils. Kingsolver and Koehl discovered from the fossil record the structure of these early insects and this, combined with their ingenuity, enabled them to extract information concerning the salient outputs of early proto-wings. But of course we are in the very same kind of evidential situation in the case of psychological traits. In order to justify the attribution of any adaptive function to any psychological capacity, we need a plausible how-possibly explanation of that capacity. But such explanations require that we know in advance the structure and the salient outputs of the specified psychological traits. To know this, however, is to know in advance the very thing we are trying to discover—Marr's computational theory of the relevant psychological capacities. To pursue the evolutionary inquiries that evolutionary psychologists wish to promote, therefore, we must have accomplished already the central aim of our cognitive psychological inquiry.

I thus conclude that the methods of evolutionary psychology add nothing to our search for the architecture of our cognitive capacities. The methods of evolutionary psychology are idle. For as we have seen, step (ii) in the methodology of evolutionary psychology saddles us with substantial evidential commitments, requiring specific evidence concerning our natural selective and evolutionary history; the acquisition of such evidence, however, requires that we know in advance the functional status of the traits we are investigating; so we must have in advance the very thing we are after, in which case the search for adaptive functions is idle in our search for the mechanisms of the mind.

4 History and Cognitive Psychology

The conflict in evolutionary psychology, then, is that demonstrably barren methods appear to bear theoretical fruit. Whence this conflict? The source, I believe, is the conflation of the aims of cognitive psychology with those of evolutionary history. The apparent fruits of some evolutionary psychologists are borne not by the methods of evolutionary psychology but rather in the pursuit of certain evolutionary historical speculations. This conflation is quite clear in the work by Cummins described above.

Cummins speculates that the selective and evolutionary advantages of social alliances were significant; this is step (i*) of her strategy. On the basis of these speculations, Cummins postulates the existence of certain psychological traits with various adaptive functions; this is her (ii*). But of course the attribution of such adaptive functions to these traits is plausible only against the background of our prior knowledge of the organism's psychology. After all, certain species of bees and ants live in highly structured colonies with distinct social rankings and thus seem to fit the speculations given in (i*), yet it is an obvious mistake to attribute to bees or ants the capacity to recognize rank, recognize permissions and prohibitions, or make decisions. We know on nonevolutionary grounds that the psychology of these organisms does not include such capacities. By contrast, the attribution of these functions in the case of primates is plausible because we know on nonevolutionary grounds that their psychology includes such capacities.[12]

It thus is clear that the move from (i*) to (ii*) in Cummins's strategy is plausible only to the extent that we already understand a significant portion of the psychology of the relevant organisms. Indeed Cummins begins with rather detailed information concerning primate psychology derived from observations of the organisms in their current environments.[13] And this shows that Cummins is *not* employing her own evolutionary speculations to investigate primate psychology. She is *not* reasoning from the speculations in (i*) to the adaptive functions in (ii*), as the methods of evolutionary psychology recommend. On the contrary, she is employing antecedent knowledge of primate psychology and speculating about the evolution of that psychology. She appeals to a prior account of psychological functions, and on the basis of that account, she then speculates about the evolutionary origins of those functions. Crucially, the functions with which she begins are *not* adaptive functions of any sort, but simply the sorts of functions recommended by Marr: functions discovered by observing the organism in its current environment. Obviously this is not an objectionable form of inquiry, but neither is it an inquiry guided by the methods expressed in (i)–(iii). This is not evolutionary psychology but merely the evolutionary history of cognitive capacities the functions of which are known already.

The conflict in evolutionary psychology, then, derives from the conflation of two distinct research projects, cognitive psychology and evolutionary history. Advocates of evolutionary psychology claim that we can discover how we are built, how our minds are structured, by first discovering the evolutionary processes that built us. But that is wrong; in fact it is backwards. We cannot discover which of several evolutionary processes built and sustained us unless we first know how we are built, unless we first know the structure and salient outputs of traits the history of which we are trying to trace. The conflict in evolutionary psychology comes from imagining that there is theoretical unity when all there is is the confusion engendered by this conflation.

Acknowledgments

This essay was prepared while I was a Charles P. Taft Postdoctoral Fellow at the University of Cincinnati. I am grateful to the Taft Foundation for their generous support of my work. An ancestor of this paper was presented to the University of Cincinnati and to the 1997 meeting of the International Society for the History, Philosophy, and Social Studies of Biology. I am grateful to both audiences, especially my philosophy colleagues at the University of Cincinnati, for helpful comments.

Notes

1. This section and section 3 below are drawn from Davies (1996). For a fuller account of the methods of evolutionary psychology, see sections 1 and 2 of that paper; for a fuller critique of those methods, see sections 3, 4, and 5.

2. See Tooby and Cosmides (1992, p. 68) and Cosmides and Tooby (1987, pp. 284–285) for textual support. For further discussion, see Davies (1996, pp. 563–564).

3. See note 2 for citations.

4. Cosmides (1989) and Cosmides and Tooby (1992), for example, claim that humans today possess information-processing algorithms for detecting cheaters. The hypothesis is that the capacity to detect cheaters in social exchange situations—situations in which two or more parties agree to swap benefits for payments—would have conferred an advantage in terms of fitness. It thus is suggested that humans today have inherited algorithms that underwrite the capacity to detect cheaters. Cosmides and Tooby assert that the existence of such algorithms is confirmed by data generated in a series of experiments first reported in Cosmides (1989). But there is little here that is plausible. At least four general points tell against this social exchange theory. (1) The evolutionary speculations are weak (Davies, Fetzer, and Foster 1995). There is no reason to accept that ancestors endowed with an algorithm for detecting cheaters would have enjoyed greater fitness than those not so endowed. It all depends upon the presence or absence of other ancestral perceptual and cognitive capacities. (2) Cosmides and Tooby seem to assume that the Pleistocene environment was uniform with respect to cost-benefit interactions, and that is a dubious assumption (Sterelny 1995). (3) The evolutionary speculations and associated experiments are conceptually muddled (Davies et al. 1995). (4) Cosmides's experiments fail to rule out competing hypotheses and thus do not support her evolutionary hypothesis (ibid.).

5. See Richardson's (1996) assessment of Nozick's (1993) work on rationality and evolutionary psychology, and of Pinker and Bloom's (1992) work on language learning.

6. While Cummins's speculations appear plausible, they raise rather difficult questions. It is unclear, for example, whether or not Cummins is claiming that the evolution of dominance hierarchies involves a form of group selection. She seems to suggest a two-step progression. The first step involves selection at the level of individual organisms. Organisms that tended toward greater sociality, we may presume, enjoyed relatively greater expected fitness, thus giving rise to more individuals with the propensity to form alliances. The second step involves selection acting on these newly emerging alliances. Within a given population, we may presume, some alliances were more effective and thus enjoyed greater expected fitness than others. On the assumption that the properties that distinguish the various groups are group-properties, and on the further assumption that such group-properties are heritable, this latter selection is plausibly an instance of group selection. Assessing the plausibility of Cummins's evolutionary speculations depends at least in part on the status of her commitments to group selection. Nevertheless, for the sake of argument I propose to take her evolutionary speculations at face value.

7. I am unmoved by Cummins's appeal to deontic relations. None of the research she reports requires the postulation of such elements. Appeals to considerations of self-interest and prudence suffice. Or so it seems to me. And as Davies, Fetzer, and Foster (1995) argue, genuine deontic rules are demanding in terms of cognitive capacities, which suggests that the capacity to reason according to such rules is well beyond those possessed by our distant primate ancestors.

8. As, for example, suggested by R. Cummins (1975) and by Amundson and Lauder (1994).

9. For doubts about this second burden—doubts concerning the evolutionary approach to functions—see R. Cummins (1975), Amundson and Lauder (1994), Davies (1994), and Davies (ms1) and (ms2).

10. I adapt Sober's (1984) felicitous distinction between *selection for* and *selection of*. I believe that more than Sober's *selection for* is required to justify the attribution of adaptive functions. The specified trait must have been selected for and then *evolved for* as well—*selectively evolved for*, as I prefer to put it. I attempt to clarify and extend Sober's original distinction in Davies (ms3).

11. I do not wish to commit myself to any strong distinction between how-possibly explanations and how-actually explanations. All that matters for present purposes is that the evidential demands be relaxed and nevertheless enable us to distinguish compelling from uncompelling explanations.

12. We know this from the work, for example, of Cheney and Seyfarth (1990)—one of the many references cited by Cummins.

13. See the many citations throughout Cummins's well-informed discussion. It is clear that the wealth of information to which she refers was acquired with little or no concern for evolutionary history, but rather from observations and manipulations of organisms in their present environments.

References

Amundson, R. and Lauder, G. V. (1994) "Function without Purpose: The Uses of Causal Role Function in Evolutionary Biology," *Biology and Philosophy* 9:443–470.

Barkow, J., Cosmides, L., and Tooby, J. (1992) *The Adapted Mind: Evolutionary Psychology and the Generation of Culture*. New York, NY: Oxford University Press.

Brandon, R. (1990) *Adaptation and Environment*. Princeton, NJ: Princeton University Press.

Buss, D. (1995) "Evolutionary Psychology: A New Paradigm for Psychological Science," *Psychological Inquiry* 6:1–30.

Cheney, D., and Seyfarth, R. (1990) *How Monkeys See the World*. Chicago, IL: University of Chicago Press.

Cosmides, L. and Tooby, J. (1987) "From Evolution to Behavior: Evolutionary Psychology as the Missing Link," in J. Dupre (ed.), op. cit., pp. 277–306.

Cosmides, L. (1989) "The Logic of Social Exchange: Has Natural Selection Shaped How Humans Reason? Studies with the Wason Selection Task. *Cognition* 31:187–276.

Cosmides, L. and Tooby, J. (1992) "Cognitive Adaptations for Social Exchange," in J. Barkow et al. (eds.), op. cit., pp. 163–228.

Cosmides, L., Tooby, J. and Barkow, J. (1992) "Introduction: Evolutionary Psychology and Conceptual Integration." In J. Barkow, L. Cosmides, J. Tooby (Eds.) *The Adapted Mind: Evolutionary Psychology and the Generation of Culture*, pp. 3–15. New York: Oxford University Press.

Cummins, D. D. (1996) "Dominance Hierarchies and the Evolution of Human Reasoning," *Minds and Machines* 6:463–480.

Cummins, R. (1975) "Functional Analysis," *The Journal of Philosophy* 72:741–765.

Davies, P. S. (1994) "Troubles for Direct Proper Functions," *Nous* 28:363–381.

Davies, P. S. (1996) "Discovering the Functional Mesh: On the Methods of Evolutionary Psychology," *Minds and Machines* 6:559–585.

Davies, P. S. (ms1) "Evolutionary Malfunctions."

Davies, P. S. (ms2) "Why Evolutionary Functions are Causal Role Functions."

Davies, P. S. (ms3) "On the Causes and Conditions of Natural Selection."

Davies, P. S., Fetzer, J., and Foster, T. (1995) "Domain Specificity and Social Exchange Reasoning: A critique of the Social Exchange Theory of Reasoning," *Biology and Philosophy* 10:1–37.

Dupre, J. (ed.) (1987) *The Latest on the Best: Essays on Evolution and Optimality*. Cambridge, MA: The MIT Press.

Kingsolver, J. and Koehl, M. (1985) "Aerodynamics, Thermoregulation, and the Evolution of Insect Wings," *Evolution* 39:488–504.

Marr, D. (1982) *Vision*. San Fransico, CA: W. H. Freeman and Company.

Nozick, R. (1993) *The Nature of Rationality*. Princeton, NJ: Princeton University Press.

Pinker, S. and Bloom, P. (1992) "Natural Language and Natural Selection," in J. Barkow et al. (eds.), op. cit., pp. 451–493.

Richardson, R. (1996) "The Prospects for an Evolutionary Psychology: Human Language and Human Rationality," *Minds and Machines* 6:541–557.

Sober, E. (1984) *The Nature of Selection*. Cambridge, MA: The MIT Press.

Sterelny, K. (1995) "The Adapted Mind," *Biology and Philosophy* 10:365–380.

Tooby, J. and Cosmides, L. (1992) "The Psychological Foundations of Culture," in J. Barkow et al. (eds.), op. cit., pp. 19–136.

5 Presence of Mind

Lawrence A. Shapiro

1 Evolutionary Psychology and the Behavior-to-Mind Inference

In the introduction to their influential anthology *The Adapted Mind: Evolutionary Psychology and the Generation of Culture*, Barkow, Cosmides, and Tooby observe that until recently psychologists have appealed to evolutionary theory only to advance hypotheses about the adaptive function of known phenotypic traits. Given that we have color vision, for instance, they have asked what adaptive function might this have served? For what reason did language evolve? Why are men more strongly attracted to women with clear and smooth skin than they are to women with mottled and wrinkled skin, and why do women prefer men who are powerful and well respected to those who are meek and lowly? Answers to these questions involve an inference from an observed trait to an unobserved history of evolution. With the advent of optimality theory, researchers can make this inference with an impressive degree of precision. We build a model describing that trait which, under given constraints, would be optimal and compares this optimal trait to the actual trait of the organism. Depending on how closely the actual trait matches the optimal trait, we can assess the contribution selection made in its evolution, thereby deepening our understanding of what, if anything, the trait was selected for.

But there is another kind of inference that evolutionary psychologists often try to make. In addition to inferring adaptive function from observed behavior, Barkow et al. claim that "[o]ne can use theories of adaptive function to help one discover psychological mechanisms that were previously unknown" (1992, p. 10). This inference, from adaptive function or behavior to psychological mechanism, is important enough to deserve a name, and henceforth I will refer to it as the behavior-to-mind (or "BTM") inference. The BTM inference, as Barkow, Cosmides, and Tooby view it, is central to evolutionary psychology, and, fittingly, evolutionary psychologists are not shy about making it. Some examples will help illustrate the BTM inference at work.

In "Origins of Domain Specificity: The Evolution of Functional Organization," Cosmides and Tooby infer the existence and nature of a psychological kin-recognition mechanism from facts about constraints under which kin selection is adaptive. According to Cosmides and Tooby:

Hamilton's kin selection theory raises—and answers—questions such as: How should the information that X is your brother affect your decision to help him? How should your assessment of the cost to you of helping your brother, versus the benefit to your brother of receiving your help, affect your decision?... In general, how should information about your relatedness to X, the costs and benefits to you of what X wants you to do for him, and the costs and benefits to X of your coming to his aid, affect your decision to help X? (1994, p. 97)

Having stated the problem the kin selection mechanism has been designed to solve—the function it serves—Cosmides and Tooby proceed with the BTM inference:

As these questions show, an organism's behavior cannot fall within the bounds of the constraints imposed by the evolutionary process unless it is guided by cognitive programs that can solve certain information-processing problems that are very specific. To confer benefits on kin in accordance with the evolvability constraints of kin selection theory, the organism must have cognitive programs that allow it to extract certain specific information from its environment: Who are its relatives? Which kin are close and which distant? What are the costs and benefits of an action to itself? To its kin? The organism's behavior will be random with respect to the constraints of kin selection theory unless (1) it has some means of extracting information relevant to these questions from its environment, and (2) it has well-defined decision rules that use this information in ways that instantiate the theory's constraints.   (1994, p. 97)

So, from facts about the behavior a kin recognition mechanism produces, Cosmides and Tooby conclude that organisms, human or otherwise, in which kin selection evolved as an adaptation, must employ a cognitive mechanism that can (1) detect one's degree of relatedness to others in the group and (2) apply decision rules to this information so as to derive an optimal solution.

Janet Mann, in "Nurturance or Negligence: Maternal Psychology and Behavioral Preference among Preterm Twins," discusses a mother's investment strategies when caring for high-risk infants. When faced with a high-risk infant, under what conditions should the mother invest extra resources to care for it rather than simply let it perish? After a careful analysis of the circumstances that favor one or the other strategy, Mann makes the BTM inference:

This evidence leads me to suggest the following set of parental care algorithms. First, mothers have a template (or prototype) for normal, healthy infants. This template is sensitive to acoustic, physical, and some behavioral features of infants. This template is probably refined through experience, direct or indirect, with infants. Second, this template, if not matched closely, tells the mother that something is wrong with the infant or, if matched closely, tells the mother that the infant is fine. Third, subsequent decision rules concerning parental care are responsive to relevant environmental inputs. If the template is not closely matched (abnormal infant), I would expect the psychological mechanisms of care allocation to be more sensitive and reactive to social, demographic, and ecological information than if the infant was normal (matched the template). The greater the mismatch between her infant and the template, and the poorer the social and economic conditions, the greater the maternal feelings of despair, detachment, and negative infant attributions ... and subsequently the more likely she will be to terminate or minimize investment. If the template is not closely matched, but all the social and economic conditions are favorable for investment, the greater the maternal feelings of dedication and the more positive the infant attributions ... and subsequently the more likely she will be to increase maternal investment.   (Mann 1992, p. 373)

Like Cosmides and Tooby, Mann follows her discussion of a behavioral capacity, in this case parental-care strategies, with a quite specific description of the psychological mechanism that must underlie such a capacity. By her lights, a mother's adaptive behavior toward high-risk infants requires a mechanism that employs both templates and decision rules.

David Buss, in "Mate Preference Mechanisms: Consequences for Partner Choice and Intrasexual Competition," writes:

Imagine a state in which human males had no mate preferences aside from species recognition and instead mated with females randomly. Under these conditions, males who happened to mate with females of ages falling outside the reproductive years would become no one's ancestors. Males who happened to mate with females of peak fertility, in contrast, would enjoy relatively high reproductive success. Over thousands of generations, this selection pressure would, unless constrained, fashion a psychological mechanism that inclined males to mate with females of high fertility over those of low fertility. (Buss 1992, p. 250)

And Bruce Ellis makes a similar point about female sexual preferences in "The Evolution of Sexual Attraction: Evaluative Mechanisms in Women":

Consider, for example, a woman who can choose between two husbands, A and B. Husband A is young, healthy, strong, successful, well liked, respected by his peers, and willing and able to protect and provide for her and her children; Husband B is old, weak, diseased, subordinate to other men, and unwilling and unable to protect and provide for her and her children. If she can raise more viable children with Husband A than Husband B, then his "mate value" can be said to be higher. Over evolutionary time, ancestral females who had psychological mechanisms that caused them to find males of high mate value more sexually attractive than males of low mate value, and acted on this attraction, would have outreproduced females with opposite tastes. (Ellis 1992, p. 267)

While Buss and Ellis are silent about the details of the mate preference mechanisms, they agree that such mechanisms must be psychological in nature. So, although their BTM inferences may not be as elaborate as Cosmides and Tooby's or Mann's, they are nonetheless making a substantive claim about the kind of process from which mate preferences emerge—it is psychological rather than, for example, merely physiological.

If I have belabored the point that the BTM inference is accepted currency among evolutionary psychologists it is only because I find it absolutely remarkable that no defense is offered for such an inference (Shapiro, forthcoming; Shapiro and Epstein, forthcoming). Consider, for instance, Cosmides and Tooby's application of the BTM inference. Kin selection in any species, they believe, implies the existence of sophisticated cognitive programs that do things like compute degrees of relatedness and apply decision rules to the products of these computations. Surely, however, this inference overlooks many cases in which a noncognitive explanation is more likely. Krebs and Davies (1987, p. 255), for instance, note that female Belding's ground squirrels are able to distinguish full sisters from half sisters via phenotype matching. Such matching, they suggest, requires only sensitivity to particular odors. Why attribute to female Belding's ground squirrels complex cognitive programs when the differential behavior they exhibit toward group members might be triggered by nothing more than odor preferences?

Similarly, why does a mother need a template to tell her that her infant is normal or abnormal? Why can't the cries of distressed babies simply cause a mother to react in ways that normal cries do not, just as the sound of fingernails on a chalkboard causes us to cringe whereas the sound of a brush moving through hair does not? Why must psychological mechanisms produce our mate preferences? Perhaps nubile females release pheromones that draw a male's attention in the same way that carbon dioxide attracts mosquitoes. "But they don't!" you might say. "But we can't rule out the possibility a priori!" I say back. If there are cognitive or psychological mechanisms directing adaptive behavior, this itself must be a matter of empirical discovery. How we should decide when to apply psychological explanations to the mechanisms that produce adaptive behavior will be the main topic of what follows. In particular, I will characterize circumstances under which the BTM inference is justified; in doing so I will also defend a particular conception of what minds are for.

2 Two Claims about What Minds Do

We must now turn to the difficult task of saying when the BTM inference is warranted, when some bit of behavior is likely to implicate the presence of mind. I should say at the outset that I am not looking for necessary or sufficient conditions for the presence of mind—I think it unlikely that any claim of the form "X has a mind if and only if C" would stand up to prolonged scrutiny. Rather, my goal is to find a reason to believe that some type of behavior has a mental rather than merely physical cause. We are, in short, looking for a reason to believe we should look to psychology to understand the proximate causes of some adaptive behavior, rather than, for example, molecular biology.

A natural approach to the job at hand is to try to conceive of tasks that minds could perform but that mindless biological mechanisms could not. Then, given that an organism has a capacity whose exercise requires something more than mindless biology, we can infer that the organism possesses the relevant psychological mechanisms. Basically, the BTM inference increases in strength as nonpsychological mechanisms appear less able to produce the behavior in question. In pursuing this strategy it will pay to look at Godfrey-Smith (1996) and Fodor (1986). Though neither Godfrey-Smith nor Fodor are concerned precisely with the question of when it is legitimate to infer the presence of mind from adaptive behavior, they are interested in issues that promise to shed light on our question. Specifically, Godfrey-Smith is concerned with developing a theory of why minds evolved. This account is useful to us insofar as it tells us what minds are good for, why an organism may benefit from having a mind. Likewise, Fodor's project is to draw a rough line between responses that require representational capacities for their production and those that don't. To the extent that representational capacities are psychological capacities, our tracks run in parallel. As we shall see, Godfrey-Smith is right to emphasize the significance of complexity in the evolution of minds, but his characterization of complexity must be refined. Fodor, similarly, succeeds in pointing to kinds of selective responses that biological and physical sciences by themselves seem poorly suited to explain. However, Fodor misidentifies what it is about these responses that makes them likely candidates for psychological explanation.

In *Complexity and the Function of Mind in Nature*, Godfrey-Smith defends what he calls the environmental complexity thesis (ECT). According to this thesis, the mind evolved as an adaptation to a complex environment. Psychological mechanisms can perform better in very complicated situations than can mindless biological mechanisms. Hence, presumably, selection would favor those organisms which, when confronted with a complex environment, evolved a mental means for responding to it. If true, ECT could be used to fill the gaping hole in the BTM inference. When seeking to determine if adaptive behavior B in organism O indicates the presence of psychological processes, the evolutionary psychologist should, given the truth of ECT, examine the environment in which organisms of O's type evolved. If the environment to which behavior B is an adaptive response is suitably complex, we have, Godfrey-Smith believes, reason to believe the proximate causes of B include psychological processes.

Godfrey-Smith's suggestion has great intuitive appeal. The idea that minds become useful (perhaps even necessary?) as the complexity of one's environment increases is not far beneath the surface of Cosmides and Tooby's confidence that kin selection behavior requires a mind. There are just so many variables that enter into successful kin recognition behavior that it seems impossible an organism could "figure out" whom to help and by how much without aid of a mind. Simple biological mechanisms work fine for solving simple problems, but when the going gets tough and the adaptive problems grow more complex, selection will favor mental solutions over their alternatives.

As Godfrey-Smith construes environmental complexity, it is a measure of the heterogeneity present in the organism's environment. "Complexity is changeability, variability; having a lot of different states of modes, or doing a lot of different things" (p. 14). To call an environment complex is to say it "is in different states at different times, rather than the same state all the time; a complex environment is different in different places, rather than the same all over" (p. 14). Consider, for instance, the environment of bryozoans, or sea moss (Harvell 1986). Godfrey-Smith turns to this example in the course of describing mathematical models of the evolution of flexible behavior. In the model Godfrey-Smith explores, he assumes the bryozoan's normal environment has just two states that make a difference to its survival. In one state there are no predatory sea slugs and in the other there are. Godfrey-Smith explores scenarios in which it would be adaptive for the bryozoan to exhibit flexible rather than rigid behavior in response to its binary environment. Depending on how frequently bryozoans encounter sea slugs, how reliably they detect the presence of sea slugs, and what the costs of such encounters are, the bryozoan may benefit from a strategy in which it never grows defensive spines, always grows spines, or grows spines only when it detects sea slugs.

Though Godfrey-Smith is careful not to attribute minds to bryozoans, he does believe that flexible behavior in response to a heterogeneous environment does, when the environment is heterogeneous enough, entail the existence of minds. In dealing flexibly with a two-state environment, the bryozoan is exhibiting behavior with a "proto-cognitive character" (p. 114). As the environment grows in complexity and conditions favor flexible behavior, the more likely it is that a mind guides this behavior. However, rather than finding Godfrey-Smith's discussion of flexible behavior in bryozoans suggestive of minds, I'm inclined to see it as casting doubt upon the claim that environmental heterogeneity, at least without further specification, increases the likelihood that minds will evolve. As Godfrey-Smith notes, the bryozoans detect the presence of sea slugs by "making use of a water-borne chemical cue" (p. 114). The detection of such a chemical cue, I shall assume, requires no psychological equipment on the part of the sea moss. But if no mind is necessary for adapting bryozoans to an environment with a complexity of two (i.e., an environment in which two states matter to the survival of bryozoans), then why should a mind be necessary for dealing with an environment with a complexity of three? Indeed, we can imagine an environment in which the bryozoan is at risk from two hundred distinct kinds of predators, each with its tell-tale chemical cue. Perhaps the bryozoan has at its disposal two hundred distinct (but compatible) strategies that defend it against each of these predators:

when detecting the cue associated with sea slugs, grow spines;

when detecting the cue associated with sea cucumbers, emit noxious chemical Q;

when detecting the cue associated with starfish, grow spines and emit noxious chemical R;

etc.

Our super-bryozoan is now superbly adapted to an extremely heterogeneous environment, but because the flexibility of its behavior requires only the detection of chemical cues, there is no more reason to attribute psychological processes to it than there was to our run-of-the-mill bryozoan. Moreover, it's hard to see how adding to this kind of environmental complexity will make it any more likely that the super-bryozoan will evolve a mind (see also Sober 1997).

One might suspect that my criticism of the environmental complexity thesis relies unfairly upon a kind of sorites paradox. One twig does not make a pile of twigs, nor do two, but this does not mean that we can never build a pile of twigs. Likewise, the objection goes, responses to simpler environments do not require minds, but this does not imply that minds are not needed to deal with vastly more complex environments. Yet, the analogy between building a pile of twigs and evolving a mind is not a good one. A pile of twigs is nothing more than a lot of twigs. In contrast, a mind is more than a lot of mindless detectors. If natural selection favors an organism with a mind over an organism with a lot of dumb detectors, it must be in virtue of something minds can do that platoons of detectors cannot. This, presumably, is why we need psychology rather than biology or physics or chemistry to understand minds.

Of course, it may appear as if our super-bryozoan has a mind. In particular, if we didn't know that each of its predators released an idiosyncratic chemical cue that caused an appropriate defense, we might be tempted to ascribe a sort of craftiness to the super-bryozoan (look how it changes its strategy when the sea slug gives up and the snail has a go—it must have certain beliefs about how to defend against snails. Perhaps it even has second-order intentions: beliefs about what snails dislike). Daniel Dennett (1978; 1987) has suggested that we are driven to take an intentional stance toward an organism just when predicting its behavior becomes too difficult otherwise. However, whether something has a mind should not, in my view, be determined by our epistemological limitations. If an organism's interactions with its environment are directed by a million detectors, each one as simple and tropistic as the next, I do not see why, despite what might be our superior ability to predict its behavior from an intentional stance, we should conclude the organism has a mind.

The reflections of the last few paragraphs suggest that environmental complexity, at least without further specification, is not what we should be looking for when trying to say under what conditions a mind is likely to have evolved. The problem with the unrefined complexity answer is that an organism equipped with a mind has no obvious selective advantage over an organism equipped with mindless detectors for each salient condition in the environment. But appreciation of this problem does suggest another kind of answer: rather than looking at the number of obstacles an organism confronts, we should look at the kind of obstacles. If some features of an environment to which an organism responds cannot be detected in the simple manner assumed above, then we have good reason to think the organism's responses are controlled by more than simple detectors. What simple detectors can't detect, perhaps minds can.

Before making this last suggestion more precise, it is worth motivating it with an example. Consider Ellis's list of properties that women look for in a mate. Included on this list are properties like "successful," "well-liked," "respected," and "able to provide for children." Ellis claims that sensitivity to these properties requires psychological processes. One reason this may be so, however, is not that this list of properties is vast in number, but that the properties that identify a good mate are not properties to which a simple, mechanistic detector can be hardwired to respond. What, for instance, constitutes a "successful" man? Success, as a factor in mate preference, presumably is closely associated with material wealth. But how does a woman know a wealthy man when she sees one? In the USA in 1997, a man who drives a Mercedes Benz, wears Ralph Lauren suits, dines at chic restaurants, has a high-status profession, etc., is likely to stand out as successful. But it is important to make two observations about this set of success-indicators. First, women have no reflexive responsiveness to things like Mercedes Benz automobiles and chic restaurants. Accordingly, if a woman identifies a successful man by use of these indicators, she's relying on more than mindless detectors of properties that are coextensive with success. So, unlike the stickleback fish that identifies a rival by sensing red, or a fruit fly that keeps to humid areas by staying away from bright areas, a woman does not identify a successful man by mindlessly sensing expensive automobiles. Second, while this set of indicators is suggestive of success, it is not coextensive with success. As evolutionary psychologists frequently note, *Homo sapiens* evolved during the Pleistocene period. Surely the marks of a successful man today differ from the marks of a successful man in the Pleistocene. More generally, what constitutes a successful man undoubtedly differs through time and across culture, country, and community. The task women face in identifying a successful man cannot, therefore, be solved by any mindless detection device. It must be a device that can spot the successful Pleistocene man, the successful Victorian Englishman, and the successful late-twentieth-century Japanese man.

Fodor (1986) tries to sharpen the above intuition. Fodor distinguishes nomic properties from nonnomic properties as follows:

[S]ome properties are such that objects fall under laws in virtue of possessing them, and some properties are not. For convenience, I shall observe the following convention, if a property is such that objects fall under laws in virtue of possessing it, then that property is ipso facto nomic. All and only nomic properties enter into lawful relations; and, since not all properties enter into lawful relations, not all properties are nomic. (pp. 9–10)

Fodor's idea, then, is that an organism's ability to respond to a nonnomic property (in those cases where it is not relying on the detection of a coextensive nomic property) strongly suggests that the organism employs psychological processing in order to derive from those nomic properties it can detect a representation of the nonnomic property as that nonnomic property. So, to keep to the example I introduced above, a twentieth-century American woman can represent a successful man as a successful man in virtue of psychological operations over those states she possesses as a result of detecting whatever nomic properties successful American men possess. Psychological processes infer the presence of a successful man from the presence of certain nomic properties that women can detect without the need of any psychological processing.

Before scrutinizing Fodor's proposal more closely, we should note that is does provide the kind of help we've been looking for to strengthen the BTM inference. In particular, Fodor's proposal recommends the following methodology: before attributing to an organism a psychological mechanism, check to see whether the features in its environment to which its behavior is directed are nonnomic. If they are nonnomic and are not coextensive with features that are nomic, then the BTM inference gains credibility. If, for example, women are attracted to successful men, but "successful" is a nonnomic property and is not coextensive with any nomic properties, then a woman's capacity to spot a successful man is plausibly under the control of a psychological process.

Notice further that Fodor's suggestion does not appeal to the sort of complexity central to Godfrey-Smith's proposal. We might attribute a psychology to an organism that responds to a single state of an environment if this state is nonnomic. Likewise, we might find no need to attribute a psychology to an organism that lives in an environment with two hundred states, if each of these states is nomic. What matters from Fodor's perspective is the kind of things to which an organism responds rather than the number of things. And this, I think, is a smart perspective to take when trying to justify the use of psychological explanation. Surely biological, chemical, and physical explanations deal successfully with interactions involving large numbers of factors. So if psychological behavior is to be distinguished from biological, chemical, and physical behavior, it must be in virtue of the kinds involved in this behavior.

In my view, Fodor has given us an important insight for understanding when the BTM inference is justified. However, as it stands, Fodor's proposal has two flaws. First, it is incomplete. Second, and more seriously, the concept of "nomicness" Fodor employs is too poorly understood to serve as the keystone of a proposal. Regarding the first problem, Fodor has told us that sensitivity to nonnomic properties requires a mind, but this condition leaves open the question of what to say about an organism that is responding to a nomic property. Surely some responses to nomic properties employ psychological processes (e.g., the phenomenon of shape constancy I describe later), and if we are to use Fodor's proposal to help us understand when a kind of behavior implicates the presence of mind, it would be nice to have in hand some characterization of what, if anything, there is in common between those cases when an organism responds to a nonnomic property and those in which it responds via psychological processing to a nomic property. Until we have such a characterization, Fodor's proposal may be of some help in justifying the BTM inference, but it won't help in those cases where we're wondering whether an organism's response to nomic properties in its environment is a result of mental processes.

Turning to the second problem, we find that the distinction between nomic and nonnomic properties on which Fodor depends is too problematic to carry much weight. How, for instance, are we to categorize complex conjunctive or disjunctive properties? Classical ethologists give the label "sign stimuli" to complex conjunctive properties that cause reflexive responses in certain organisms. The ethologist Robert Hinde provides an example of such a sign stimulus:

The laughing gull chick obtains regurgitated food from the parent by pecking at the latter's beak. The most effective stimulus is an oblong rod, about 9 mm in width, held vertically, and moved horizontally about eighty times a minute. It must contrast with and be darker than its background; if coloured, red and blue are more effective than intermediate wavelengths. (1982, p. 39)

The relevance of sign stimuli for our discussion of Fodor's proposal is this: Responsiveness to sign stimuli seems, as Hinde himself recognizes (1982, p. 41), to rely on mindless physiological processes. Natural selection (presumably) has designed a mechanism in the laughing gull that causes the gull chick to peck at a stimulus of the sort Hinde described. However, Fodor can agree that the gull chick's response is the product of a nonpsychological mechanism only if he grants nomic status to the following property: being an oblong rod 9 mm in width, oriented vertically while moving horizontally about eighty times a minute, contrasting with and darker than its background, and, if colored, reflecting wavelengths at the red or blue portions of the spectrum. But is this property nomic? Note that even if each of the singular properties mentioned in this complex property were nomic, this would be of no help to Fodor, for the gull chick's pecking response is not a function of each singular property by itself, but of only the complex property.

Similarly, suppose an organism responds to a nomic property like "being above 25 degrees Celsius." If there is a law such that when an organism exhibits behavior B then it is above 25 degrees Celsius, then there is also a law such that when an organism exhibits behavior B then either it is above 25 degrees Celsius or X, where "X" could be any property at all. Because such a law exists, properties like "being either above 25 degrees Celsius or being next to Jerry Fodor" must count as nomic, which is absurd. Of course, Fodor would presumably deny that the necessary consequences of laws are themselves laws, but without an account of why this is so we are left resting the BTM inference upon Fodor's intuitions about what sorts of properties are nomic, what constitutes a law, etc.

Finally, Fodor's insistence that it is for the detection of nonnomic properties that psychological processes are necessary has a rather embarrassing consequence. Consider Fodor's example of a nonnomic property, *being a crumpled shirt*. Fodor claims that our ability to recognize an object as a crumpled shirt requires psychological operations, But now suppose that the following regularity is true: when a person sees a crumpled shirt they tend to think "that's a crumpled shirt." This regularity may have all the tell-tale signs of a law: it supports counterfactuals, etc. But if it is a law, then, because laws tell us what properties are nomic, *being a crumpled shirt* is a nomic property. On the other hand, if Fodor can't stomach the possibility that *being a crumpled shirt* is a nomic property, then he can't allow that there are laws of the sort I just described. He can't, in short, allow that there are psychological laws relating organisms to the very kinds of things he thinks psychology has evolved to provide us access to.

In this section we have considered two proposals about the conditions that favor the presence of mind. The first had to do with complexity, with the heterogeneity in environmental features to which an organism has to be responsive, and the second had to do with the kind of environmental features to which an organism has to be responsive. I have argued that both these proposals face difficulties that prevent them from offering any sort of guarantee of the presence of mind. Yet, despite the troubles with these proposals, each does speak to some intuition about why minds might be present. Human beings do have minds and our environments are apparently far more complex than the environments of a paramecium or an earthworm. Can this be just coincidence? Moreover, we have the ability to respond to kinds we are unequipped to detect in any mindless way. Can we say something general about what kinds lend themselves to psychological rather than mindless detection?

3   Justifying the BTM Inference

We are seeking justification for inferring mind from behavior. Under what circumstance are we licensed to say behavior X is evidence for the presence of mind? For Godfrey-Smith what matters is the heterogeneity in environmental conditions an organism encounters, and for Fodor what matters is whether these conditions are nomic or not. Both proposals encounter difficulties for essentially the same reason: they fail to motivate the need in an organism for anything other than mindless detectors. I propose to approach the problem as follows. I will describe a condition under which simple detectors of the sort we have been imagining cannot do the job. As we'll see, complexity of a sort is a good measure of when to expect the presence of psychological processes, so Godfrey-Smith is right about that. However, the complexity that matters to psychology is not one of mere numerical diversity, but of variation in the kinds of cues by which an adaptively significant feature is detected— a point Fodor is after with his insistence that it is responsiveness to nonnomic stimuli that requires psychology.

Here, then, is the picture. Let X be an adaptively significant feature in organism O's environment. X is, for instance, a mate, a predator, a parent, or food. Assume that O has no mindless detector for X and instead tracks X by means of cues for which it does have mindless detectors. But now assume further that these cues vary over time or place to such an extent that the class of cues by which O obtains information about X is essentially open-ended. In such an event, a mindless detector of any one of the cues for X will be of limited or no use to O. So, for instance, consider again the bryozoan that relies on a chemical cue to inform it of the presence of a sea slug. Surely, I claimed, the bryozoan's ability to detect this chemical cue involves no mental processes, and so the bryozoan's responsiveness to sea slugs is not a subject for psychological explanation. But now suppose that the bryozoan manages to track sea slugs despite the fact that the sea slug, though unable to prevent itself from leaking some chemicals, is able to change in nearly limitless ways the kinds of chemicals it does emit. In this case, the bryozoan equipped only with a set of mindless chemical detectors would be no match for the sea slug. If the bryozoan is going to avoid sea slugs not just today but also tomorrow, not just over here but also over there, it will need something other than mindless detectors of chemical cues.

It's important to the above example that the sea slug's capacity to change the kind of chemical it emits remains nearly boundless. To understand why a bryozoan might be better off with a mind requires that we see the contest between the bryozoan and the sea slug as involving more than a simple arms race. If the sea slug's ability to vary its chemical signature were strictly limited, then we might expect that the bryozoan would answer with an ability to detect each of the chemicals in the sea slug's repertoire. Each time the sea slug added a new chemical signature to its set, the bryozoan would evolve a new detector. It is when there is practically no limit to the number of chemical kinds a sea slug releases that the strategy of evolving new detectors becomes unwieldy.[1]

So it seems Godfrey-Smith is right that environmental complexity makes likely the evolution of minds. However, it is not heterogeneity of any sort that increases the probability of mental evolution. There are cases of heterogeneity to which mindless mechanisms produce perfectly adequate responses, as the earlier example of the super-bryozoan illustrates. The kind of heterogeneity that matters, or at least a kind of heterogeneity that matters, is heterogeneity of the cues that provide information about the adaptively significant condition in the environment. When there is tremendous variance in these cues, selection will favor those organisms that evolve a psychological means of identifying the relevant condition.

Similarly, though Fodor is on the right track when looking for those properties that cannot be detected in a simple, mechanistic fashion, it is not because a property is nonnomic that it cannot be mindlessly detected. Many properties with a nonnomic "feel" seem suitable candidates for mindless detection; correlatively, there may be many properties with a nomic "feel" that, because one has access to them only through a limitlessly varied set of cues, require psychological operations for their recognition.

To sum up, the BTM inference that evolutionary psychologists like to make is on safest ground when the behavior being studied involves responses to an environmental condition that can be detected only through an extremely varied set of cues. This consideration explains why Ellis is no doubt right that a woman's mate preferences depend on psychological processing. There is no fixed set of cues—no set of cues that has remained constant from the time our Pleistocene ancestors roamed the earth, no set of cues that has remained invariant across geographic and cultural borders—that mark the successful man. In contrast, the cue by which the laughing gull chick knows its mother, though nonnomic, is regularly correlated with its mother across time and place. Hence the BTM inference in the case of the laughing gull would be quite precarious. As these cases reveal, assessing the BTM inference requires extensive empirical inquiry on a case by case basis. Making the inference requires familiarity not only with the cues that direct responses of a species in one locale, but in many locales; not only today, but over the course of the species' history.

4   The Nature of Psychological Processes

Before closing, I should say something about why, in those situations where I have claimed the BTM inference is sound, it is psychological processes that are doing the work, that are directing the organism's response. How can psychology help an organism respond to X when the cues by which the organism can detect X vary limitlessly? What is the nature of psychological processes such that they are useful in these kinds of situations?

Again, the example of a woman's mate preferences is instructive. Let X be a successful man. Our question is how does a woman know an X when she sees one? The problem facing women is that there is no fixed set of cues that indicate the presence of an X. This is why, presumably, even if a mutation provided a woman with a means for mindlessly detecting nonnomic properties like Mercedes Benz automobiles and Ralph Lauren suits, this mutation would not spread through a population. Who knows what cars and what clothes will distinguish the successful man a generation from now? So, mindless detection of cues for X is not an option for women.

Rather than providing women with mindless detectors of cues for X, selection has favored women who possess general sensory systems—the same as those which all (normal) human beings possess. At bottom, these systems consist of the kind of detectors from which we have been withholding psychological explanations: neural circuitry composing receptive fields that respond selectively to specific forms of stimulation. How we characterize the output of these sensory systems is a controversial matter, but let's allow that these sensory systems provide human beings with perceptions of shapes, surfaces, colors, etc. Our question now is "How does the perception of shapes, surfaces, and colors provide a woman with the means for detecting a successful man?" The answer is: by combining with things like memories and beliefs. The woman remembers that things shaped and colored so are Mercedes Benz automobiles, and she believes that such automobiles cost a lot of money, and if one can afford a car like that then one is successful. Psychological processes are what extend our knowledge of the world beyond what mindless detectors can tell us about, and they do so by way of inferential patterns. It is in virtue of such inferential patterns drawing on beliefs and memories that Pleistocene women, Victorian women, and contemporary Japanese women, all equipped with the same sensory apparatuses, can spot a successful man.

One must be cautious not to read too much into my mention of inferential patterns. In saying that psychological processes rely on inferences, all that's really being said is that they are ampliative: they build on the information they receive from receptors. Thus it is consistent with this understanding of psychology that the sensory systems I mentioned above, which produce perceptions of shapes, surfaces, and colors, are themselves proper objects of psychological investigation. But this is so not because these systems draw on memories and beliefs—for presumably they do not. However, as in the case of perceptual constancies, we perceive an object as being fixed in shape and size despite variations in our viewing angles and viewing distances because of processes that combine information from multiple sources. By drawing on multiple sources of information, our perceptual systems describe for us the world as it is rather than as it appears on the retina.

This view of psychological processes as ampliative fits nicely with my claim about when the BTM inference is justified. In essence, we should expect an organism to have evolved a psychological solution to some problem when the problem requires the organism to have more information about a feature of its environment than cues in its environment provide.[2] When the information an organism can mindlessly detect falls short of the information the organism actually possesses it is because psychological processes are present to span the gap.

Notes

1. I'm of course simplifying matters here. It's possible that even with a small repertoire of chemical emissions it's no longer cost effective for the bryozoan to keep pace with the sea slug. However, this fact does not affect the proceeding point about what counts as good evidence for the BTM inference.

2. I'm simplifying again. Incomplete information may be good enough, depending on the costs and benefits associated with encountering the feature. I'm assuming here that the costs of missing the feature, or the benefits of finding it, are very high.

References

Barkow, J., Cosmides, L., and Tooby, J. (eds.) (1992) *The Adapted Mind: Evolutionary Psychology and the Generation of Culture*. New York: Oxford University Press.

Buss, D. (1992) "Mate Preference Mechanisms: Consequences for Partner Choice and Intrasexual Competition," in Barkow, Cosmides, and Tooby (eds.), op. cit.

Cosmides, L. and Tooby, J. (1994) "Origins of Domain Specificity: The Evolution of Functional Organization," in L. Hirschfeld and S. Gelman (eds.), *Mapping the Mind*, pp. 85–116. New York: Cambridge University Press.

Dennett, D. (1978) *Brainstorms*. Cambridge: The MIT Press.

Dennett, D. (1987) *The Intentional Stance*. Cambridge: The MIT Press.

Ellis, B. (1992) "The Evolution of Sexual Attraction: Evaluative Mechanismsin Women," in Barkow, Cosmides, and Tooby (eds.), op. cit.

Fodor, J. (1986) "Why Paramecia Don't Have Mental Representations," *Midwest Studies in Philosophy* X:3–23.

Godfrey-Smith, P. (1996) *Complexity and the Function of Mind in Nature*. New York: Cambridge University Press.

Harvell, D. (1986) "The Ecology and Evolution of Inducible Defences in Marine Bryozoan: Cues, Costs, and Consequences," *American Naturalist* 128:810–823.

Hinde, R. (1982) *Ethology: Its Nature and Relations with Other Sciences*. New York: Oxford University Press.

Krebs, J. and Davies, N. (1987) *An Introduction to Behavioural Ecology*. Oxford: Blackwell Scientific Publications.

Mann, J. (1992) "Nurturance or Negligence: Maternal Psychology and Behavioral Preferences among Preterm Twins," in Barkow, Cosmides, and Tooby (eds.), op. cit.

Shapiro, L. (1998) "Do's and Don'ts for Darwinizing Psychology," in C. Allen and D. Cummins (eds.), *The Evolution of Mind*; pp. 243–259. New York: Oxford University Press.

Shapiro, L. and Epstein, W. (1998) "Evolutionary Theory Meets Cognitive Psychology: A More Selective Perspective," *Mind and Language* 13: 171–194.

Sober, E. (1997) "Is the Mind an Adaptation for Coping with Environmental Complexity?" *Biology and Philosophy* 12:539–550.

6 DeFreuding Evolutionary Psychology: Adaptation and Human Motivation

David J. Buller

Symons claims that the "potential contribution of Darwinism to psychology does not lie merely in assigning ultimate causes to psychological mechanisms" (1987b, p. 143); rather, by understanding the evolution of the human mind, we will "be aided in understanding its nature" (p. 121). Elsewhere he writes that "an evolutionary view of life can shed light on psyche, which *eludes us because it is us*" (1979, p. vii; emphasis added). This goes well beyond the idea that understanding the evolution of the mind will enable us to infer its internal dynamics. It also conveys the Freudian legacy that our "manifest" image of human motivation is largely a veneer of illusion concealing the truth about the "latent" motives that actually cause us to behave as we do. Indeed, Symons claims, "Darwinism's most significant contribution to psychology may lie in its potential to shed light on these goals, wishes, purposes and desires—these mechanisms of feeling that motivate human action" (1987b, p. 131). For, although common sense "has proved to be a reliable guide to reasoning about the design of perceptual mechanisms," when it comes to the nature of human motivation "the most fertile hypotheses are likely to come from imaginations informed by Darwinism" (p. 131). Informed by these Darwinian imaginations, evolutionary psychology will penetrate the veneer of our manifest image of human motivation and reveal a latent image of a psyche consisting in "emotional/motivational mechanisms, to recognize and look after ... reproductive 'interests'" (1979, p. 308).

This picture of evolutionary psychology coexists uneasily in the literature with occasional explicit denials that evolutionary psychology reveals the nature of human motivation. As Daly and Wilson put it (1988, p. 7):

Evolutionary psychology is not a theory of motivation. No one imagines that genetic posterity (fitness) is a superordinate "goal" in any direct sense.... The concept of natural selection explains behavior at a distinct level complementary to the explanations afforded by motivational theories.

Such denials are peculiar; for, like this passage from Daly and Wilson, they all deny that evolutionary psychology trades in motivational theories by denying that it supposes humans to have a *single* unconscious motive with *inclusive fitness* as its *direct* goal (see also Barkow 1989, p. 112 n. 4; Ellis 1992, p. 284; Buss 1995, p. 10).[1] But it is not at all uncommon to find evolutionary psychologists postulating unconscious motives whose goals pertain to specific aspects of social competition and reproductive success—that is, a *group* of unconscious motives with inclusive fitness as their *indirect* goal. Indeed, in this respect, explanatory practice in evolutionary psychology frequently violates the explicit denials that evolutionary psychology is a source of motivational theories. And, when it does violate the denials, evolutionary psychology presents itself as offering a picture of human motivation that will replace our commonsense manifest image and provide us with a deeper understanding of ourselves.

I will argue that evolutionary psychology is not, and should not be seen as, a source of insight into the latent motives that drive human behavior—that its explanatory function is not, and should not be seen as, that of replacing our manifest image of human motivation with a latent image. To use evolutionary considerations to infer the hidden dynamics of the mind in this way is, to turn a phrase of Symons's (1992) against him, a "misuse of Darwinism in the study of human behavior." But I will *not* be claiming that there are *no* hidden dynamics in the mind, that our commonsense manifest image is a complete account of the mind. Rather, I will only be arguing against a particular way of portraying the mind's hidden dynamics. So let me clarify this issue before moving on.

Consider first what I will call *personal psychology*, which explains behavior by appeal to the full range of thoughts, motives, knowledge, and emotions that we attribute to *persons* in our daily lives.[2] The paradigmatic use of personal psychology is to explain behavior in terms of conscious states; but when we insist on a personal psychological explanation of someone's behavior even though they sincerely disavow possessing the motives and beliefs that feature in the explanation, we postulate what I will call the *personal unconscious*. For example, we may insist that Pat is jealous of Kim's flirtation with Leslie although Pat is not aware of it; here the jealousy is postulated as an unconscious state that is causally efficacious in controlling Pat's behavior. Explanations in terms of the personal unconscious trade in precisely the same types of state, with precisely the same types of content, as explanations in terms of conscious personal psychology; both trade in state attributions such as "Sue is afraid of the dark" and "Sam wants more attention from his father." Because of this, explanations appealing to the personal unconscious make it appear that there is another *person* "in there," with their own motives and beliefs, for whom "I" am merely a front. And when people come to accept at explanation of their behavior in terms of personal unconscious motives, they come to see themselves as having been mistaken about their *true* motives—to see those other persons "in there" as who they *really* are. Acceptance of attributions of personal unconscious motives thus leads to a new sense of self, a closer identification with that "other person" than with the previously avowed motives.

Contrasted with personal psychology is *subpersonal psychology*, which is concerned not with explaining a person's behavior, but with explaining the functioning of the individual psychological mechanisms that compose the human mind, such as those involved in parsing sentences of our native language (see Dennett 1987). Of course, none of us is consciously aware of the subpersonal cognitive processes involved in, for example, estimating depth from binocular two-dimensional retinal displays or parsing a sentence of our native language. So these processes are in a straightforward sense unconscious, constituting what I will call the *subpersonal unconscious*. The information deployed in subpersonal unconscious processes, however, is not the sort of information that figures in the content of personal psychological states (e.g., *that* the orange juice is in the refrigerator); rather, it is information about texture gradients, the angles subtended by the retinas, and so forth. Such information is *subdoxastic*, "inferentially isolated" from the sorts of information that figure in the contents of the conscious motives and beliefs of personal psychology (Stich 1978). Thus, although explanations in terms of the subpersonal unconscious do provide us with insight into the hidden workings of the mind, they never rival or lead to a revision of the attributions of conscious motives that figure in the explanations of personal psychology. Rather, subpersonal psychology complements explanations of personal psychological by revealing the substructural functioning of a mind of which personal psychology is true (Dennett 1987).

There are two ways, then, in which psychological theories can claim to offer us insight into the hidden dynamics of human motivation. On the one hand, a psychological theory can provide us with insight into the subpersonal processes underlying human motivation. In this way, for example, we can learn that increases in female sexual desire around ovulation are a function of peaking levels of estrogens (Adams et al. 1978). Such subpersonal explanations, however, do not involve postulations of *additional* motives; they merely illuminate the mechanisms underlying the motives of which we are antecedently and independently aware. On the other hand, a psychological theory can construct an image of *personal* unconscious motives that allegedly drive our behavior, which is then offered as an alternative account of human motivation intended to *replace* our commonsense manifest image. This type of explanatory project is designed to show that human motivation is not what it appears, that our manifest image is mistaken and that we are actually driven by motives that are not apparent to us. A classic example of this is psychological egoism, according to which all manifestly altruistic motives are just a disguise for the latent selfish motives that are the true causes of behavior. When evolutionary psychologists employ "Darwinian imaginations" to "shed light on" human motivation, they are engaged in this latter type of explanatory project, and it is this to which I will object in what follows. Thus my arguments will not be directed against any explanatory appeals to subpersonal processes underlying human motivation; for the subpersonal unconscious is a respectable staple of the cognitive and brain sciences, and there is every reason to expect that these sciences will teach us a great deal about the subpersonal unconscious processes underlying human motivation. But this is radically different from teaching us that our manifest image of human motivation is mistaken. In what follows, then, I will argue that evolutionary psychology is not, and should not be seen as, a source of theories about *personal* unconscious motives. With this in mind, I will now proceed to use the expression "unconscious motives" as a abbreviated reference only to *personal* unconscious states.

To see how some evolutionary psychological theorizing has constructed a latent image of human motivation by postulating unconscious motives, it is important first to distinguish levels of description in evolutionary psychology (see Buss 1995, pp. 2–5). At the most general level, psychological mechanisms are described as adaptations that enhanced fitness in the *environment of evolutionary adaptedness*. At a slightly less general level, the ecology of the environment of evolutionary adaptedness is taken into consideration in order to identify adaptive problems that psychological mechanisms are described as having evolved to solve. These adaptive problems are, to put it roughly, the *components of fitness* for the human species. For example, one component of fitness in the environment of evolutionary adaptedness was that of successful intrasexual competition, "besting members of one's own sex to gain access to desirable members of the opposite sex" (Buss 1991, p. 465). At an even more specific level, adaptive problems are analyzed into subtasks, the solutions to which constitute solving the adaptive problems identified at the level above. Solving the problem of intrasexual competition, for example, required "acquisition of resources required by a potential mate" and "successful courtship of the potential mate" (Buss 1991, p. 465). Finally, specific descriptions can be derived from the task analysis about how evolved psychological mechanisms function under specified input conditions, where the input conditions characteristic of contemporary environments are of particular interest.

It is true that no evolutionary psychologist postulates that humans have a desire for inclusive fitness per se (the most general level of description above). But some frequently employ evolutionary considerations to postulate unconscious motives with *components of fitness* as their goals. For example, Nesse and Lloyd write: "Psychodynamic psychologists and psychiatrists may find in evolutionary psychology new possibilities for a theoretical foundation in biology" (1992, p. 601). "For instance," they continue, "Freud's emphasis on the sexual origins of human motivation as reflected in the concept of 'libido' is remarkably congruent with the evolutionary psychobiologist's recognition of the crucial importance of reproductive success to human motivation" (p. 619). Barkow concurs that "your conscious *and* unconscious goals presumably are linked to the kinds of activities that would have tended to enhance the fitness of your ancestors" (1989, p. 112; emphasis added).

Such goals are to be defined as "evolutionary functions" (p. 113) that "are rather clearly linked to inclusive fitness" (p. 110), and they include "not merely having sex, but attracting mates who, in terms of the differing fitness interests of the sexes . . . , are likely in effect to enhance one's genetic fitness" (p. 110). And in their luridly fascinating work on human sperm competition, Baker and Bellis write, "unconscious programming is actually likely to be a more accurate indicator of adaptive function than is conscious rationalization" (1995, p. 185). Thus, when they discovered that women are more likely to engage in extra-pair copulations during ovulation than at any other time in their menstrual cycle (pp. 160–166), they were led to interpret the desire for sex with the extra-pair partner as merely the conscious manifestation of the unconscious motive to harvest an ejaculate with better genes than those contained in the primary partner's ejaculates (pp. 184–185; see also Baker's [1996] popularization of their work, which is rife with such attributions of unconscious motives).

Even Daly and Wilson, in violation of their own claim, appeal to unconscious motives. In a critical discussion of Freud, they fault Freud not for his theory of the dynamic unconscious, but for having gotten its dynamics wrong. A young boy's conflicts with his father are due not to unconscious sexual jealousy over his parents' sexual relationship as per Freud's theory of the Oedipus complex, they argue, but to a desire to postpone as long as possible the addition to the family of a sibling competing for parental resources (1988, p. 115). They derive this conclusion from Trivers's (1974) theory of parent-offspring conflict—a clear case of using evolutionary theory to formulate motivational hypotheses.

Another telling instance in Daly and Wilson occurs in their discussion of infanticide (1988, chap. 3). On the basis of a cross-cultural analysis, they argue that infanticide is virtually always committed under circumstances that enhance parental fitness—for example, the murdered infant is not viable (so it would be an undue drain on parental resources that would more efficiently be invested in, or saved for, other offspring), the parent lacks the resources to rear the child (so attempting to rear it might jeopardize future reproductive potential), or a husband kills a child his wife bore from an extra-pair copulation. Daly and Wilson claim to find just one clear violation of the rule that infanticide enhances parental fitness. In Yanomamö culture, it is taboo for a woman to have sex from the time of discovered pregnancy until the completion of weaning; so, in order to sooner resume a regular sex life, some young Yanomamö couples have committed infanticide. Finding it so paradoxical that the desire for sex itself could counter reproductive interests in this way, Daly and Wilson claim that this motive is actually inconsistent with predictions of evolutionary theory (p. 58). But this can only appear paradoxical if it is assumed that the desire for sex is a subordinate motivational goal to the superordinate motivational goal of reproduction. If, instead, sex is seen as a *motivational end* in itself, which merely *functions on average* to promote reproductive success without (always) being motivationally subordinate to it, it is not surprising that people will pursue the motivational end wholly detached from its evolutionary function. Indeed, once the motivational genie is let out of the bottle, it has the power to call the shots for itself.

Unconscious motives work their way into evolutionary psychological theorizing in other subtle ways. The concept of a *Darwinian algorithm*, as typically articulated, requires the tacit postulation of unconscious motives whose goals are components of fitness. According to the standard articulation, the theory of evolution by natural selection "defines adaptive information-processing problems that the organism must have some means of solving," and psychological mechanisms evolved to deal with these problems (Cosmides and Tooby 1987, pp. 284–285; cf. Tooby and Cosmides 1989, pp. 40–41). The input to an evolved psychological mechanism "specifies to the organism the particular adaptational problem it is facing," and a Darwinian algorithm is a decision procedure that transforms this input into a behavioral output that "solves a particular adaptational problem" (Buss 1991, p. 464 n. 2; cf. Cosmides and Tooby 1987, p. 286).

Discussions of Darwinian algorithms tend to ignore motivational goals and to focus exclusively on the *cognitive* processes involved in solving adaptive problems (the extraction and processing of information from the environment). But the mental causation of behavior requires not only cognition, but also motivation. Neither information-bearing states nor motivational states cause behavior in the absence of the other; for, to put it crudely, an organism with information but no motivation has nowhere to go, and one with motivation but no information has no clue how to begin getting there. In addition, the mental causation of behavior requires that information-bearing states be *about motivational goals*, that is, that the contents of information-bearing states match up appropriately with the goals of motivational states. If one has reliable information about how to obtain X, but lacks motivation to obtain X, having instead only motivation to obtain Y, the information about how to obtain X will fail to cause X-directed behavior (unless one also believes that obtaining X will result in obtaining Y). So, for a Darwinian algorithm to produce an actual behavioral solution to an adaptive problem, an individual must possess some motivational state whose goal is (roughly) the solution of that adaptive problem.

To illustrate this point, consider the following. Research has shown that on average across cultures heterosexual women prefer as long-term mates men who are loyal and have access to resources essential to child rearing (Buss 1994, chap. 2). But, once in a stable relationship with such a man, women appear to employ a single, different criterion in choosing an extra-pair sex partner: a low degree of bodily asymmetry, which is a sign of heritable developmental stability (Gangestad and Thornhill, 1997). And, it is typically assumed, women have extra-pair copulations with men with low asymmetry during fertile periods of their cycles in order to obtain superior genes for a child that can then be reared on the secured resources provided by the cuckolded long-term mate with inferior genes (Buss 1994, p. 90). So suppose that, in choosing partners with low asymmetry, women employ a Darwinian algorithm for extracting information about male bodily asymmetry, performing cross-male comparisons of asymmetry and arriving at a decision about which potential partner has "good genes." For such information to result in actual behavior, it must be accompanied by a motivational state directed toward achieving *the very thing* that the information is about—in other words, by a motive to have a sex partner with "good genes." If a woman has no motivation to have sex with a man with "good genes," no amount of information about the genetic quality of potential partners will result in sexual behavior directed toward them; for, in the absence of an appropriately contentful motivational state, the information-bearing states will not interact with motivational states in such a way as to cause behavior.

A few passing comments can be found that seem to acknowledge the postulation of such unconscious motives driving Darwinian algorithms. For example, Buss writes that "the major goals toward which humans direct action" are "problems that historically had to be solved to enable reproductive success" (1991, p. 484). He continues: "Although there exists substantial variability in how individuals *frame* their goals," evolutionary theory can illuminate "the underlying species-typical goal structure" within the human mind (p. 485; emphasis added). In a Freudian vein he adds, "nothing in an evolutionary perspective requires that humans be aware of" this underlying goal structure (p. 470). But, if the evolutionary perspective leads to postulating motives to secure a mate with "good genes," for example, it is trading almost exclusively in goals of which humans are *not* aware. Symons provides an example of this when he describes "one mental mechanism for assessing sexual attractiveness as a rule that specifies, 'detect *and prefer* the population composite'" (1987a, p. 118; emphasis added).

Postulations of unconscious motives such as these paint a picture of humans as motivated to achieve maximal reproductive success and as continually calculating the best means of achieving it. But such explanatory appeals to unconscious motives directed at reproductive success (or other components of fitness) are problematic, and for the very reasons that evolutionary psychologists have criticized sociobiology. Sociobiology tended to view humans as "inclusive fitness maximizers." There are two problems with this view. First, it is empirically inadequate; for human behavior too often fails to promote inclusive fitness. As Buss says: "If men had as a goal the maximization of fitness, then why aren't they all lined up to give donations to sperm banks, and why do some individuals decide to forego reproduction entirely?" (1995, p. 10). Such examples, of course, can easily be multiplied. But this gross empirical inadequacy is merely a symptom of a second problem, which Buss calls the *sociobiological fallacy*; for viewing humans as fitness maximizers "conflates a theory of the origins of mechanisms (inclusive-fitness theory) with a theory of the nature of those mechanisms" (p. 10). Evolutionary psychology has presented itself as free of these difficulties in virtue of seeing humans as *adaptation executors*, rather than fitness maximizers, and seeing psychological mechanisms as adaptations to specific problems in ancestral environments. Behavior that fails to maximize fitness is thus explained as produced in response to contemporary environments by psychological mechanisms that are adapted to ancestral environments. So, the story goes, human sexuality was formed by natural selection well before the appearance of sperm banks and the wealth of contraceptives available at the local pharmacy; and that is why human sexuality doesn't necessarily lead directly to reproductive success within contemporary cultural environments.

But, when evolutionary psychology appeals to unconscious motives whose goals are components of fitness, it succumbs to the same objections that Buss levels against sociobiology. Simply replacing the single goal of fitness maximization with a group of goals related to components of fitness does not make the problems disappear. For, if we have unconscious motives for reproductive success (for lots of children by mates with "good genes," for example), then *given our unsurpassed intelligence* we should be more effective in achieving our unconscious goals than we are. At the very least, we should be *unconsciously* achieving our unconscious goals—as per the suggestion by Baker and Bellis that a "man who 'accidentally' slips out of a condom, leaving it in the vagina full of sperm" is following unconscious programming in pursuit of the unconscious goal of reproduction (1995, p. 185). If there are such unconscious programs for pursuing unconscious motives, however, it is a mystery why they do not lead to more consistent achievement of our unconscious goals of reproductive success. Why don't men "accidentally" slip out of condoms more often than they do? Why don't women forget to take their pills, or improperly place their diaphragms, more often than they do?

Since the standard view is that evolved psychological mechanisms are *domain-specific modules*, responsive only to highly specific forms of environmental information (see Cosmides and Tooby 1987), it could be argued that, when the environment does not provide sufficient unambiguous cues of the sort to which our psychological mechanisms are responsive, they will not function effectively in achieving our unconscious goals. But there are well-known reasons that the mind cannot consist solely of domain-specific mechanisms; there must be some domain-general "central systems" that interface with the modules (see Fodor 1983, pp. 101–103). Thus the failure of a module to achieve the goal it's evolved to achieve wouldn't exclude the possibility of domain-general processing taking over to achieve that goal.

Consistent failure to achieve our unconscious goals, then, requires that the mind be constructed in such a way that unconscious goals are concealed from domain-general processing. But then it's a mystery why such a mental architecture would have been favored by natural selection over an architecture in which domain-general processing had greater access to information about the success rate of the functioning of the modules and that passed control to domain-general processing when modules failed to achieve their goals. Evolutionary psychologists have argued persuasively that there are definite adaptive advantages to having domain-specific modules dedicated to stable and recurrent problems (Cosmides and Tooby 1987; Tooby and Cosmides 1989). But the most those arguments show is that an advantage would accrue to an architecture in which the *default assignment* of solving adaptive problems fell to dedicated domain-specific mechanisms. If those mechanisms ever persistently failed to achieve the goals for which they were designed, an architecture that passed control to domain-general processing to figure out how to achieve those goals would surely have been favored by natural selection. Of course, one could argue that the necessary mutations never arose to build such an effective mental architecture. But this response would be grossly ad hoc given the strong adaptationism that informs *all* theorizing in evolutionary psychology; so this response is not available to the typical evolutionary psychologist.

Besides, we needn't be driven to such speculative lengths by the failure of our behavior to satisfy more successfully our putative unconscious motives for reproductive success. For there is a far simpler explanation of this failure: We possess *no such unconscious motives*. The far simpler explanation is that human motivation is what our commonsense manifest image takes it to be. That is, we are motivated by desires for sex and love, loyalty to loved ones, and a powerful concern for the welfare of our children—not by desires for reproductive success. A woman engaging in extra-pair copulation during ovulation is motivated by desire for her lover, period—not *at any level* by a desire to harvest "good genes." Not only is this explanation simpler, but it avoids the fallacy involved in postulating motives directed at reproductive success. For to attempt to infer the latent nature of human motivation on the basis of evolutionary considerations is to recommit the sociobiological fallacy of misconstruing a theory of the *origins* of the human mind as a theory of its *nature*. Simply replacing sociobiology's single motivational goal of fitness maximization with a group of goals related to reproductive success does not immunize evolutionary psychology against this fallacy; it merely relocates the source of infection. So, to avoid this fallacy, I suggest that we take literally Daly and Wilson's (explicit) claim that evolutionary psychology is not a theory of motivation, which in turn involves rejecting the idea that evolutionary psychology can inform us of the latent psychology beneath our manifest psychological image of ourselves.

I have on offer both a diagnosis of and a cure for the temptation to view evolutionary psychology as a source of insight into latent motives directed at reproductive success. I will begin with the diagnosis, of which there are two aspects. First, the evolutionary half of evolutionary psychology views us through the lens of evolutionary theory, which conceives of organisms as "striving" to achieve the adaptive goals of successful survival and reproduction. Nothing in evolutionary theory, however, takes these adaptive goals to be "internal" motivational goals of organisms that cause the "striving." Rather, adaptive goals are "external," imposed on organisms by the standards of success in the competition for survival and reproduction without their needing to be aware that they are engaged in that competition. It just so happens that some of them do things that make them more likely than others to be crowned "winners" of the competition by the process of natural selection. The psychology half of evolutionary psychology views us, independently of evolutionary considerations, as having minds comprised of "goal-directed mechanisms" (Symons 1987b, p. 121). When these two halves are joined, and our goal-directed psychological mechanisms are seen as having been "necessarily designed by natural selection" (p. 121), it becomes easy to think that the goals that direct psychological mechanisms from the inside (motivational goals) are the goals of success in reproductive competition (adaptive goals), since natural selection both sets the adaptive goals and designs the psychological mechanisms that achieve them. But it doesn't follow that psychological mechanisms contribute to reproductive success in virtue of having internalized adaptive goals. We would never be tempted to think that livers contribute to success in the game of reproductive competition in virtue of having internalized adaptive goals; and the fact that psychological mechanisms themselves contain goals should make us on more tempted to suppose that the reason they contribute to reproductive success is that their internal goals are adaptive goals.

Second, all talk of goals involves the teleological expression "in order to" or its equivalents, both in psychological contexts (I went to the freezer in order to get the ice cream) and adaptive contexts (the male lion killed the cubs in order to bring their mother into estrus). One of the functions of the expression "in order to" is to link goals in a hierarchy—as in "I went to the freezer in order to get the ice cream in order to satisfy my desire for something sweet" or "the male lion killed the cubs in order to bring their mother into estrus in order to mate with her." Since the expression "in order to" cuts across both motivational and adaptive contexts *and* serves to link goals in a hierarchy, it is tempting to link motivational with adaptive goals in a common hierarchy so that motivational goals appear to be subordinate to, or the *intentional means* to achieving, the superordinate adaptive goals. This kind of mixed hierarchy is explicit, for example, in Barkow (1989, pp. 106–116). Once this point is reached, the Freudian legacy easily comes into play: If individuals are consciously aware only of their subordinate (motivational) goals, their superordinate (adaptive) goals must be unconscious; so their conscious goals are pursued only *in order to* achieve their unconscious goals. In this way, adaptive goals get construed as having a *psychological* reality.

But at this point the sociobiological fallacy has been committed: An (adaptive) account of the origins of the mind has been confused with an account of the (motivational) nature of the mind. Or, as Ghiselin once said more simply, "motives have been confused with functions" (1973, p. 965). This confusion stems from a failure to recognize that "in order to" has a different meaning in motivational contexts than in adaptive contexts. Very roughly, saying that I went to the freezer in order to get the ice cream means that a representation of the goal state of getting the ice cream was among the *causes* of my going to the freezer. In motivational contexts, then, "X in order to Y" means that a representation of Y was a causal antecedent of X. But, saying that the male lion killed the cubs in order to bring their mother into estrus means that killing the cubs has the *function* of bringing their mother into estrus, and this in turn means something roughly like the following: Killing cubs contributed to the reproductive success of ancestral male lions because it had the effect of bringing the mother into estrus so that the killer male could mate with her. So, in adaptive contexts, "X in order to Y" is functional, meaning that Y is a *effect* of X because of which X contributed to ancestral reproductive success. The idea that evolutionary psychology will inform us of the latent psychology beneath our manifest psychological image is thus a product of the fallacy of conflating the adaptive function (or effects) of a particular behavior with the motives that cause it.

With the fallacy now diagnosed, I turn to its cure. At a general level, the cure is simple: Keep motivational goals and adaptive goals distinct. But this requires revising the evolutionary psychological conception of psychological *strategies*. Currently, the model of theory construction in evolutionary psychology is as follows: Identify an adaptive problem, assume the solution of that problem to be the goal of an evolved psychological mechanism, and then work up a Darwinian algorithm for solving the problem (which it is then hypothesized the psychological mechanism executes). Goals, in this picture, are adaptive goals and strategies are procedures for achieving them. But, since Darwinian algorithms can only be in the service of *motivational goals* and do not operate to directly achieve adaptive goals, they can contribute to achieving adaptive goals only insofar as motivational goals themselves *function* to achieve adaptive goals. Thus we should see our *motivational goals* as our evolved psychological strategies and Darwinian algorithms as the procedures for pursuing them. In relation to achieving adaptive goals, then, so-called Darwinian algorithms are at best loose *heuristics*.

In this revised picture, humans possess all the familiar motivational goals, which are not merely means to the achievement of other goals, but are the *ends* that we pursue in our actions. Having one of these motivational goals is itself a strategy—a *motivational strategy*—in the competition to achieve adaptive goals (which are not among the goals that *motivate* us). Some motivational strategies, however, no doubt led to greater success than others in achieving adaptive goals in our ancestral population. In such cases of differential success between competing motivational strategies, there will be an adaptationist explanation of why one of the strategies was more successful than its rivals in achieving some evolutionary goal. And, once in possession of that explanation, we can say that people now pursue that motivational strategy *in order to* promote their reproductive interests. But this is not a motivational "in order to." It is a purely *functional* "in order to," indicating *only* that among individuals in the ancestral population pursuing that motivational strategy, the *average ratio* of beneficial *effects* to costs of playing that strategy was greater than the same average ratios among the groups of individuals playing alternative strategies. Thus evolutionary explanations of our motivational strategies do not complement our ordinary motivational explanations of someone's behavior by postulating additional motives with adaptive goals. Rather, they simply explain why the motives that we manifestly possess constituted more adaptive motivational strategies than any alternative motivational strategies in our ancestral population. Thus the evolutionary part of evolutionary psychology does not lead us to replace our conception of human motivation with an alternative picture of what "truly" motivates us to behave as we do; it simply informs us of how our manifest motives contributed to the fitness of our ancestors. So, contra Symons, an evolutionary view of life sheds light only on the *ecological* area *around* psyche; it does not illuminate the previously dark area *within* psyche. This is all evolutionary psychology *can* do if we take literally Daly and Wilson's other explicit claim that evolutionary explanations account for "behavior at a distinct level complementary to the explanations afforded by motivational theories" (1988, p. 7).

To illustrate this alternative picture of evolutionary psychology, consider two contrasting motivational strategies, *jealousy* and *insouciance*. Those playing jealousy suffer extreme emotional duress whenever they detect signs they perceive as indicating that their mate is engaging in or contemplating extra-pair copulations, and this duress motivates "mate retention tactics," actions "typically intended either to cut off a rival or to prevent the mate's defection" (Buss 1994, p. 126). Such tactics may include "dissuading potential competitors, luring one's mate with positive inducements, or even rendering one's mate less attractive or evocative to competitors" (Buss 1988, p. 292). Those playing insouciance, in contrast, show total indifference to signs that their mates might be engaged in extra-pair involvements. Other things being equal, the jealousy strategy will result *on average* in more mate retentions than the insouciance strategy, and this in turn will result in the continued investment by the retained mate in current offspring or future reproductive endeavors. So jealousy should lead to greater reproductive success than its rival motivational strategy insouciance. When this is the case, we can say that people are motivated by jealousy *in order to* protect their reproductive interests. But it is important that we not construe this "in order to" motivationally; that is, it does not function to link jealousy to the superordinate motivational goal of protecting reproductive interests. Rather, it is a purely functional "in order to," indicating only that the reason jealousy is a successful motivational strategy is that *on average* it has the *effect* of retaining one's mate and, consequently, promoting one's reproductive interests. Thus the evolutionary account of jealousy merely demonstrates why jealousy was favored by selection over its rival motivational strategies among our ancestors.

This picture of the role of adaptationist explanations should be boringly familiar from all of the nonpsychological domains of evolutionary theory. From the standpoint of psychologists that have been looking to evolutionary theory for insights about the nature of the human mind, however, it will appear disappointingly deflationary. But there are three benefits of this deflationary picture of evolutionary psychological explanation, which I would like to sketch in concluding.

First, it brings evolutionary explanations of psychological traits into line with evolutionary explanations of nonpsychological traits. In the nonpsychological case, we would never explain how some observable trait contributes to fitness by postulating that it has internalized, and now pursues, goals related to reproductive success. For example, when we discover that the kidneys filter metabolic wastes from the blood, we do not explain how this contributes to fitness by attempting to link it with some additional, underlying operation of the kidneys; rather, we simply show the fitness benefits that accrue due to *that very process* of filtering metabolic wastes. It is the Freudian legacy of the dynamic unconscious that tempts us in the psychological case to internalize adaptive goals into the unconscious and then view them as the hidden driving force behind our behavior and reproductive success. But the Freudian legacy should be resisted in favor of a non-psychologically-mediated explanation of how observable psychological traits succeeded in contributing to fitness in ancestral populations. That is, we should not attempt to explain how some manifest psychological trait contributed to fitness by postulating additional, "hidden" psychological traits or processes; rather, we should explain it in terms of the operation of extra-psychological, ecological factors alone, just as we do in nonpsychological cases. The picture of explanation in evolutionary psychology that I have presented here precisely fits this bill.

Second, this deflationary picture of explanation in evolutionary psychology succeeds in avoiding the sociobiological fallacy, since it sharply distinguishes adaptive goals (which figure in an account of the *origins* of the human psyche) from motivational goals (which figure in an account of the *nature* of the human psyche). So there is no special problem posed by the persistent failure of our behavior to satisfy our unconscious motives to achieve adaptive goals, for we simply have no such motives. And the motivational strategies that we do possess need not be perfect or even highly reliable in achieving reproductive success; they need only to have had a higher ratio of beneficial effects to costs than any competing motivational strategies in our ancestral population. This allows a motivational strategy a great deal of room for failure to achieve reproductive success in specific instances.

Third, in spite of these facts, the deflationary picture offered here is nonetheless fully compatible with all existing evolutionary psychological explanations. For, whenever such an explanation postulates an unconscious motive, it will be linked to a manifest motive by the expression "in order to" (or some equivalent), as in "Jane desired an extra-pair copulation during ovulation *in order to* satisfy her desire for an ejaculate rich in 'good genes.' " Such explanations can be systematically transformed into explanations of the type I have urged simply by reinterpreting this motivational "in order to" functionally. Thus, rather than taking such explanations as linking a manifest motive with a superordinate motive in a common motivational hierarchy, we should take them as providing a (partial) explanation of the *effects* because of which the manifest motive was favored by selection over alternative motivational strategies in ancestral environments. So, the above example should be interpreted as an explanation sketch of the following sort: The desire for extra-pair sex during ovulation *on average* had the *effect* of promoting sperm competitions in which the sperm with "good genes" won; so, among women in our ancestral population, the motivational strategy of desiring extra-pair sex during ovulation had a higher average ratio of beneficial effects to costs than any alternative motivational strategies, and thus led to the greatest reproductive success. In this way, all references to unconscious motives can be purged from evolutionary psychological explanations, while still leaving these fully evolutionary explanations of psychological traits. Then it's only a matter of determining whether such adaptationist evolutionary psychological explanations are *true*.

Acknowledgments

I would like to thank Anjay Elzanowski, Peter Godfrey-Smith, Valerie Gray Hardcastle, Karen Neander, Thomas Polger, and Kim Sterelny for their comments on an earlier version of this essay; and I would like to thank Elliott Sober for very helpful comments and an ensuing volley of e-mails.

Notes

1. Oddly, Symons himself offers such a denial, but related to a specific explanatory case: "it would be inaccurate to infer that a furious, cuckolded husband only imagines himself to be angry at his wife's sexual peccadilloes when, in some more profound sense, what he is 'really' doing is promoting the survival of his genes" (1979, pp. 306–307).

2. One might be tempted to think of personal psychology as what is commonly called "folk psychology." But I want to resist this identification. For folk psychology is typically taken to consist of a set of lawlike generalizations formulated over the intentional states of *belief* and *desire*. But our everyday explanations of behavior are very seldom formulated in terms of beliefs and desires. Indeed, we generally explain behavior in terms of motives, emotions, character traits, and moods. I want to include all these in personal psychology—to include the full range of explanations in terms of love, lust, envy, jealousy, irascibility, and so forth—regardless of whether such states are fully analyzable in terms of the intentional states of belief and desire. If they are, then personal psychology is just folk psychology; if they are not (and I think they are not), then folk psychology, if there is such a thing, forms only a proper part of what I am calling personal psychology.

References

Adams, David B., Ross Gold, Alice, and Burt, Anne D. (1978) "Rise in Female-Initiated Sexual Activity at Ovulation and Its Suppression by Oral Contraceptives," *New England Journal of Medicine* 299:1145–1150.

Baker, Robin. (1996) *Sperm Wars: The Science of Sex*. New York: Basic Books.

Baker, R. Robin and Bellis, Mark A. (1995) *Human Sperm Competition: Copulation, Masturbation, and Infidelity*. New York: Chapman & Hall.

Barkow, Jerome H. (1989) *Darwin, Sex, and Status: Biological Approaches to Mind and Culture*. Toronto: University of Toronto Press.

Buss, David M. (1988) "From Vigilance to Violence: Tactics of Mate Retention in American Undergraduates," *Ethology and Sociobiology* 9:291–317.

Buss, David M. (1991) "Evolutionary Personality Psychology," *Annual Review of Psychology* 42:459–491.

Buss, David M. (1994) *The Evolution of Desire: Strategies of Human Mating*. New York: Basic Books.

Buss, David M. (1995) "Evolutionary Psychology: A New Paradigm for Psychological Science," *Psychological Inquiry* 6:1–30.

Cosmides, Leda and Tooby, John. (1987) "From Evolution to Behavior: Evolutionary Psychology as the Missing Link," in J. Dupré (ed.), *The Latest on the Best: Essays on Evolution and Optimality*, pp. 277–306. Cambridge, MA: The MIT Press.

Daly, Martin and Wilson, Margo. (1988) *Homicide*. New York: Aldine de Gruyter.

Dennett, Daniel C. (1987) "Three Kinds of Intentional Psychology," in D. Dennett, *The Intentional Stance*, pp. 43–68. Cambridge, MA: The MIT Press.

Ellis, Bruce J. (1992) "The Evolution of Sexual Attraction: Evaluative Mechanisms in Women," in J. H. Barkow et al. (eds.), *The Adapted Mind*, pp. 267–288. New York: Oxford University Press.

Fodor, Jerry A. (1983) *The Modularity of Mind*. Cambridge, MA: The MIT Press.

Gangestad, Steven W. and Thornhill, Randy. (1997) "The Evolutionary Psychology of Extrapair Sex: The Role of Fluctuating Asymmetry," *Evolution and Human Behavior* 18:69–88.

Ghiselin, Michael T. (1973) "Darwin and Evolutionary Psychology," *Science* 179:964–968.

Nesse, Randolph M. and Lloyd, Alan T. (1992) "The Evolution of Psychodynamic Mechanisms," in J. H. Barkow et al. (eds.) *The Adapted Mind*, pp. 601–624. New York: Oxford University Press.

Stich, Stephen P. (1978) "Beliefs and Subdoxastic States," *Philosophy of Science* 45:499–518.

Symons, Donald. (1979) *The Evolution of Human Sexuality*. New York: Oxford University Press.

Symons, Donald. (1987a) "An Evolutionary Approach: Can Darwin's View of Life Shed Light on Human Sexuality?" in J. H. Geer and W. T. O'Donohue (eds.), *Approaches and Paradigms in Human Sexuality*, pp. 91–125. New York: Plenum.

Symons, Donald. (1987b) "If We're All Darwinians, What's the Fuss About?" in C. Crawford et al. (eds.), *Sociobiology and Psychology: Ideas, Issues, and Applications*, pp. 121–146. Hillsdale, NJ: Lawrence Erlbaum Associates.

Symons, Donald. (1992) "On the Use and Misuse of Darwinism in the Study of Human Behavior," in J. H. Barkow et al. (eds.), *The Adapted Mind*, pp. 137–159. New York: Oxford University Press.

Tooby, John and Cosmides, Leda. (1989) "Evolutionary Psychology and the Generation of Culture, Part I: Theoretical Considerations," *Ethology and Sociobiology* 10:29–49.

Trivers, Robert L. (1974) "Parent-Offspring conflict," *American Zoologist* 14:249–264.

III INNATENESS

7 Innateness Is Canalization: In Defense of a Developmental Account of Innateness

André Ariew

Lorenz proposed in his (1935) articulation of a theory of behavioral instincts that the objective of ethology is to distinguish behaviors that are "innate" from those that are "learned" (or "acquired"). Lorenz's motive was to open the investigation of certain "adaptive" behaviors to evolutionary theorizing. Accordingly, since innate behaviors are "genetic," they are open to such investigation. By Lorenz's lights, an innate/acquired or innate/learned dichotomy rested on a familiar Darwinian distinction between genes and environments. Ever since Lorenz, ascriptions of innateness have become widespread in the cognitive, behavioral, and biological sciences. The trend continues despite decades of strong arguments[1] that show, in particular, the dichotomy that Lorenz invoked in his theory of behavioral instincts is literally false: No biological trait is the product of genes alone. Some critics[2] suggest that the failure of Lorenz's account shows that innateness is not well defined in biology and that consequently the practice of ascribing innateness to various biological traits should be dropped from respectable science.

Elsewhere (Ariew 1996) I have argued that despite the arguments of critics, there really is a biological phenomenon underlying the concept of innateness. On my view, innateness is best understood in terms of C. H. Waddington's concept of "canalization," that is, the degree to which a trait is innate is the degree to which its developmental outcome is canalized. The degree to which a developmental outcome is canalized is the degree to which the developmental process is bound to produce a particular endstate despite environmental fluctuations both in the development's initial state and during the course of development. The canalization account differs in many ways from the traditional ways that ethologists such as Konrad Lorenz originally understood the concept of innateness. Most importantly, on the canalization account the innate/acquired distinction is not a dichotomy, as Konrad Lorenz had it, but rather a matter of degree—a difference that lies along a spectrum with highly canalized development outcomes on the one end and highly environmentally sensitive development outcomes on the other end. Nevertheless, I justified the canalization account on the basis of a set of desiderata or criteria that I suggested falls out of what seemed uncontroversial about Lorenz's account of innateness. Briefly: Innateness is a property of a developing individual; innateness denotes environmental stability; and innate-ascriptions are useful in certain natural-selection explanations (more below). From that same set of desiderata I argued (1996) that neither the concept of heritability nor of norms of reactions—two concepts from population genetics—suffices to ground innateness.

In this chapter, I wish to provide further support of the canalization account in two ways. First, I wish to better motivate the desiderata by revisiting a debate between Konrad Lorenz and Daniel Lehrman over the meaning and explanatory usefulness of innate ascriptions in ethology. Second, I wish to compare my canalization account of innateness with accounts proposed by contemporary philosophers, one by Stephen Stich (1975), another by Elliott Sober (1998), and a third by William Wimsatt (1986).

1 Lorenz's Theory of Instincts

Within Lorenz's theory of instincts, the theoretical significance of identifying innate traits is threefold: it is useful in taxonomy, in individual explanation, and in evolutionary explanation (Richards 1974). Roughly:

(1)   Taxonomy involves characterizing species-specific traits as opposed to traits that pick out differences between individuals in a species or a population.

(2)   Individual explanation: Citing innate behavior may explain why individual organisms act as they do in certain circumstances. For example, why does a stickleback attack a wax model of a fish that lacks all structural resemblance to a rival stickleback? The answer is that the model has a red underbelly, which serves in sticklebacks to release an innate attacking behavior (Richards 1974).

(3)   Innate traits are genetically determined and hence are subject to evolutionary investigation and explanation. Lorenz sought to provide natural-selection explanations for certain "adaptive" species-specific behaviors.

According to Lorenz, field observations and "isolation experiments" constitute the empirical investigation of behaviors. In an isolation experiment, the organism under study is deprived of the opportunity to "learn" or acquire the candidate behavior from environmental cues. If the organism undergoing isolation develops the trait "normally" then the trait or behavior is said to be innate. For example, Lorenz observed that female mallards raised to reproductive age in exclusive company of pintail ducks show no sexual affinity for the pintail drakes. But upon seeing a male mallard for the first time, the females immediately engage in the sexual courtship behavior particular to their species. Remarkably, a mallard expresses courtship behavior even if it has had no opportunity to learn it. That is, mallards that are naturally or experimentally deprived of the opportunity to observe courtship behavior tend to develop it nonetheless. Determining that the mallard's courtship behavior is innate allows the investigator to (1) include the behavior as part of the taxonomy of mallards, (2) explain why mallards exhibit courtship behavior, and (3) open the study of courtship behavior to evolutionary investigation.

As Lorenz's critics have demonstrated, however, Lorenz's account is unsatisfactory. To attribute behavior to the genes is to misunderstand how biological traits are manifested. No biological trait is the instantaneous product of genes alone or of the environment alone. Biological traits are the product of a developmental process that depends on the actions and interactions of *both* genes and environment. Lorenz thought that the results of deprivation experiments showed that some traits appear "out of a vacuum" (Lehrman 1953). Yet, even behaviors manifested in deprivation depend on *some* environmental interaction, namely the environment of the deprivation device (assuming that the creature is kept alive). It follows from this truism that no phenotype is the product of genes alone.

2   The Lorenz-Lehrman Debate

Lorenz devoted a small book to respond to his critics (Lorenz 1965). In his response, Lorenz makes some concessions though he maintains that the category of innateness is useful to ethology. Lorenz's main concession is to acknowledge that attributing behavior to the genes is misleading in the context of a developmental explanation of behavior. Nevertheless, Lorenz asserts that the categories of innate and acquired (or learned) are explanatorily useful (and hence ought not be dropped) as they serve to articulate the "adaptiveness" of certain behaviors. I think that Lorenz's response is significant. It reveals that the exchange between Lorenz and his critics is in part a clash of divergent explanatory strategies or approaches to explain the nature of organic life.

What is clear is that Lorenz's interest in the category of innateness is to explain an evolutionary phenomenon, adaptiveness. What does it mean to be interested in explaining adaptiveness? Adaptiveness is a relation between an organic form and its habitat. Within a particular habitat, an adaptive trait is one that confers relative survivability and/or reproductive success to those individuals who possess it. Darwinists explain adaptiveness by appeal to the theory of natural selection. More precisely, since evolution is a population-level phenomenon, what Darwinists explain is the origins and prevalence of adaptive traits in a population of organisms.

In contrast, Lorenz's critics are in large part developmental biologists, interested not in the relation between organic form and its habitat but in the organic form itself. Further, developmentalists are not necessarily interested in prevalence of organic form, but rather in its origins in an individual embryo.

Given this distinction between what evolutionary theorists and developmentalists seek to explain, I think we can clarify Lorenz's response to his critics. He admits that attributing behavior to the genes is an error if the point is to explain how a trait comes to develop in an individual (the developmentalist project). However, if the point is to explain the characteristics of adaptiveness (the evolutionist project), then attributing behavior to the genes is part of a standard Darwinian evolutionary explanation. Let me briefly explain why Lorenz might have thought this.

There are many points of intersection between evolution and development, but one difference is that evolutionary explanations often "black-box" development. That is, evolutionary theorists often talk about the genetic transmission of phenotypes without regard to what might be a complex correspondence between genotypes and phenotypes. From the evolutionary theorist's point of view, what is important is not knowledge of the causal factors that lead individuals with a given genotype to *develop* particular phenotypes, but that genotype and phenotype are correlated in a way that allows investigators to examine an evolutionary phenomenon (Sober 1984; Amundson 1994). In other words, the developmental story that links genes to phenotypes is not the central issue for evolutionary theorists, though it is important that that there is some developmental story to tell.

Although both explanatory strategies, the developmentalist's and the evolutionary theorist's, appeal to the concept of "gene" and "environment," in the case of genes what counts as the significant causal role diverges for each strategy, and in their uses of the term "environment," their meanings (and purported references) differ. For evolutionary theorists (especially adaptationists), genes serve necessarily as the unit of inheritance and sometimes serve as the unit of selection. The laws that dictate how genes play that role (e.g., Mendelian and non-Mendelian laws of segregation and assortment) are insensitive to the molecular composition of genes (i.e., the fact that genes are bits of chromosomes composed of DNA in the form of a double-helix, etc.). But for the developmentalist, genes play a crucial biochemical role in the determination of organic form. Genes code for amino acids that determine the 3-D folding structure of protein molecules, and so on. Further, in regard to what counts as the "environment," what distinguishes modern-day epigeneticism from pre-nineteenth-century preformationism is the tenet (truism) that genes alone do not produce biological traits. Development involves complex interactions between a few or many genes, between genes and other components of a cell, etc. Everything outside of a particular gene is "the environment" for developmentalists.

The evolutionary theorist (again, especially the adaptationist) regards the environment as part of the "selective regime" including all conditions external to the organism itself (or outside the unit of selection): resources (food, shelter), existence of predators, climatic conditions, topography, location of mates, etc. Typically, they gloss over the role of the environment in the development of individuals. Are they justified in doing so?

Consider how natural selection explains the prevalence and maintenance of biological traits. Selection operates over *variants* in a population. Certain organisms, because of particular traits they possess, are better suited to survive and reproduce than their conspecifics in the environment they all inhabit. The key is that an adaptationist is interested in trait differences between organisms exhibited in an environment where selection is going on. In a particular selective regime, the factors of the environment that instigate selection (climate, resources, etc.) are *fixed* as far as the evolutionary theorist is concerned. The phenotypic differences are all due to genetic differences (much of those genetic differences are likely responsive to selection). This is not to say that interactions between gene and environment do not occur. In other words, it is not to say that the development of the organism is due solely to the genes. That would violate the central tenet of epigenetics and make evolutionary theory incompatible with the best theory of development. Rather, it is *assumed* that these various ways in which environment affects development does not account for the phenotypic variation. It is this sense in which, vis-à-vis the evolutionary project, the environmental role in development is said not to matter and is often taken for granted.

If the evolutionary project were the extent of Lorenz's methodology, I think that Lorenz's response to his critics is relatively uncontroversial. For example, it might interest an evolutionary theorist to consider the selection pressures that might have been present to favor individuals that possessed innate behavior capacities rather than individuals who had to learn their behaviors. In the spirit of adaptationism a model may be constructed to speculate on possible answers (see Sober 1994a).

But Lorenz fails to distinguish the question of explanatory significance from the question of what makes a trait innate. Recall that Lorenz originally identified innateness with the ability of an individual organism to *develop* a trait in isolation of environmental cues. In response to his critics, Lorenz asserted the value of innateness from an evolutionary theorist's point of view. But his response fails to answer the original question: What sort of *developmental* phenomenon is innateness?

Lorenz recognized this problem and provided an answer consistent with his evolutionary approach. His answer went roughly as follows: For adaptive behavior to develop in an individual, the organism requires *information* about its environment. There are two possible sources of this information. Either it is provided by the individual's interaction with its environment, or it is provided by the evolutionary process of natural selection, in which case it is encoded in the organism's genes. When the source of the information is provided by natural selection, Lorenz argued, the trait is said to be "innate." So, the category of innateness is important not only to the study of evolutionary biology, but also to the study of *development*. It allows one to distinguish traits whose developmental information is due to natural selection from those traits whose developmental information is in the individual organism's environment.

Setting aside Lorenz's associations between innateness and "information," let us focus on his views of the explanatory role of natural selection. I take it the key point is that knowing that a trait is a product of natural selection helps explain its development in an individual. But as psychologist Daniel Lehrman (1970) pointed out, Lorenz's theory about the role of natural selection in development is mistaken. According to Lehrman, knowing that a trait is a product of natural selection does not settle any questions about the developmental process by which the phenotypic characteristic is produced in an individual organism.

Lehrman writes, "The clearest possible genetic evidence that a characteristic of an animal is genetically determined in the sense that it has been arrived at through the operation of natural selection does not settle any questions at all about the developmental processes by which the phenotypic characteristic is achieved during ontogeny" (1970, pp. 25–28). Lehrman's argument rests on an example of a trait, species-specific mating preferences, that is widely under selective control. Compare the development of this trait in doves and cowbirds. For both the trait is a product of natural selection, yet individual doves develop their mating preferences through "learning" whereas the cowbird does not.

I think the most perspicuous way of illustrating Lehrman's point is to consider an example used by Elliott Sober (1984), which I alter to fit our purposes. Imagine that gaining entrance to a school requires that individuals read at the third-grade level. Suppose that Sam, Aaron, Marisa, and Alexander pass the test and so are admitted, whereas other students do not pass, and so are excluded. The selection process explains why the room contains only individuals that read at the third-grade level. But the selection process does *not* explain why *those individuals* in the room (Sam, Aaron, and so on) read at the third-grade level. As Lehrman put it, the selection process favors certain *outcomes*, not the processes by which those outcomes are achieved. Notice that it is quite possible that the story about *how* each individual came to read at the third-grade level varies. For example, suppose Sam and Aaron have parents who taught them how to read at a young age, Marisa took a pill which enhances her reading ability, and Alexander reads well despite living with parents who did nothing to aid his progress. These individual differences are not part of the selection explanation for the prevalence of children in the room who read at the third-grade level. But they are part of the developmental explanation for how each child came to read at the third-grade level.

Lorenz's mistake was to misapply natural-selection explanation. Selection explanations are tailor-made for the investigation of evolutionary phenomenon. Yet they are inappropriate for the investigation of what is essentially a developmental phenomenon, the "innateness" of biological traits. Further, by defining innateness as a product of natural selection (via genetic transmission), we are left without an adequate definition of innateness as a *developmental phenomenon.*

Let me put the point in a different way. Attributing biological traits to genes as Lorenz would have it does not answer the central question that innateness ascriptions were meant (by Lorenz) to address.[3] The issue is, broadly, why does an individual have the traits that he or she does? (Or, how does an individual come to manifest the traits that he or she has?) To say that a trait is innate, that is, to say that it is genetically determined, is only to deny that it is, in some unclear sense, not environmentally acquired. Identifying traits as innate as opposed to acquired says nothing of the developmental *process* involved in the production of the trait (Lehrman, p. 344). If we want to know why an individual has the traits he or she has we need to ask *how* environment and genes interact to produce the trait in question. Attributing a trait "mostly" to the genes says nothing about the developmental process that gets us from genes to the trait in question.

3   Apportioning Casual Responsibility between Genes and Environments

Is there another way to preserve Lorenz's genetic account and still have it that innateness is a feature of development? Some authors, most notably Robert Richards (1974) and Michael Levin (1994), have suggested that the distinction between innate and acquired traits is in fact useful to the developmental project of *decomposing* phenotypes into genetic and environmental components. The thought is that the fact of development does not preclude the possibility that the causal role of each may be distinguished. As both Richards and Levin point out, the decomposition of phenotypes is consistent with standard practice in physics and chemistry. Levin (1992) writes that "the causal role of genes can be isolated, just as the causal role of any single chemical in a reaction can be" (p. 50). Richards (1974) concurs, "The discernment of the sources of behaviour, a little like the discernment of spirits, is rarely easy. . . . The situation is no different, of course, in any of the other sciences. . . . [T]he problem of holding constant or eliminating factors other than those of interest is one which . . . all experiments in natural science share" (p. 127).

Unfortunately, Levin's and Richards's approach won't work. The method of decomposition to which they refer is not applicable to the investigation of phenotypes. Let us see why.[4]

In chemistry and physics, one can meaningfully ask, "How much did two causal factors contribute to an event?" We can ask this about chemical or physical systems because events of physics and chemistry both obey J. S. Mill's principle of the composition of causes. Consider a Newtonian example where the event is a particle's acceleration, one causal factor is an effect of gravity, and the other is an effect of electricity. By Mill's principle, the result of the two forces is just the sum of what each would have *achieved had each acted alone* (Sober 1994b, p. 184). To answer the question, "How much did gravity and electricity contribute to the particle's acceleration?" we ask two further questions: (1) How much acceleration would there have been, had the gravitational force been present, but the electrical force been absent? and (2) How much acceleration would there have been had the electrical force been present, but the gravitational force been absent?

Such questions have no analogue in developmental biology. The nature and behavior of phenotypes do not obey Mill's principle. It follows from the fact of development that one cannot ask what effect genes or the environment would have had on the production of a biological trait if genes or environment had acted *in isolation*. For example, one cannot ask of Sally how tall would she have been if genes acted in isolation or if the environment acted in isolation.

Interestingly, Sober (1994b) argues that Mill's principle is inessential to the issue of whether it is possible to ask how much genes and environment contribute to a biological trait. If this is right, then the truism of the phenotype does not automatically yield the negative conclusion. What is essential to the conclusion is that the contributions of genes and environments to the production of a biological trait are incommensurable; as Sober puts it, "their contributions are not made in a common currency" (Sober 1994b, p. 193).

Consider the following example to illustrate the notion of commensurability between two causal factors and to demonstrate why the truism of the interaction of genes and environments is inessential to the conclusion.[5] Suppose two brick-layers are responsible for building a wall. Each begins work on one end of the wall and works her way to the middle, both involved in the task of laying down the mortar and the bricks. Since both perform the same task, it is possible to ask which brick-layer contributed more to the job. To answer, we simply count the number of bricks each worker contributed to the job. There is a way to determine the relative causal contribution of each worker for each makes her contribution in a common currency.

Notice that to ask how much each brick-layer contributed to the wall, it is not necessary to ask how much would each have contributed *had she acted alone*. That is, to answer the question about causal contribution we do not need to appeal to Mill's principle of contribution of causes. We just need to compare the number of bricks each brick-layer laid.

One might object that counting each brick-layer's contribution just amounts to asking how much each would have contributed had she acted alone.[6] A slightly different example throws some doubt on that intuition. Suppose the masons coordinate their workload such that each takes turns laying down one brick before the other lays down hers. So, one mason sets down brick number 1, 3, 5, etc., while the other sets down brick number 2, 4, 5, etc. After the wall is built, we cannot answer the question how much did each contribute had each acted alone, because each worker's contribution by itself does not constitute a wall at all. A wall with successive bricks missing is not a wall, but a heap of bricks. Again, if we want to know the relative contribution of each worker we simply count the bricks each mason laid down.

Now suppose that the division of labor is such that each brick-layer has a specific task: one lays down the mortar and the other sets the bricks. Is it meaningful to ask which brick-layer contributed more to the task of building the wall? Since, in this scenario, each brick-layer does not make her contribution in a common currency, this question has no answer. The interaction between genes and environments is akin to this second brick-laying example, in which the contributions the factors make are incommensurable.

What would it be like for biological traits to be decomposable and for their relative contributions to have a common currency? Suppose height were the result of an accumulation of height particles whereby both genes and environments contributed a certain number of particles to make up a person's height. If so, then we could compare the number of particles genes and environments contributed to, say, Jane's six-foot stature (Sober 1994b, p. 193). To return to innateness, if the term is meant to denote a trait that owes a great deal of its structure to genes, and if the contributions genes and environments made were commensurable in this way (e.g., if there were height particles), then innateness ascriptions would be helpful in apportioning causal responsibility among genes and environments. We could determine the relative number of particles genes and environments contributed to the trait by elaborating various deprivation experiments. But alas, there are no height particles; the contributions that genes and environments make to biological traits are not commensurable; we cannot answer the question, *how much* each factor, genes and environments, contribute to a trait. Hence innateness cannot be ascribed in answering such questions.

The nature/nurture case is not unique. The distance a cannonball travels depends on both the angle of the cannon's barrel and the muzzle velocity. But the two factors do not make their contributions in a common currency; there is no telling how much muzzle velocity and barrel angle contribute to the distance the cannonball travels (Sober 1994b, p. 197).

I conclude that Levin's and Richards's suggestion intended to save Lorenz's genetic account of innateness fails. Does it follow, as Lehrman and his modern-day counterparts (Oyama, Johnston, Griffiths, Gray, and others) would have us believe, that the gene/environment dichotomy upon which the innate/acquired dichotomy rests is unexplanatory? Is there no refinement of the notion of innateness that is useful for any field of inquiry?

No. Lorenz's naive account is not the only one available. There are other biological definitions of innateness, the best of which are not falsified by the fact that genes alone do not produce biological traits. I will offer one and criticize three others in this essay. But first I need some criteria of evaluation. I take it that the best source of these criteria comes from what remains fairly uncontroversial about Lorenz's theory.

(a) Development. An account of innateness should make it a feature of development. In the birdsong literature, for example, you often find a contrast between birds that develop their songs "innately" and those that don't. Recall that part of Lorenz's project was to pick out certain adaptive behaviors that particular animals exhibit even if they had no opportunity to "learn" or "experience" the behavior in the wild. Ethologists are still interested in such behaviors. For example, ethologists of birdsong have discovered, via the use of deprivation experiments, various ways in which species-specific birdsong develops in different species of birds. Certain birds develop their species-specific song regardless of whether or not they hear another bird sing that song. Other birds seem to require experiencing the song before they themselves can sing it. Here we have a contrast in developmental requirements: some birds require a specific environmental cue for the development of their own song, whereas others do not require that specific environmental cue.

(b) Environmental Stability. Innateness should denote an environmentally stable trait. In the ethology and biology literature, innateness seems to have something to do with what environment does not do to influence development in an individual. Evidence from deprivation experiments and observations in the wild suggests that some traits develop normally in a range of environments, including impoverished and abnormal ones. In such cases, the environment does not prevent the trait from being manifested. Ethologists since Lorenz sometimes associate innateness with environmentally stable, as opposed to labile, traits (Hinde 1982, p. 86).

(c) Natural Selection Explanandum. An account of innateness should make clear how natural selection can effect the prevalence of some adaptive traits. Recall that Lorenz was interested in applying natural selection to explain the prevalence of certain highly adaptive species-specific traits. For example, he was interested in how selection might explain the prevalence of ducklings that display courtship behavior without any need of experiencing other ducklings displaying that behavior. I object only to defining innateness solely as a product of natural selection. I have no in-principle objection to natural-selection explanations of the prevalence of innate features. The natural-selection project seems promising although it is an open question whether all such traits are explainable via natural selection.

First, I'll propose my own account of innateness, which stems from C. H. Waddington's concept of "canalization." After demonstrating how the canalization account of innateness captures the desiderata I'll compare it with three other accounts, one put forward by Stephen Stich, another by Elliott Sober, and a third by William Wimsatt.

4   The Canalization Account

In 1936, developmental biologist C. H. Waddington along with his associates (Joseph Needham and Jean Branchet) made a remarkable discovery (see Gilbert 1991). They had set out to determine the specific substance that induces competent ectoderm tissue to develop neural plates. Waddington had thought that the substance was a particular steroid and that only that steroid could induce neural-plate development from competent ectodermal tissue. Waddington's crew discovered, however, that a large variety of natural and artificial compounds rather than a specific steroid induce neural-plate development. Waddington proposed that since lots of different substances serve as an inducer and not all cells can form neural plates, there must be something special about the competent tissues of the ectoderm that allows it to respond to the range of the inducing chemicals. What is special, Waddington conjectured, is that competency to respond to a range of inducing agents is a genetically controlled feature of certain tissues. The wide-ranging competency of the ectoderm allows neural plates to develop in a wide range of environments.

If competency is genetically inherited, it might be susceptible to forces of natural selection. For instance, neural plates develop out of ectoderm in a wide range of developing environments, even in environments lacking steroids. For certain selective regimes the capacity to develop in a range of environmental conditions might confer reproductive success to individuals possessing them, say, in unstable environments.

In such a case, Waddington said that the pathway became *canalized* by natural selection. Canalization denotes a process whereby the endstate (the product of development) is manifested despite environmental perturbations (Waddington 1940).

According to Waddington, canalization explains why developing organisms tend to produce a number of distinct and well-defined body types despite environmental variation between the individuals. Waddington envisioned development as "an epigenetic landscape," a branching out of various developmental pathways, each leading to the production of a distinct endstate. Once development starts (e.g., in the egg) any number of a range of inducing agents force the developing organism down one or another canalized pathway (Waddington 1957).

Further, the canalization model could account for why, for some quantitative characters, there appears to be a "normal" range such that deviant morphologies are difficult to produce. Waddington writes, "For most animals there seems to be a 'normal' size, to which the adult individual often closely approximates even though it may have suffered various enhancements or retardations of growth during its lifetime as a consequence of factors such as the number of its litter-mates, the level of its feeding and so on" (Waddington 1975, p. 99).

For our project of providing an adequate biological account of innateness, canalization fits the bill. Recall the three desiderata for an adequate account of innateness. An adequate account ought to (a) make innateness a feature of development, (b) contrast traits that are environmentally stable rather than labile, and (c) be part of natural-selection explanations for the prevalence of certain traits. Waddington provided an account of the development of environmentally stable traits that occur in individuals and whose prevalence can be effected by natural selection, satisfying the three desiderata. I propose the following as an adequate account of innateness:

The degree to which a biological trait is innate for individuals possessing an instance of a genotype (or set of genotypes) is the degree to which the developmental pathway for individuals possessing an instance of that genotype (or set of genotypes) is environmentally canalized.

The degree to which a developmental pathway is canalized is the degree to which development of a particular endstate (phenotype) is insensitive to a range of environmental conditions under which the endstate emerges.

The concept of canalization is useful because there often exists a high degree of constancy ("robustness") of phenotypes over a fairly well-defined "normal" range of environmental conditions.

Several features of the canalization account of innate traits are worth noting. First, the canalization account preserves the idea that traits are innate with respect to certain genotypes. It may turn out that, for example, some of the genotypes that typically express blue eyes are canalized whereas others are more sensitive to environmental fluctuations.

Further, although Waddington's account of canalization was intended to provide an illustration of the genetic control of phenotypic characters, it is important to note that Waddington recognized that development is an interactive process involving both genes and environmental conditions. Certain environmental conditions are essential for the development of certain traits. No developmental pathway is resistant to all possible environmental perturbations. As Waddington states, "Even the most well-canalized character is, of course, never entirely invariant" (Waddington 1975, p. 100). That there is a limit in the environmental ranges to which a developmental pathway produces a single phenotypic endstate should come as no surprise, given the fact of development.

To reflect the significance of development, on the canalization account innateness is treated as a matter of degree: The greater the environmental range against which a developmental pathway is buffered, the more canalized is the developmental pathway. The steepness of an entrenched pathway (or "chreode") in Waddington's representation of development represents the degree to which a pathway is canalized. The steeper the sloping sides of the pathway, the more resistant the pathway is to disturbances.

On the canalization account, development too is a matter of degree. Loosely speaking, limb development in many organisms is highly canalized; limbs develop in all but only the harshest environments. However, a trait requiring several very specific environmental cues to develop in an individual, for example, the ability to speak French (as opposed to linguistic ability in general), is not highly canalized.

Finally, consider that Waddington's comparisons are made within a specific environmental range. This leads to the question of what counts as an appropriate environmental range to determine innateness. I doubt that there is a uniquely correct answer to this. We probably cannot avoid having to determine the appropriate range on pragmatic considerations, say, depending on the interests of the biologist. As Waddington himself recognized, as a consequence of the fact that no biological trait develops independently of biological factors, no trait is canalized simpliciter. Mutatus mutandis for innateness. Recall that Lorenz's mistake was to think innateness denoted traits that appear in vacuo. An adequate account of innateness should not make the same error.

However, biologists and ethologists tend to ascribe innateness to traits whose development are to some surprising degree insensitive to certain environmental conditions (we made this environmental stability condition as one of our desiderata of an adequate account of innateness). Let us look at three examples illustrating that what constitutes the appropriate range depends on the trait in question.

First, Lorenz was interested in understanding the development of goslings who exhibit a "follow-mother" behavior even in the absence of mother geese. For Lorenz's purposes, the exhibition of a gosling's follow-mother behavior is insensitive to whether mothers are present or not. Lorenz was not interested in how fluctuating embryonic temperatures affect the development of the behavior although it may be an open question as to whether temperature does affect the outcome. Lorenz was impressed that a behavior that is specifically designed to allow an individual to follow mother (that is, a well-adapted activity) is exhibited without the need of the actual mother or any other gosling present. According to Lorenz, there is something interesting about the behavioral development of goslings that would allow such a trait to emerge. In the terms of the canalization account, what's interesting is that the endstate is canalized within the range of environments whereby mothers are present or not. (Note that it may be that the development of the follow-mother behavior is canalized or innate within environments where mothers are either present or not, but not canalized within environments in which, say, embryonic temperature is fluctuating.)

Second, birdsong ethologists are interested in the contrasting developmental pathways of individuals of different sparrow species. Some sparrows produce their characteristic song even if they are reared in silence, whereas in other species, a sparrow produces its song only if it first hears that song performed (Gould and Marler 1991). Birdsong ethologists are interested in contrasting the developmental pathways of individuals representing different songbird species, that is, ones that develop when reared in silence with ones that require their specific song as a recondition for development.

Third, Waddington expressed interest in traits whose development appeared to be insensitive to "normal" environmental conditions. Waddington sought to provide selection explanations for why certain traits persist in "normal" environments and express variants only in "stressful" ones. What counts as a "normal" and "stressful" environmental range is vague, but it probably denotes the range of environments a population of individuals typically shared throughout their selective history.

To reflect the varied interests of biologists, what counts as an appropriate range might change depending on the trait in question. But at least one caveat is in order: At minimum biologists should restrict themselves to environments in which the organism can *develop*. This is an important condition vis-à-vis an account of innateness. For example, one does not prove that hair color is not innate (for individuals possessing a genotoken) by showing that there are environments in which the individual does not develop hair at all (Sober, personal communication).

5   Innateness and Natural Selection

In accepting that Waddington's notion of canalization applies well to the idea of innateness, we open the door to a range of parallels between the work of Waddington and the work of ethologists like Lorenz. A striking example will further strengthen the case for thinking that innateness is canalization. In this instance, we find Waddington solving a problem that Lorenz failed to solve: How can we invoke natural selection to explain the prevalence of highly adaptive traits, such as sexual affinity in mallards, that are seemingly environmentally acquired yet turn out upon further investigation (e.g., by performing a rearing in isolation experiment) to manifest in isolation? For Waddington, the key is the concept of canalization.

Recall Waddington's notion of competence. Competent tissues are inducible by a range of compounds. Most importantly, the range can extend to compounds found either to be internal or external to the developing organism. Waddington hypothesized that once a developmental pathway is canalized, the competent tissue may transfer its responsiveness from one inducing agent to another (Gilbert 1994, p. 851). The possibility of transfer of competency is significant for evolutionary biology. For example, callouses are typically environmentally induced by friction between skin cells and some external surface. Here, the genes play a role in proliferating cells to form the callous. Suppose both that the competence of the skin cells to form a callous structure when induced by friction is a product of natural selection and that the pathway initiated by the friction that leads to the callous structure is canalized. The canalization of callous formation depends on an external agent in this instance. But if the skin cells are competent for a range of agents, it is possible for an internal agent to substitute for the external agent. For instance, suppose a mutation appears that enables the skin cells to respond to a stimulus within the developing embryo. Then the ability to form a callous due to friction may become part of the "genetic heritage" of the organism. That is, what starts as an externally induced trait undergoes a mutation that in effect transfers its competence to internal inducers. The end result is an organism that develops a callous in the absence of friction. In some selective regimes the transfer of competence from external to internal inducer may confer fitness to organisms possessing the capacity. Waddington called the transfer of competence "genetic assimilation." He hypothesized that selection aiding in the genetic assimilation of the capacity of ostriches to develop callouses explains why ostrich *embryos* possess callouses (Waddington 1975). If the transfer of competence is due to a mutation, genetic assimilation is said to be an instance of what is known as the "Baldwin effect."

Unfortunately, Waddington did not support his hypothesis concerning ostrich callouses with experimental evidence. However, he did find evidence of genetic assimilation in experiments on thorax development in *Drosophila*. To illustrate the phenomenon, Waddington (1975) managed to induce an extreme environmental reaction in the developing embryos of *Drosophila*. In response to ether vapor, a proportion of embryos developed a radical phenotypic change, a second thorax. At this point in the experiment we should say that bithorax isn't innate; it is a kind of chimera induced by an unusual environment. But then Waddington continually selected for *Drosophila* with the developmental capacity to respond to the environmental stress. After about twenty generations of selection, some *Drosophila* were obtained that developed *bithorax without being exposed to ether treatment*. What happened, according to Waddington, is that selection favored a particular pathway that led to the production of the optimal (in this case desired) effect. Eventually the pathway became canalized, and hence the endstate, bithorax, appeared regardless of environmental conditions.

6 Critique of Stich's, Sober's, and Wimsatt's Theories of Innateness

Now I wish to compare the canalization account of innateness with three other developmental accounts, one by Stich (1975), one by Sober (1998), and a third by Wimsatt (1986). Using certain counterexamples, I hope to demonstrate the superiority of the canalization account over these. Further, in demonstrating why the canalization account is superior, I will highlight its key features. For example, I have emphasized that canalized development is (relatively) insensitive to developmental environments. Resiliency, not interaction with environmental conditions, accounts for the developmental invariance of the phenotype in question. This feature is not clearly a part of either Stich's or Sober's developmental account of innateness.

Stich's aim is to provide an account of innate *ideas*. Following Descartes, his strategy is to provide a more general account of innateness (actually, one of innate diseases) and then apply it to the case of ideas: "In calling ideas innate, Descartes tells us, he is using the same sense of the word we use when we say certain diseases are innate. So let us launch our analysis of innateness by pursuing Descartes's hint and asking what it is to be afflicted with an innate disease" (p. 3).

Stich's account resembles Descartes's own; both are "disposition accounts." On Descartes's view, individuals that suffer an innate disease "are born with a certain disposition or liability to acquire them" (Stich 1975, p. 6). Stich's account unpacks the disposition to provide the following account:

A person has a disease innately at time t, if and only if, from the beginning of his life to t it has been true of him that if he is or were of the appropriate age (or at the appropriate stage of life) then he has or in the normal course of events would have the disease's symptoms.   (1975, p. 6)

The Cartesian approach (of which Stich's is a piece) is sensitive to the intuition that "inborn" or "present at birth" is not a necessary condition for "innateness." Muscular dystrophy is, intuitively, an innate disease, yet the symptoms do not appear until later in a child's life. Stich's Cartesian account preserves this intuition: A child born with muscular dystrophy has the disposition to experience the associated symptoms, and hence muscular dystrophy is an innate disease. Or, more precisely, it is true of the diseased child that at the appropriate age she will begin to experience the symptoms (Wendler 1996, p. 91).

Stich's account does not fare well with diseases associated with parasites that are acquired in the normal course of one's development, as Wendler (1996) demonstrates. Humans typically possess an abundant supply of a particular species of bacteria *clostrium difficile* (*c. diff.*) in our intestines. Humans are not born with *c. diff.*; we typically acquire it by ingesting food and water. *C. diff.* is not harmful to healthy humans, but it can make sick people on antibiotic treatment sicker by colonizing in areas which the antibiotic treatment (to which *c. diff.* may be resistant) is killing off harmful bacteria. Unabated colonization of *c. diff.* produces toxins that lead to diarrhea and associated symptoms. Is the possession of *c. diff.* in one's intestines innate?

Innateness is generally and intuitively contrasted with environmentally acquired (albeit as a matter of degree). *C. diff.* is acquired from ingesting food and water, so, intuitively, the possession of *c. diff.* is not innate. However, in the normal course of events humans eat and drink water and hence acquire *c. diff.* As Descartes would say, humans have a disposition to suffer the symptoms of *c. diff.* in the intestines. It follows, on Stich's Cartesian account, that *c. diff.* is an innate disease of the intestines. This is an unfortunate consequence of Stich's account.

Stich recognizes the loophole in his account: "There are commonly a host of necessary environmental conditions for the appearance of the symptoms of a disease. If these conditions all occur naturally or in the normal course of events, the symptoms will be counted as those of an innate disease" (p. 7). Stich's response amounts to hoping for vagueness in one's intuitive judgments per case: "it is often unclear whether the occurrence of a certain necessary [environmental] condition is in the normal course of events. So it will often be unclear whether a person is afflicted with an innate disease or is, rather, susceptible to a (noninnate) disease" (p. 7). But Wendler's counterexample is a paradigm. Clearly *c. diff.* is, in the normal course of events, at least part of the normal task of eating and drinking.

It must be admitted that the problem that Stich alludes to is difficult for any account of innateness. Stich's problem can be generalized as follows: For the development of any biological trait, disease, or otherwise, there will be a host of necessary environmental conditions for the appearance of that trait. This just follows from the fact of development, that genes do not operate in vacuo.

The canalization account handles this fact in two ways, first by making innateness a matter of degree, and second by noticing that for the development of some (special) traits, once a certain environmental (or genetic) condition is met, development becomes resilient to further environmental perturbations—in other words, development becomes canalized. For instance, suppose a developmental system is found to correspond to the developmental "rule of thumb": "Develop X no matter what the state of the world happens to be." This exemplifies an extreme form of canalization. Less extreme are systems that conform to the conditional rule: "Develop X if experience is C" or "Develop X if experience is C, develop Y if experience is D." The more sensitive to the environmental condition the developmental system is, the less inclined we are to say that the developmental endstate is canalized. Let's see why this is an improvement over Stich's account.

Stich's account fails to take into account the *causal dependency* of development on the environmental conditions that constitute the normal course of events. Sober's account suffers from a similar problem. For Sober, "a phenotypic trait is innate for a given genotype if and only if that phenotype will emerge in all of a range of developmental environments" (1998). In other words, innateness amounts to phenotypic invariance across a range of environmental conditions. Both Stich and Sober fail to recognize that there are two ways in which a trait emerges (invariantly) in a (e.g., normal) course of development, namely (both conditions come from Johnston 1982, p. 420)[7]:

(1)   By means of strict genetic control over development so that the outcome of development is *insensitive* to the conditions under which it occurs. Such outcomes are said to be strongly canalized against environmental perturbation.

(2)   By means of a developmental sensitivity *only* to environmental factors that are themselves invariant within the organism's (e.g., normal) developmental environment.

The second outcome is not canalized in Waddington's sense but is invariant under the normal conditions of development. The *c. diff.* case is a paradigm example of the second case. Contracting the symptoms associated with *c. diff.* (or even contracting *c. diff.* itself) is part of one's normal course of environment. Further, acquiring the disposition to suffer the symptoms associated with the possession of *c. diff.* emerges across quite a range of (normal) human environments. So we would say the outcome is invariant and hence innate on Sober and Stich's account. But acquiring *c. diff.* and having the disposition to suffer the consequences are both normal and developmentally invariant *because* humans are susceptible to an environmental condition that is itself invariantly part of human development. *C. diff.* is in the food we eat and the water we drink, and we obviously need to eat and drink to develop at all. Although the outcome is invariant it is acquired, not innate and not canalized.

Next, I turn to Wimsatt's "generative entrenchment" account of innateness. I reject it because generic entrenchment is not at all an essential condition for innateness.

The "crucial feature" of Wimsatt's account is the idea that "features that arise early in development have a higher probability of being required for features that appear later ... and tend to have a larger number of downstream traits depending on them" (1986, p. 198). Accordingly, Wimsatt defines "generative entrenchment" of traits "to the degree that they have a number of later developing traits depending on them" (p. 198). There appear to be two central claims here. One is that innate traits tend to appear early in development. The second is that innate traits have a number of later developing traits depending on them. I'm not altogether sure how Wimsatt conceives of the relationship between these two concepts, though it does seem that the concept of generative entrenchment is most central to Wimsatt's notion of innateness judging from his quip, "on this analysis, if it is generatively entrenched, it is 'innate'" (p. 200). Nevertheless I think neither are *necessary* features of innate traits. To see what's wrong with Wimsatt's theory, consider the innate development of pubic hair in adolescents. Because (i) pubic hair appears late in development and (ii) pubic hair has no further trait (at least prima facie) depending upon it, Wimsatt must say that pubic hair is not innate. Whether or not a trait is innate has nothing to do with how late in development it appears or what further trait depends on it.

Wimsatt might retort that pubic hair is in fact not innate. But that would contradict his motivation to unify many philosophical and ethological accounts of innateness under the concept of generative entrenchment. The idea that pubic hair is innate is supported by several of these accounts, including this one: "Innate behavior for a given species is universal among normal members of that species in their normal environment either: (a) because the behavior has a genetic base [or] (b) because the behavior is 'canalized' or homeostatically regulated in development" (p. 187).[8]

According to Wimsatt, an interesting consequence of his developmental model is that some environmental experiences count as being innate. For example, Wimsatt claims that "not only is the imprinting mechanism of the greylag goose at birth 'innate' ... but the object of imprinting is also 'innate'" (p. 200). Take one of the minimal requirements for an environmental condition to be called innate: "the acquisition of that kind of information at that stage of development is deeply generatively entrenched with respect to subsequent behavior" (p. 200). Accordingly, imprinting is innate because a lot of later developing traits depend on it: "there is a very high probability that the young goose will properly imprint on its mother and will, in short order, learn to distinguish her cries and her appearance from that or other female greylag geese nearby" (p. 201).

To see what's wrong with this account, try applying it to the case of the fetal deformities resulting from the mother's taking thalidomide during pregnancy. For those who wish to preserve the contrast between innate and acquired, the effects of thalidomide are not at all innate when we consider that the presence of thalidomide, an environmental condition, makes all the difference between a child being born with a limb or without. Whether or not a lot of later traits depend on the taking of thalidomide does not sway our intuition that the child's deformities are not innate. I conclude that whether or not a trait is innate has little to do with whether it is generatively entrenched.

7 Conclusion

Taking my cue from critics of Lorenz, I have presented an account of innateness that avoids the fallacy of claiming that traits can develop by genetic causes alone. Innateness, on the canalization account, is a property of a developing individual. At the same time, the proposal captures what is thought to be distinguishing features of innateness: satisfying the ontogeny condition, referring to the capacity to produce environmentally stable traits, and making sense of the idea that natural selection can install innate traits in a natural population.

Acknowledgments

Parts of this essay were read at ISSHP in Seattle, the University of Arizona, the University of Rhode Island, Arkansas State University, and Virginia Tech. I wish to thank the audiences for helpful comments. In addition I wish to thank Denis Walsh, Elliott Sober, Chris Stephens, Robert Cummins, Tom Bontly, Joel Pust, Paul Bloom, and Henry Byerly for extensive comments on earlier drafts.

Notes

1. E.g., Kuo (1921), Lehrman (1953), Oyama (1985), Johnston (1988), Griffiths and Gray (1994).

2. Especially Oyama (1985), Johnston (1988), and Griffiths and Gray (1994).

3. Johnston writes on behalf of Lehrman: "The main point of Lehrman's argument was not to take a stand on the issue of whether, or how much, behavior is due to the environment as opposed to the genes, but rather that any such stand simply reflects a misunderstanding about the nature of development" (1987, 153).

4. My demonstration is borrowed heavily from Sober (1994b).

5. The example is taken from Lewontin (1974).

6. I thank Joel Pust and Tom Bontly for pointing this out to me and helping me with a solution.

7. Note that the distinction between the two means of invariant outcomes is not a dichotomy but a matter of degree.

8. I should point out that both intuitively and on my canalization account, the universality condition is neither central nor necessary to identify innateness. Consider those unfortunate persons who possess a gene that increase one's chances of developing Tay-Sach's syndrome. It appears that Tay-Sachs syndrome is canalized to a large degree for those members possessing the gene. Neither possessing the gene nor developing the syndrome is universal among humans.

References

Amundson, R. (1994) "Two Concepts of Constraint: Adaptationism and the Challenge from Developmental Biology," *Philosophy of Science* 61:556–578.

Ariew, A. (1996) "Innateness and Canalization," *Philosophy of Science* (63 Proceedings), pp. S19–S27.

Gilbert, S. (1991) "Epigenetic Landscaping: Waddington's Use of Cell Fate Bifurcation Diagrams," *Biology and Philosophy* 6:135–154.

Gilbert, S. (1994) *Developmental Biology*, 4th edition. Sunderland, MA: Sinauer Associates, Inc.

Gottlieb, G. (1991) "Experiential Canalization of Behavioral Development: Theory," *Developmental Psychology* (27)1:4–13.

Gould, J. and Marler, P. (1991) "Learning by Instinct," in D. Mock (ed.), *Behavior and Evolution of Birds*, pp. 4–19. San Francisco: Freeman.

Griffiths, P. and Gray, R. (1994) "Developmental Systems and Evolutionary Explanation," *Journal of Philosophy* 91(6):277–304.

Hinde, R. A. (1982) *Ethology*. New York: Oxford University Press.

Johnston, T. D. (1982) "Learning and the Evolution of Developmental Systems," in H. C. Plotkin (ed.), *Learning, Development, and Culture*, pp. 411–442. Chichester: Wiley.

Johnston, T. D. (1987) "The Persistence of Dichotomies in the Study of Behavioral Development," *Developmental Review* 7:149–182.

Johnston, T. D. (1988) "Developmental Explanation and the Ontogeny of Birdsong: Nature/Nuture Redux," *Behavioral and Brain Sciences* 11:617–629.

Kuo, Z. (1921) "Giving Up Instincts in Psychology," *Journal of Philosophy* 18:645–664.

Lehrman, D. S. (1953) "A Critique of Konrad Lorenz's Theory of Instinctive Behaviour," *The Quarterly Review of Biology* 28:337–363.

Lehrman, D. S. (1970) "Semantic and Conceptual Issues in the Nature-Nuture Problem," in L. R. Aronson, E. Tobach, D. S. Lehrman, and J. S. Rosenblatt (eds.) *Development and Evolution of Behavior*, pp. 17–50. San Francisco: Freeman.

Levin, M. (1994) "A Formal Treatment of Gene Identity, Genetic Causation, and Related Notions," *Behavior and Philosophy* 22(2):49–58.

Lewontin, R. (1974) "The Analysis of Variance and the Analysis of Causes," *American Journal of Human Genetics* 26:400–411.

Lorenz, K. (1935) "Der Kumpan in der Umwelt des Vogels." *Journal für Ornithologie* 83:137–213.

Lorenz, K. (1957) "The Nature of Instincts," in C. H. Schiller (ed.), *Instinctive Behavior*, pp. 129–175. New York: International University Press.

Lorenz, K. (1965) *Evolution and Modification of Behavior*. Chicago: University of Chicago Press.

Oyama, S. (1985) *The Ontogeny of Information: Developmental Systems and Evolution*. Cambridge: Cambridge University Press.

Oyama, S. (1988) "How Do You Transmit A Template?" *Behavioral and Brain Sciences* 11:644–645.

Richards, R. J. (1974) "The Innate and the Learned: The Evolution of Konrad Lorenz's Theory of Instinct," *Philosophy of Social Sciences* 4:111–133.

Sober, E. (1984) *The Nature of Selection*. Cambridge, MA: The MIT Press.

Sober, E. (1994a) "The Adaptive Advantage of Learning and *A Priori* Prejudice," in E. Sober, *From a Biological Point of View*. New York: Cambridge University Press.

Sober, E. (1994b) "Apportioning Causal Responsibility," in E. Sober, *From a Biological Point of View*. New York: Cambridge University Press.

Sober, E. (1998) "Innate Knowledge," in *Routledge Encyclopedia of Philosophy*, pp. 794–797, Routledge.

Stich, S. (1975) "The Idea of Innateness," in S. Stich (ed.), *Innate Ideas*. Los Angeles: University of California Press.

Vance, R. (1996) "Heroic Antireductionism and Genetics: A Tale of One Science," *PSA 1995* 63:S36–S45.

Waters, C. (1994) "Genes Made Molecular," *Philosophy of Science* 61:163–185.

Waddington, C. H. (1940) *Organizers and Genes*. Cambridge: Cambridge University Press.

Waddington, C. H. (1957) *The Strategy of the Genes*. London: Allen and Unwin.

Waddington, C. H. (1975) *The Evolution of an Evolutionist*. Ithaca: Cornell University Press.

Wendler, D. (1996) "Locke's Acceptance of Innate Concepts," *Australasian Journal of Philosophy* 74(3): 467–483.

Wimsatt, W. (1986) "Developmental Constraints, Generative Entrenchment, and the Innate-Acquired Distinction," in W. Bechtel (ed.) *Integrating Scientific Disciplines*. Dordrecht: Martinus Nijhoff.

8 Generativity, Entrenchment, Evolution, and Innateness: Philosophy, Evolutionary Biology, and Conceptual Foundations of Science

William C. Wimsatt

This essay is part of a larger project to develop an account of the evolution of phenotypic structure that complements the account we have of genetic structure. As developmental genetics increasingly shows, understanding the developing phenome and its environment is crucial to understanding gene action. This project turns pivotally on a strategy for reintegrating development into evolutionary theory, and it promises strategies for a variety of problems in the biological and human sciences that are not solvable or not easily accessible on the genetic approach. These include the question of how to generate an adequate model of cultural evolution—which must include or interface closely with a theory of cognitive development in the broadest sense, applying to all cognitive, conative, and affective skills, and the domains of their employment—and a new and importantly different approach to the phenomena that have motivated the innate-acquired distinction. Section 1 provides an orientation to the approach and to why what I call generative entrenchment is so important to a theory of evolving systems. In Section 2, I focus on the classical innate-acquired distinction and provide a new and richer account in terms of generative entrenchment of the phenomena invoked in its support. The new analysis is compared with traditional accounts of innateness as genetic or canalized. I conclude that the old innate-acquired distinction should be retired, but its conceptual niche is not dispensible and is filled fruitfully by the new concept.

1 The Evolution of Generative Systems

Optimal Design and Historical Contingency

History matters to evolution. It's not far wrong to say that everything interesting about adaptation is a product of selection for improvements in design, or of history, or of their interaction.[1] Gould has emphasized the role of contingency in evolutionary processes, arguing that minor unrelated "accidents" or "incidents" can massively change evolutionary history.[2] It seems plausible—indeed almost inescapable—to believe that a successively layered patchwork of contingencies has affected not only the detailed organic designs we see and variations between conspecific organisms, but also much deeper things—the very configuration and definition of the possible design space and the regions they occupy in it. Deep accidents from the distant past not only define the constraints of our current optimizations, but constraints on these constraints, and so on, moving backward through a history of the deposition of exaptive[3] dependencies, which become framing principles for the design of successively acquired and modified adaptations. Ecological past and present run on similar tracks: genetic events—mutations, segregation events, independent assortment, and recombinations in inheritance are the most commonly cited sources of contingency, but no less important are chance ecological events—meetings leading to matings, migrations, symbioses, parasitisms, and predations. As George Williams quipped in 1966, "To a plankton, a great blue whale is an act of God." (Better design as a plankton cannot save it if it happens to be in the wrong place at the wrong time.)

But accidental or contingent events *needn't* leave visible historical traces much later. Most do not. Many are (1) not heritable, or even if locally heritable, are (2) averaged out—lost in multiple intersecting entropic processes. Also (3) "Noise tolerant" design—both of phenotype and of genotype—damps out many fluctuations, and biological processes are designed to be noise tolerant through diverse adaptations, ranging from third-position synonymies in the genetic code, diploidy, and alternative metabolic pathways for many critical functions, up through developmental canalization, growth allometries, bilaterally symmetric organs, and macroscopic regulatory features of individual physiology, to various mechanisms in social groups, breeding populations, ecosystems, and trophic levels. (4) Optimization also erases history on the evolutionary time scale: over time, changing adaptations to changing circumstances gradually erode and reconfigure everything which is changeable (this point can be argued in different ways: See e.g. Lewontin 1996 and Van Valen 1973). The closer you approach an adaptive peak and the longer you remain there, the less obvious is the path you took to get there—or at least so it seems.

To mark history, an event must cause cascades of dependent events which affect evolution. Some contingent events are massive, immediately marking diverse biotic and geophysical processes, like the "large body" impact or impacts recorded at the K-T boundary that extinguished the dinosaurs and most other species on earth and gave small mammals a chance. No surprise: such a massive cause *should* have had far-reaching effects.

Most "contingencies" start small–single-base mutations that initiate selective cascades of layered exaptations with divergent consequences. Oxygen production as a metabolic by-product in ancient plants presumably started small, but hardly any contingency has had broader or greater consequences for evolution, which it had by spreading as these plants succeeded and becoming a much larger process. As oxygen rose in concentration, this atmospheric poison was initially adapted to, and then eventually utilized by nearly all creatures throughout the animal kingdom, thereby driving an energetically richer metabolism. Small contingencies that leave a mark in evolutionary history do so by becoming larger—amplified by recurrent processes—for organisms through the process of reproduction.[4]

Cascading sequential dependencies also occur in each individual during development. These reflect an evolutionary history of contingencies, of exaptation layered upon exaptation, a history unique to, characteristic of, and divergent in different lineages. This is the architecture of adaptation. This creation of layered dependencies, the structure of the product, and measures of the degree of such dependencies I call "generative entrenchment," or GE. GE is not limited to contingent events. Probabilistically common or inevitable "generic" events (Kauffman 1993) or even unconditional laws of nature can become generatively entrenched—deeply utilized or "presumed" in the design of adaptive structures (Wimsatt 1986; Schank and Wimsatt 1988). But generative entrenchment can also happen to arbitrary contingencies, rendering them more or less context-dependent adaptive necessities—all the more striking for their seeming arbitrariness. *GE is essentially the* only *way that such smaller contingencies have a reasonable chance to be preserved over long stretches of macroevolution. Drift alone will not suffice* (Wimsatt forthcoming). *We see evolution as a contingent process* because *of generative entrenchment.*

So these systmeatic effects—the structure of dependencies—can also have systematic evolutionary consequences. Developmental processes are the source of these patterns—they are what allows these contingencies to persist. Models of them can be used alone and in concert with evolutionary genetics to expand the compass of an evolutionary account. Their scope is nothing less than the dependency structure of our adaptations and heuristics. This approach can also inform cognitive development, cultural, technological, and scientific change, and aspects of the structure of scientific theories. An account of generative entrenchment should get at and facilitate theorizing about these contingencies and how they are preserved, elaborated, or modified in evolution and in development.

Generative Entrenchment and Developmental Perspective on Evolutionary Dynamics

Development was left out of the evolutionary synthesis. This is now increasingly perceived as a fundamental and crippling omission. In the so-called synthesis of the 1930s and 1940s genetics called the tune, and that omission partly reflects the then primitive state of developmental genetics. But accelerating progress there since the early 1980s is now articulating, with interests of paleontologists and macroevolutionists, an emerging multidisciplinary convergence within biology richly delineated (and in part engendered) by Raff (1996). Developmental geneticists look for invariant features of development in search of broadly important mechanisms. As molecular geneticists they also focus on the micromachinery of the expression of genes, both as these are articulated in development and at their patterns of distribution across phylogeny. These mechanisms and patterns of distribution will reveal important things about how the detailed contingencies structure the architecture of our developmental programs, though not about why contingencies—rather than just *these* contingencies—should matter.

The very existence and *structure of dependencies* in developmental programs is the metacondition that makes contingencies (and history) important in evolution—these contingencies, or those of other interacting lineages. *Features (whether contingent or not) that accumulate many downstream dependencies become deep necessities, increasingly and ultimately irreplaceably important in the development of individual organisms. This causes them to be increasingly conservative in evolution, and this together with the inheritance of features down taxonomic lineages leads them to become taxonomic generalities of increasing scope, broadly represented across many organic types.*[5] This basic fact has broad implications: models of these processes can go far in explaining the structure of change in adaptive generative structures, and they may often be able to do so without detailed dependency on the micromachinery if they can capture more abstractly the degree and character of these dependencies. As a measure of the magnitude and character of these downstream dependencies, generative entrenchment becomes a tool for theorizing about them.[6]

This need for integrating development into accounts of evolution is even greater for cultural evolution: there, it may be the easiest thing to get a handle on. In biology, the genetics is straightforwardly combinatorial and relatively accessible, has a stable architecture through successive generations, and is traditionally (though increasingly problematically) treated as inherited through a single channel—the germ line. The whole genome is inherited in one bolus at the beginning of the life cycle, so it seems that we can clearly specify the genetic complement at the start of life. Developmental interactions, by contrast, seem complex: they depend on a constantly changing context both within and external to the developing and increasingly large and complex organism, and they are hard to analyze. Thus we seek to trace the *genetic* architecture of evolution and development, hoping to bring its apparent clarity and stability to bear on the complexity of phylogenetic and ontogenetic processes. Thus the rising promise (and promises!) of developmental genetics.

But for culture, the glass is reversed: (1) despite the common talk of "memes," there is no intra- or inter-organismal Mendelism for ideas, practices, norms, or any other of our artifacts, and (2) there is no "memome": (a) no bolus of significant ideas transmitted at the start of life, and (b) no standard size and (c) no form for the cultural "memotype," and (d) no standard "memetic" units. (3) The means of transmission for memes are varied and baroque, involving multiple complementary and conflicting channels, which (4) are acquired and act sequentially throughout the development and life span of the individual. Moreover, unlike the biological case, (5) the transmission channels used for a given idea can change from one generation to the next.[7] Worse still, (6) acquisition of specific ideas or practices modulate later receptivity to others, so (7) heredity and selection are interwoven throughout the periods of development and learning—that is for humans, throughout virtually all of the life cycle. They are not separable as we can presuppose when we construct population genetic models of evolution. The "internal" memetics for the combination of ideas—confounding as they do processes of cognitive development, selection, and heredity—are too complex for a simple algebra.[8]

Cultural evolution is supposed to be different because it is "Lamarckian," but *these* problems—all embedded within that uninformative label (also critically reviewed by Hull 1988, pp. 452–458)—are what makes cultural evolution so profoundly difficult to theorize about. If we try to imagine constructing a multigenerational evolutionary computer simulation for culture, we find that almost everything we can safely treat as constants in biological models needs to be treated as variables—even from generation to generation. The resultant explosion in the necessary complexity of the model is daunting.

The situation is not totally impossible, however. We can learn a lot from various idealized limiting cases, if we look not for a single model but for a family of related models which together explore the behavior in question. (See e.g. Boyd and Richerson 1985.) We need to be aware in such modeling that some questions are likely reasonable ones to pursue at that level of coarse detail, but that many are not. The more recently explored individual-based modeling approaches, which allow both for complex behavioral rules and intrapopulational variability, should be increasingly widely employed in this area—as they have already in artificial-life models incorporating social interactions.

But there are other saving graces for culture, which I want to exploit: cognitive developmental interactions, whether invariant or context-dependent, are more accessible and public than their biological counterparts, since in mediating learning, and because they are culturally transmissible, they must interface richly with the external world. (1) The richness of our socially and technologically developed language, visual representations, skills, disciplinary knowledge, and our means for communicating these, provide much more sensitive probes for assessing the ontogeny of our knowledge and practices than we have for virtually any aspect of biological development. Every skill or discipline that has sequential dependencies for the order in which its components must be acquired provides potential tests for unraveling characteristics of our conceptual ontogeny. This is not seriously compromised if the necessities are contextual so long as their contextual dependencies can be specified. (2) Moreover, for the most part these probes are accessible, and their effects understandable, in readily understood terms—the cognitive skills through which these developing skills are exercised. (Of course, one must understand and possess these cognitive skills to analyze the dependencies, so "good old-fashioned internalism" still has an important place in *this* socialized account of scientific change.) (3) And those elements of our skills and knowledge that are significantly anchored by their downstream dependencies should possess a stability over time that could allow them to serve the structuring functions for theory and observation that genes did in biology but which "memes"—given the complexities discussed above—cannot do for culture. These strategies for exploring the developmental structure of our knowledge, capacities, tools, and regulative norms have been applied only in cognitive developmental psycholinguistics and some other areas of early cognitive development, and even there not as richly as they can be. (Compare Rasmussen 1987 for rich and varied applications in biology.) There is much more to be done to expand and increase the effectiveness of their use.

Such advantages suggest that to further understand cultural evolution, we should get as much as we can out of modeling the structure of developmental processes. We should continue to study ideas using traditional narrative internalist methods for detailed accounts of the development of particular ideational lineages, both within and across individuals (ontogenies and phylogenies, respectively). Only with these methods can we deal with the conciliations and interferences between ideas sequentially acquired through individual ontogenies, and also with the phylogenies of socially held cognitive structures, be they norms, scientific theories, or practices. Or sometimes, the memetics of how individuals construct their various conceptual schemes may be left as a neuro-psycho-sociological "black box" for general models that emphasize transmission while exploring the epidemiological consequences of different social structures. We will need a variety of models that look at different aspects of the whole system and its parts with varying resolutions. Occasionally, as with our studies of Weismann diagrams (Griesemer and Wimsatt 1988), or Punnett squares (Wimsatt forthcoming-a), there are cultural systems of well-individuated units for which we can track the hereditary, developmental, and evolutionary details of the same case with relative ease and clarity. Both of these papers discuss the methodological advantages of such studies. If we are going to make progress in analyzing cultural evolution, we need to look for more of these sorts of systems.

Models that integrate development and evolution promise other analogies between biological evolution and scientific change (Wimsatt forthcoming-c). They can suggest foci where scientific changes are more (or less) likely; and they predict a pattern of scientific revolutions qualitatively reminiscent of Kuhn's model, though postulating different causes. They can also provide a molar dynamics for such change and offer a new dynamical but generically Quinean, naturalistic account of the analytic-synthetic distinction. They give productive handles on the phenomena for which the innate-acquired distinction is invoked, and they also provide strategies for relating scientific and cognitive processes to biological development and provide broader views of the interaction of developmental and evolutionary processes. A developmental model of evolutionary change emerges both from observations common to the generation of any complex adaptive structure and from dependency relations arising from that process. *These processes are thus of great generality and must be features of any account integrating development and evolution.* I illustrate their import with an example from a different source—which can be taken either as concretely or as metaphorically as you like—in the next section. I return to provide a general theoretical account in the following section.

Engineering a Dynamical Foundationalism

Consider a common observation from everyday engineering. Trying to rebuild foundations after we have already constructed an edifice on them is demanding and dangerous work. It is demanding: unless we do it just right, we'll bring the house down and not be able to restore it on the new foundations. It is dangerous: the probability of doing just that seems very great, and there are seldom strong guarantees that we *are* doing it right. We are tempted to just "make the best of it" and do what we can to fix problems at less fundamental levels. Given the difficulty of the task this is not just back-sliding temptation, but usually well-founded advice. These are the phenomena to be explained and exploited. They are extremely general and have many and diverse consequences.

This is as true for theories or any complex functional structures—biological, mechanical, conceptual, or normative—as it is for houses. *This is why evolution proceeds mostly via a sequence of layered kluges.* It is why scientists rarely do foundational work, save when their house threatens to come down about their ears. (Philosophers like to mess around with foundations, but usually when working on someone else's discipline!) Neurath's image of this activity as one of rebuilding the boat while we're in it is heroic. (That's why the image is so powerful—it is not an activity one recommends lightly!) Actual revisions are preceded by all sorts of vicarious activity, and if we must we fiddle at all levels to make it work. We'd all prefer to redesign a *plan*, rebuild only after we're satisfied with the revisions, keep in touch during the reconstruction to deal with problems that inevitably come up, and move in only after the rebuilding is done or nearly so—complete with all sorts of local patches and in-course corrections. Jokes about construction projects executed by theoreticians or by architects who never visit the building site are legion, a rich source of folk wisdom about the difficulties involved in foundational revisions, or in getting from new foundations to the finished product.[9]

This bias against doing foundational work if one can avoid it is a very general phenomenon. Big scientific revolutions are relatively rare for just that reason—*the more fundamental the change, the less likely it will work, and the broader its effects; so the more work it will make for others, who therefore resist it actively.* The last two facts are institutional, social, or social-psychological in character, but the first two aren't. They are broad, robust, and deeply rooted logical, structural, and causal features of our world—unavoidable features of both material and abstract generative structures.

It is also rare for individuals to undergo major changes in conceptual perspective later in life.[10] It is common folk-wisdom that we get more conservative as we age. Behavioral and mental habits "build up," "increase in strength," and increasingly "channel" our possibilities and our choices. These marked metaphors delimit proto-theories likely to leave their issue in any future theoretical accounts because they reflect deep truths about the common architecture of our behavior. Related features permeate biological evolution even more strongly and deeply than the cognitive or cultural realms. Von Baer's "law" that *earlier developmental stages of diverse organisms look more alike than later ones* is broadly true (with some revealing exceptions). It is most fundamentally an expression of the evolutionary conservatism of earlier developmental features. More usually depend on them, and so mutations affecting them are more likely to be strongly deleterious or lethal. So they persist relatively unchanged.

Differential dependencies of components in structures—causal or inferential—are inevitable in nature. Their natural elaboration generates foundational relationships. *The rise of generative systems in which some elements play a generative or foundational role relative to others has been a pivotal innovation in the history of evolution*, as well as—much more recently—in the history of ideas. There are eminently good reasons that mathematics, foundational theories, generative grammars, and computer programs have attracted attention as ways of organizing complex knowledge structures and systems of behavior. *Generative systems would occur and be pivotal in any world*—biological, psychological, scientific, technological, or cultural—*where evolution is possible.* Generative systems came to dominate in evolution as soon as they were invented for their greater replication rate, their fidelity, and their efficiency. We must suppose that even modest improvements in them spread like wildfire.

The spread of the informational macromolecules, RNA and DNA, has been one of the most irreversible reactions in the history of life. It was followed by others.

Skipping to the cultural level, the invention of language, the advent of written and alphabetic languages, printing and broader literacy, and other means of improving the reliable transmission and accumulation of increasingly complex information should have spread rapidly, under suitable circumstances. Essentially from this perspective, Diamond (1997) attempts a global and integrated explanation of the rise and character of civilizations, providing a mix of contingencies and autocatalytic and hierarchially dependent processes virtually designed for analysis via generative entrenchment. Adaptations (beginning with agriculture) that allow and support growing population densities, cities, and role differentiation (and consequent interdependencies) play a central role and make GE inevitable. Many levels of adaptation, mind, and culture have yielded similar inventions—new generative foundations. Campbell (1974) distinguishes ten levels of "vicarious selectors"— all prima facie suitable candidates. In each case, a similar dominance and irreversibility, products of runaway positive feedback processes, result in a contingent—but once begun, increasingly unavoidable—freezing in of everything essential to their production.

Similar phenomena emerging across importantly different disciplines suggest a new attitude toward foundations and foundationalisms that has broader philosophical merit: traditional foundationalisms are too static and poorly adapted to our constant state of acquiring, confirming, and infirming plausible and usually effective beliefs. I'm interested in construction (of knowledge and adaptations), but less with traditional static issues of *in principle* constructibility than with dynamical issues of how differential *rates* of construction and reconstruction affect what we will find in the world and its stability over time. There are generalizeable and important things one can say about such processes. The generative role of entities is important to whether we keep them around. Golden or not, if you like eggs, protect the goose! This generative role is central to explaining what changes and what remains the same, the magnitude and rates of change, and ultimately, the basic features of all of our generative structures.

Classical foundationalists had it half-wrong: generative foundations are *not* architectonic principles for a static metaphysics, epistemology, or methodology. But they also had it half-right: generative foundations *are* the deepest heuristics for a dynamic evolutionary foundation. Once in place, generative elements can be so productive and become so rapidly buried in their products that they *become* foundational and de facto unchangeable. I suggest that *foundations and foundationalisms everywhere— logical, epistemological, cognitive, physicalistic, cultural, or developmental and evolutionary—owe their very existence, essential form, motivation, and power to the invention and evolution of generative structures.*

Darwin's Principles Embodied: The Evolution and Entrenchment of Generative Structures

To appreciate the generality of this approach, consider an abstract characterization of evolving structures. We can look at entrenchment in static contexts, but we find that more theoretical power arises from the generative role of entities in the adaptive structure of replicating and evolving systems. This is because ultimately, in evolving systems, it *is* the generative role of elements that causes resistance of their changing. The anchoring against change in structures of widely used elements requires few assumptions

(1)   structures that are generated over time so they have a developmental history (*generativity*), and

(2)   some elements that have larger or more pervasive effects than others in that production (*differential entrenchment*).

Different elements in the structures have downstream effects of different magnitudes. The *generative entrenchment* (GE) of an element is the magnitude of those effects in that generation or life cycle. Elements with larger degrees of GE are *generators*. This is a property of degree. If we are going to consider evolving systems (where GE really begins to have bite), the structures must

(3)   have descendents that differ in their properties (*variation*),

(4)   some of which are heritable (*heritable* variation), and

(5)   have varying causal tendencies to have descendents (heritable variation *in fitness*).

With the addition of conditions (3)–(5), these structures satisfy the requirements for an evolutionary process—conditions baptized by Lewontin (1970) as "Darwin's Principles."

So why add conditions (1) and (2) to the widely accepted (3)–(5)? The reason is that *any nontrivial physical or conceptual system satisfying (3)–(5) will do so via causal (phenotypic) structures satisfying (1) and (2)*. Condition (2) is inevitably satisfied by heterogeneous structures: try to imagine a machine or system whose breakdowns are equally severe for each kind of failure, in any part, under all conditions. (There aren't any!) Any differentiated (or nonaggregative) system—biological, cognitive, or cultural—exhibits various degrees of generative entrenchment among its parts and activities.[11] And if one could start out with a system that violates (2), with everything having effects of the same magnitude, natural and unavoidable symmetry-breaking transitions in mutation and selection processes would be self-amplifying, pushing evolution toward systems which increasingly satisfy (2) (Wimsatt and Schank 1988). It is thus unavoidable that evolutionary systems satisfy (1) and (2). They will thereby develop, and if they can reproduce and pass on their set of generators, they will have a heredity. (This order is didactic, not causal: without a minimally reliable heredity, systems cannot evolve a complex developmental phenotype, but developmental architecture can increase the efficacy and reliability of hereditary transmission. Heredity and development thus bootstrap each other—as emerging genotype and phenotype—through evolution.) This requires only that generators retain their generative powers (in context, within a tolerable range) under a sufficient fraction of accessible small changes in their structure. (Qualifications here indicate trade-offs between relevant parameters of the process.) Then they will also show phenotypic variations which inevitably (by the logic of Darwin's argument) yield fitness differences, natural selection, and evolution.

"Darwin's principles" never mention genes; neither do the new additions. *This expanded list of conditions gives heredity and development without ever introducing the usual replicator-interactor distinction.* (Due to Hull 1980, the replicator-interactor distinction has since been widely adopted as the appropriate way of characterizing the genotype-phenotype distinction for the informational generalization of "genes.") Instead, I would borrow Griesemer's account of *reproduction*, which roots a conception of heredity within an account of evolution as a lineage of developments (Griesemer forthcoming-a,b) to facilitate the central reintegration of development into our accounts of the evolutionary process. Genes are agents in these stories (at least in biology) but they are not the privileged bearers of information; rather they are co-actors with the developing phenotype and its environment as bearers of relationally embodied information.[12] I proposed in 1981b (see also Callebaut 1993, pp. 425–429) that we could employ GE to individuate genes in terms of their heterocatalytic role—as it were, by their phenotypic activity—rather than bring in talk of copying (an abstraction of their autocatalytic role) as replicator-based accounts try to do. A new "heterocatalytic" account of (biological) genes consilient with this is richly elaborated by Eva Naumann-Held (1997).

Heterocatalytic gene-like things picked out by the GE criterion in biology would include some but not all genes, some things that are not genes, and most often, heterogeneous complexes of both. In some domains (like cultural evolution) gene-like things may be picked out by GE criteria where there is arguably *nothing* picked out by autocatalytic criteria, or more commonly, where autocatalysis is such a distributed and diffuse process that there seems to be no point to try to track compact lineages through it. (How does a scientific theory make a copy of itself? *Very* indirectly!) Consider economist Kenneth Boulding's comment that "A car is just an organism with an exceedingly complicated sex life."[13] Like a technological virus, it takes over a complex social structure and redirects the resources of a large fraction of it to reproduce more of its own kind. Indeed, our economic system has fostered an environment—a "culture dish" in which the invention, mutation, and expansion of such cultural viruses is encouraged, many environmentalists would say, until it has assumed cancerous proportions. (See also Sperber 1995.) Griesemer's account dovetails naturally with GE, and it was designed to do so.

But back to the consequences of GE: If an adaptive structure meeting (1)–(5) is even minimally adapted to its environment or task, then modifications of more deeply generatively entrenched elements will have higher probabilities not only of being maladaptive but of being *more seriously* so. There probabilities become more extreme—in the simplest models, exponentially so—either for larger structures (e.g., if they grow by adding elements downstream), or as one looks to more deeply entrenched elements in a given structure. Either change increases the degree of "lock-in" of entrenched elements.[14] Crucial for cultural evolution is the large number of ways we have for modulating or weakening GE temporarily or for some purposes so we can (occasionally) make deeper modifications and get away with it (Wimsatt 1987, forthcoming-c). (Most of these ways for facilitating deep modification are not applicable for, or are present only in much weaker versions for, biology. Thus, as we already know—for this and several other reasons—cultural evolution generally proceeds much faster than biological evolution.)

Selection acts on the structures as a whole, so parts of an adaptive structure are inevitably coadapted to each other as well as to different components of the environment. So larger changes in this structure will have to meet more design constraints. Fewer changes will be able to do so. So ever-larger changes tend to have a rapidly decreasing chance of being adaptive. Mutations in deeply generatively entrenched elements will have large and diverse effects, and thus are much more likely to be severely disadvantageous or lethal. Simple analytical models (Wimsatt 1986) and more realistic structures and simulations of them (Rasmussen 1987; Schank and Wimsatt 1988; Wimsatt and Schank 1988) demonstrate that entities or their parts with greater GE tend to be much more conservative in the evolution of such systems.[15] Changes accumulate elsewhere while these deeper features appear relatively "frozen" over evolutionary time. This is the basis of von Baer's "law." It is revealed in our models to be a probabilistic generalization: there are things that appear early and are not deeply generatively entrenched, and things that appear late and are. But there are strong probabilistic associations between earliness in development and increasing probabilities of and degrees of generative entrenchment. (But these are *only* probabilistic, so Ariew's claim [this volume] that there are "innate" things late in development does not by itself provide a counterexample to this analysis. I mentioned such a case in my (1986): parental imprinting on their young complements the young's imprinting on parents in many social species, and this is clearly selectively important. Parents' divergent and complex later behavior toward their young is obviously strongly influenced by it, so it is obviously relatively deeply generetively entrenched even if occurring later in the life cycle. Such kinds of cases are dealt with particularly well on Mayr's (1974) analysis, and on this one.)[16]

So one should be able to predict which parts of such structures are more likely to be preserved and which are more likely to change—and over broader time scales, their relative rates of change—in terms of their GE. What kinds of structures? They could be propositions in a generated network of inferences, laws or consequences in a scientific theory, experimental procedures or pieces of material technology, structures or behavioral traits in a developing phenotype, cultural institutions or norms in a society, or the dynamical structures, biological and cognitive, driving cognitive development. This is a *dynamical foundationalism*, in which a larger generative role for an entity makes it more foundational (in role and properties), more likely to persist to be observed and (for some fraction of them), to grow. (And if the generatively entrenched thing is more robust than its alternatives [Wimsatt 1981a], it is more likely to have been there from the beginning, so it will appear to have an almost unconditional necessity.) Thus an element is foundational in terms of its dynamical properties; so *this* kind of foundationalism, far from being static, encourages and contributes to the study of processes of change—at all levels.

This model applies in diverse disciplines. Working in developmental genetics, Rasmussen (1987) used it to predict the broad architecture of the developmental program of *Drosophila melanogaster* form the effects of its mutants and comparative phylogenetic data. Schank and I have applied it to problems in biological evolution and development ranging from the architecture of gene control networks (1988) and the role of modularity in development (forthcoming) to the evolution of complexity (Wimsatt and Schank 1988). I reconsider below the traditional innate-acquired distinction (Wimsatt 1986). Crossing the disciplinary spectrum, Turner (1991) employed generative entrenchment to analyze the distinction between literal and figurative meaning. It also has powerful applications to the study of scientific change (Wimsatt unpublished; and partially described in: Callebaut 1993, pp. 331–334, 378–383, 425–429; Griesemer and Wimsatt 1988; Wimsatt forthcoming-a, forthcoming-c).

Since the generative form of adaptive structures affects the foci and relative rates of their evolution, and since evolution also acts on that form, one should be able to trace feedbacks from developmental pattern to evolutionary trajectory and back again, identifying pivotal points where relative stasis or elaboration can cause major changes in evolutionary direction. These are *second-order* effects. This gives "dynamical foundationalist" theories greater explanatory power than one might at first suppose. Campbell's (1974) "vicarious selectors" have demonstrated abilities to create or make possible new effectively autonomous higher-level dynamics. (See discussions of conditions for "dynamical autonomy" in Wimsatt 1981a, 1994.) Thus perception (one of his vicarious selectors) plus mate-choice have created runaway sexual selection processes and other dynamics leading natural selection in new directions (Todd and Miller 1998). Many new directions similarly become possible with cultural evolution (Boyd and Richerson 1985). And these processes build upon and interact richly with each other. They are in part products of GE, and they provide both new opportunities for its action and occasions for its use as a tool of analysis. Applied to the evolution of cognitive and cultural systems, this perspective provides new ways of extending evolutionary epistemology with new predictions and explanations. Differences between these processes can affect how the model must be developed and applied, and the kinds of results expected. These two topics—differences by subject area in how GE models should be developed and the character and consequences of elaborating these second-order effects—provide many areas for further development of this theory.

2   Generative Entrenchment vs. Innate-Acquired

Developmental Constraints, Generative Entrenchment, and the Innate-Acquired Distinction

One of these second-order effects provides a strong basis for broader connections with other disciplines (Wimsatt 1986, unpublished, forthcoming-c)—a surprising match of various features of GE with a distinction commonly made in very different terms. The innate-acquired distinction originated in philosophy, beginning with Plato's *Meno*, and it has been discussed there almost continuously since. It was exported from there to ethology as the latter emerged as a science, where it has generated as many quarrels as in philosophy and with equally inconclusive results. It has seemed both central and problematic for millennia. GE can be used to give a powerful analysis of phenomena that the innate-acquired distinction has been invoked to explain throughout its range in ethology, cognitive development, linguistics, philosophy, and elsewhere. And it does so without making problematic assumptions that have seemed inextricably linked with the distinction. Furthermore, given the distinction's traditional connections with the relations between biological and learned (or culturally fostered) inputs to behavior, this domain of phenomena occupies a critical transition zone between "strictly biological" and cognitive and cultural processes of development. If any common theory is to provide insights for both biological and cultural evolution through consideration of developmental processes, surely it should have rich applications here. If I am right, the rich uses of "innateness" in philosophy and the ease with which they could be carried over into ethology is no coincidence. The uncommon two legs of this controversial distinction have a common root—just not the one so commonly supposed.

Suppose, roughly, that to be innate is to be *deeply generatively entrenched in the design of an adaptive structure—to be a functional part of the causal expression of that system, and a relatively deeper one upon which the proper operation of a number of other adaptive features depend.*[17] I take the notion of adaptive design to be unproblematic—at least in biology and in the broadly naturalized half of psychology. My 1972 and 1997b give extensive analyses of the notion of function and of functional organization—the key ideas behind adaptive design—explicating both in terms of selective processes. I presuppose these analyses here without further comment. This move toward GE explains more of the criteria offered for innateness than any other analysis. (I have found twenty-eight criteria so far, including eight new ones predicted by this analysis.)[18] Unlike other approaches, GE also provides its explanations in a theoretically unified fashion: it explains why the criteria hang together and why they *should* be criteria for these phenomena. One might think that to do so well the new analysis must be very conservative. Not so!

I first summarize some claims made for innateness in the philosophical tradition, and I then list and discuss most of the major criteria used by ethologists. (All twenty-eight criteria I have found, and my assessments of their import for the various analyses, are included in tables 8.1 and 8.2 at the end of the chapter.) With these as data, we can begin to see the strengths of the GE analysis of these phenomena.

Claims about Innateness from the Philosophical Tradition

The philosophical tradition has provided many claims about innate knowledge. Even as ethologists eschewed talk about knowledge and substituted talk about behavior, these claims left a lasting imprint on discussions of the innate-acquired distinction. When the criteria support different interpretations, I note alternatives with parentheses. Claims explicable on the GE account are marked by a bullet. Perhaps the claim which has seen the widest range of interpretations is:

(P1)   Innate knowledge is in some sense *prior* to experience.

(P1a)   Innate knowledge exists *prior in time* to experience.

●(P1b)   Innate knowledge is a (logical, causal, epistemic, normal) *precondition for* experience.

(P2)   Innate knowledge is *independent of* experience.

(●?) (P2a)   The *justification* for or *origin* of innate knowledge is independent of or different from experience.

●(P2b)   Innate knowledge is independent of any particular experience in that it is *invariant across different experiential histories.*

(P3)   Innate knowledge often arises as an effect of or is *"triggered by"* experience.

(P4)   Innate knowledge is knowledge of *general* truths.

(P5)   Innate knowledge is *universal*—every member of a given class (usually human beings) has it. (This claim is obviously related to claim (P2b) above.)

(P6)   Innate knowledge has a *generative role* in producing other knowledge.

(P7)   Innate knowledge is often said to be different from other knowledge in being *analytic, necessary*, or *a priori*. (The first two relate to claim (P2a), above, and the last either to (P1b) or (P2a).)

    Variations of these claims have been made in many combinations by many philosophers. I skirt over subtleties, seeking only to indicate the origins of ideas that have influenced more recent linguists, psychologists, and biologists in the philosophical tradition, which has midwifed so many sciences and scientific concepts. In moving from that tradition to modern ethology, two major points are obvious and a third merits special mention:

(1) Although many ethologists ascribed a mental life to at least the more complex animals they studied, they have avoided ascribing linguistic or conceptual knowledge (or the consequences of either) to them. This reflects the power of skepticism in Western philosophy since Descartes: it was worth doing battle with skepticism (or its cousin, behaviorism) for human knowledge, but ethologists seem to have capitulated and dropped at least the fourth and seventh criteria for animals. (If animals have any knowledge, it was assumed to be too low-grade to be general. Properties ascribed by (P4) and (P7) appeared to require too rich a mental life.)[19]

(2) Almost all criteria for innateness urged by ethologists relate to one or more of the above philosophical claims, appropriately transformed for the study of animals, as we will see in the list of ethological criteria below. So *behavior* or morphology was said to be innate, but not knowledge. Criteria (P1a) and (P2b) were carried across essentially unchanged but interpreted as comments about development (E2 and E1). (P1b) has an interpretation deriving from deprivation experiments, in which deprivation of a kind of experience (particularly early in development) produces loss of a *capacity* for later acquiring or interpreting a related or broader class of experiences (E8). (P2a) has an analogue in criterion (E6) below, suggesting different sources for innate and acquired information. (E4) is a richer version of, but still related to (P3). The criterion of universality, (P5), was split into two, reflecting evolutionary taxonomy: Universality within a species was taken as central (E1), and presence of a trait in phylogentically related species as less so (table 8.1, no. 3)—an indirect indicator of a genetic basis for the trait.

(3) Surprisingly, the generative role for innate knowledge (P6) was ignored by most ethologists. It is not unconnected with the other philosophical criteria. (Thus it could give a reason for making claim (P1b), if either innate knowledge was *required* to produce other [experiential] knowledge, or [as a weaker "precondition"], if the latter were *characteristically* produced via employment of innate knowledge. As we will see, (P6) has close connections with most of the ethological criteria listed below— including some that are hard to explain without it.) This criterion was not ignored by Chomsky; generative role is central to his analysis—see his debate with Putnam (Chomsky 1967; Putnam 1967). Even though ethologists noted that some behaviors (including the ones they picked out as innate) played an important role in generating or engendering other behavior, they did not connect this directly with innateness— most likely because *innateness was seen as related to the origins or causes of knowledge or behavior, rather than as deriving from the effects of having or exhibiting that knowledge or behavior.* But generative role is the *most* powerful fulcrum in analyzing the relation between what have been called innate and acquired elements of behavior *and* knowledge. It is at the core of the analysis I give below.

Claims for Innateness in the Ethological Tradition

I now consider some of the standard claims made for innate behaviors. Each claim below is followed by comments on how (or whether) it is explained or predicted on the the two standard competing accounts of innateness, and how it fits with my GE account. The "genetic" account holds that something is innate if it is "coded in the genes"—a widespread and apparently intuitive locution, the consequences of which are quite unclear as we will see. Lorenz sometimes (e.g., 1965) speaks as if this is what he means. The "canalization" account fits better with many of the claimed criteria for innateness (including other statements by Lorenz in the same book): it holds that innate traits are developmentally buffered, so that they appear in a variety of different environments. (See, e.g., Waddington 1957; Ariew 1996 and this volume.) Not claimed in Ariew's formulation of this account (but required if it is to fit some important intuitions about innateness) is *genetic* canalization—the tendency for the trait to appear in a wide diversity (i.e., almost all) different genotypes for that species (The idea of genetic "homeostasis" or canalization was first systematically argued by Lerner 1954.)

One other point is worth noting in evaluating the accounts: does an account explain the criterion directly (as is commonly so with the GE account) or does it require additional hypotheses? (In the latter case it seems fairer to claim that the account is *consistent* with the criterion, or more charitably may even suggest it, rather than explaining it.) It is surprising how often subsidiary hypotheses are needed for either the canalization account or the "genetic" account to do their jobs. (These relationships and how they stand with the three different analyses are tabulated for the twenty-eight criteria in tables 8.1 and 8.2.)

The ethological criteria follow:[20]

(E1)   *Innate behavior for a given species is universal among normal members of that species in their normal environments.*

(E1a) On the "genetic" account, (E1) is taken to indicate that the innate behavior has a genetic basis. (But to get this universality we must also suppose that the relevant genes are fixed and that they have high penetrance—that they are virtually always expressed. These subsidiary hypotheses used to be common, but they are rarely defensible.) Talk of "high penetrance" also conceptually moves toward the canalization or GE accounts.

(E1b) On the "developmental" account, (E1) is taken to indicate that the innate behavior is "canalized" or homeostatically regulated so it appears in a range of environments (though given the genetic variability common in virtually all species, *genetic* canalization is also required for universality across the whole species. Ariew actually denies that species-typical universality is required, but he goes against many or most writers in doing so.)

(E1c) One the GE account (E1) follows directly: strong stabilizing selection produced by the number of other traits depending on the given trait's normal expression guarantees both the universality of the trait and that any species member lacking it will appear strongly abnormal. In addition, it is often adaptive for a generatively entrenched feature to be both genetically and environmentally canalized, so *we should expect that the second and third analyses will often be satisfied together.*

(E2)   *Innate behavior appears early in development before it could have been learned, or in the absence of experience.*

(E2a) On the "genetic" account this is (incoherently) taken as a basis for saying that it is "more genetic," because the environment will have had less time to act. (Note that on any realistic dynamical account of gene action, *the genes* will also have had less time to act!)[21]

(E2b) This feature is *not* predicted or explained on the developmental canalization approach, which has no resources for handling it. (Ariew denies that this feature is important, but again he thereby puts himself at odds with most writers. Indeed, this is one of the most firmly anchored intuitions that people have about innateness, and it would be nice to be able to capture it.)

(E2c) Traits expressed early in development are more likely to have high GE, so the association of (E2) follows directly, though without invoking "absence of experience." This is a characteristic tendency, but contrary to Ariew's suggestion, appearance early in development is neither necessary nor sufficient for a high degree of GE. There can be both highly generatively entrenched things late in development (recall parental imprinting on offspring), and low or even non–generatively entrenched things early in development—for examples any synonymous or neutral mutation, or "silent" gene.

(E3)   *Innate behavior is relatively resistant to evolutionary change.*

(E3) can be explained only on the GE account, from which it follows directly, or (less successfully) on the developmental fixity account (if we assume genetic as well as environmental canalization and strong stabilizing selection. But the last comes close to assuming high GE). It could also be explained on the genetic account with the (for it) arbitrary assumption of stabilizing selection.

(E4)   *Critical periods for learning certain information, or unusually rapid or "one-shot" learning, indicates the presence of an "innate teaching mechanism."*

This, with E2 above, is the basis of Chomsky's "poverty of stimulus" argument against behaviorist theories of language learning. It shows obvious generativity and so fits neatly with the GE account. Critical periods (e.g., for parental imprinting on young) can occur late in development. They seem less well accounted for by canalization (which gets less explanatory the more complex the generated behavior—as I will explain below), and not at all on the genetic account.

(E6)   *Innate information is said to be "phylogenetically acquired" (through selection) and hereditarily transmitted; acquired information is said to be "ontogenetically acquired," usually through some variety of learning.*

Cited often as a criterion, (E6) seems to be used just as a restatement of the genetic criterion. It is also explained (and becomes a distinct though derivative criterion) on the GE account, since traits with high GE are overwhelmingly likely to be phylogenetically old. One could argue similarly (but less convincingly) for a canalized trait.

(E8)   *Relatively major malfunctions occur if innate features do not appear or are not allowed to develop.*

(E8) is a direct and important consequence of the GE account, and it is not explicable on either canalization or genetic accounts without additionally supposing GE. It is actually characteristic of all paradigmatic cases of innateness, but it is surprisingly rarely mentioned as a criterion—though it does crop up in Lorenz's (1965) discussion of the deprivation experiment.

In addition, since the rise of genetics and the modern synthetic theory of evolution, two new criteria have been added, presumably because they are criteria for a trait's having a genetic basis:

(E9)   *If a trait shows simple (e.g., Mendelian) patterns of inheritance, it is innate.*

This violates E1. If a trait is showing Mendelian patterns of inheritance, it must be segregating in the population. But then it is not invariant for that species—different variations normal for that species in environments normal for that species are producing different phenotypes for that trait.

(E10)   *If a trait is modifiable through selection, it is innate.*

(E10) has the same problem as E9, though it also violates E3. Meeting this criterion would be doubly problematic for a GE trait, which would be very difficult to modify through selection because of E8. Note that on the GE account, selection is very relevant to the trait but selection is stabilizing, so for a deeply entrenched trait, that trait is not significantly modifiable through selection.

The last two criteria, (E9) and (E10), are of relatively recent provenance (since Mendel and Darwin, respectively). They were likely both added (see e.g. Mayr 1974) as criteria for a trait having a genetic basis *on the assumption that "genetic" was the appropriate gloss for innateness.* Neither fits well with either the canalization or the GE account. Both conflict with (E1), and (E10) also conflicts with (E3) and (E8). So one can't consistently maintain that all of (E1) through (E10) are criteria for the same concept. So perhaps there are two different concepts of innateness. Lehrman (1970) does essentially this—in effect individuating "genetic" and "developmental" senses of innateness, the second much as on the canalization account. But he rejects the developmental sense, interpreting the genetic (which he embraces) in terms of heritability. This move produces other problems, as Ariew (1996) shows nicely. I urge that on *any* analysis of innateness, these two criteria, (E9) and (E10), be rejected—at least in their present forms. Their intuitive fell has to do with their association with genetics. (I discuss the case for a distinct genetic concept of innateness in Wimsatt forthcoming-c: It adds nothing to saying that a trait is genetic. [I also discuss what it gives you to say a trait is genetic]. Furthermore, confusion of a supposed genetic sense of innateness with the quite distinct implications of the sense discussed here is responsible for many of the socially repugnant inferences drawn using that concept.)

A Comparison of the Generative Entrenchment and Ariew's Canalization Accounts

Many features of the GE account have already emerged. The GE analysis has numerous other interesting features, most already noted in (Wimsatt 1986). Some of them follow:

(1) GE is a degree property—widely acknowledged (since Lehrman 1970) as desireable on any acceptable analysis. (It shares this with Ariew's account.)

(2) GE captures (E2) (earliness in development), (E4) (critical periods/poverty of stimulus), and (E6) (ontogenetic vs. phylogenetic acquisition), which have been paradigmatic claims for innateness, but which the canalization account fails to do or does poorly. It captures (E2) and (E6) in radically different ways than traditional analyses, but this is a strength because it thereby avoids other serious problems with these accounts. One of these is the following:

(3) By not trying to construe innateness as something that is there before learning, GE avoids problems traditional nativists have had with trying to say what learning is and when it begins, in a way that would allow them to distinguish learning from interactions necessary for development. (Development and learning surely fall on a continuum in may respects.) This cluster of issues have probably been *the* most problematic one for the nativist tradition.

(4) Criterion (E8) (deprivation yields major malfunctions) often comes up in discussions of innateness but rarely criterially, and it is often ignored (but see Lorenz's

1965 discussion of deprivation experiments). It assumes center stage on the GE account, which actually gives a better account of disturbances of normality and why these should be relevant than any other account. On the GE account (E8) also plays a role in explaining why (E1) (species universality) and (E3) (evolutionary conservatism) are met. As Griffiths (1996) points out, through it GE also provides room for a modest and qualified essentialism. Because of the causal importance of a generatively entrenched trait, anything lacking it (even if viable) will characteristically be seriously (and deleteriously) abnormal.

(5) This new analysis of things we have thought of as innate shows that they are intrinsically relational and connected with their environments in ways not captured by prior accounts. The equation of innate with genetic is ill founded—being genetic is neither necessary nor sufficient for being innate. (Equating "innate" with "genetic" is a kind of functional localization fallacy—assuming that the function of a larger system or subsystem is realized completely in a part of that system. Writers on developmental systems theory [Oyama 1985; Moss 1992; Griffiths and Gray 1994] have argued this effectively and at length.)[22]

(6) Nor does it follow any longer that what is innate must be internal to the system. This follows from the relational character of the analysis. *Environmental* features could with equal justice be viewed as innate (Wimsatt 1986)—though neither should, properly speaking: treading carefully here would attribute innateness to the whole relationship. What is innate if anything is most fundamentally *relationships between phenotype and environment that serve to secure and increase fitness and its heritability.*

(7) The GE analysis turns traditional accounts on their heads in another way: the status of something is determined not by where it came from (the ethological distinction between phylogenetically acquired and ontogenetically acquired information) but in terms of its effects—its generative power. Generative power also allows generative linguists to invoke generativity without making claims that are developmentally and evolutionarily unsound. What is profoundly surprising is that so major a conceptual rearrangement can nonetheless capture so many of the traditional intuitions about innateness.

Ariew's analysis of innateness in terms of developmental canalization is a reasonable reconstruction of an important (arguably, the single most important) recognized theme in discussions of innateness. I agree with much of his analysis, as far as it goes. Lehrman (1970) individuates two strands in Lorenz's concept of innateness: developmental fixity (or invariance) and heritability. Unfortunately Lehrman embraces the second (Wimsatt 1986). Ariew and I would agree that fixity is more important, and we both make use of it, Ariew as the core of his analysis and I as an important consequence of mine. We agree about many other things. His adequacy conditions are sensible: an acceptable account of innateness should make it a feature of development; it should involve or explain environmental invariance or stability; and it should make it clear how innateness is relevant to selection. Our analyses both satisfy all three of these conditions, though in different ways. He later lists a fourth desideratum—that innateness should be a degree property—which canalization and generative entrenchment accounts also both satisfy.

We differ in some other important respects, however.

(1) Ariew argues that various other accounts (including mine) face counterexamples, and thus should be rejected. But he himself chooses to ignore many criteria that have been offered for innateness—without offering (at least in his 1996 or his paper in this volume) any argument for doing so. Most striking of these are (E2) earliness in development, (E1) invariance in normal members of the species, and (E4) the emergence of complex behavior from simple stimuli. These criteria could easily be used to generate counterexamples to Ariew's account. I'll discuss the last of them further below.

(2) Ariew seeks an economical definition, and he sees no reason to consider every criterion one might find for innateness, perhaps believing that many of them are redundant or mistaken. I have taken a more inductive approach to the problem. It seems more appropriate to the special features of this case: a review of the literature shows a large number of criteria offered or in use—many more than Ariew considers. They are not all mutually consistent, so no account can meet them all. But this fact compromises the value of individual counterexamples (including those Ariew provides) unless there is a more systematic way of deciding which ones *should* count. How should one proceed? One can take several considerations into account in evaluating alternative analyses:

(a) Require that the analysis be consistent with and, if possible, relate to current theories of evolution, development, and heredity. (I will suppose that Ariew's and my account fare equally well here. Traditional nativist or genetic accounts do not. In fact, traditional genetic accounts are not consistent with modern understandings of genetics.)

(b) Try to find the largest consistent subset of criteria. (But this suggests taking the larger sample of my strategy, rather than starting with just a few!)

(c) If there are to be many criteria, look for an account that integrates those that are accepted, explains why they should be criteria for innateness, and gives reasons for rejecting others. (I think that the GE account does remarkably well at this, far better than any other account, including the canalization one.)

(d) Look at how directly the accounts explain the relevant criteria, and how many subsidiary hypotheses are needed to make the explanations work. A generation ago (before the discovery of widespread genetic variation for "wild-type" traits; the ubiquity of epistasis; and the emerging complexities of gene action) the genetic account would have appeared to fare much better, but many of its presuppositions have since been falsified. And (largely because he avoids *genetic* canalization, and because many of the effects of canalization emerge from its commonly coincident generative entrenchment) Ariew's development of canalization often needs subsidiary hypotheses to meet the criteria. GE again does uncommonly well, in part because its effects are robust.

(e) Given that current "ethological" accounts of innateness are thinly veiled transformations of earlier analyses or criteria derived from the philosophical tradition, try to find an account that respects that tradition. (Again, GE· seems to score much higher on this than any of its competitors. Ignoring earliness in development and species invariance costs Ariew here, but the GE account does far better than any of the ethological accounts in capturing[23] essentially all of the claims made about knowledge—i.e., even those ignored in the move from philosophy to ethology—in circumstances where talk of knowledge is appropriate. That is, the deeply entrenched claims in a conceptual structure are those that seem most general, abstract, analytic, a priori, and necessary. (Wimsatt unpublished, forthcoming-c; Turner 1991; Griesemer and Wimsatt 1988; Griffiths 1996.) And as Griesemer and I show, these features are characteristic not only of propositional structures, but also of the canonical representation forms that develop in lineages of widely used diagrams.

I have found an account that satisfies Ariew's adequacy criteria and also meets the five preceding desiderata. It provides a single mechanistic account that exploits robust features of development and evolution to explain an unprecedented number of criteria for innateness better than—indeed *much* better than—any of the analyses that have been offered.

There is a price, however—GE violates some deeply held assumptions about innateness: characterizing a trait as innate in virtue of its consequences rather than the character of its causes; uncoupling innateness from genetics, and from being internal to the system. How fundamental can you get? Of course, none of these associations is denied simpliciter—each can be given a convincing gloss for why we should have believed it to be true. But this violation of deep assumptions alone might seem to make the inductive strategy essential: overwhelm the unintuitive character of the analysis by showing how many distinct criteria it explains. And that is surely an important power of the GE account, one which would not be apparent if we looked at just a few criteria. But the inductive strategy was not invented for this purpose; I was impressed that there were so many criteria, some of them quite diverse. I had thought it quite likely (following Lehrman 1970) that *at least* two senses were involved, and I wanted to get a sense of the whole range of claims made for innateness before attempting to construct an analysis. I was quite astounded that so many of the criteria (and now I draw from the 7 philosophical criteria (P1)–(P7) listed earlier) as well as the twenty-eight ethological and evolutionary ones listed below in tables 8.1 and 8.2) could be captured as consequences of a simple mechanistic analysis that turns on basically one criterion with a few appropriate qualifications.

The GE account is so radically different in approach that it is tempting to describe it as an eliminative account, but if so it is one with a difference. Recent discussions of eliminativism in the philosophy of mind have been associated with threats to "urban renew" (i.e., bulldoze or demolish) our ordinary conceptions of "folk psychology," replacing them with subpsychological concepts derived from theories of our neural hardware, whose conceptual basis would necessitate entirely new concepts at the macroscopic level. By implication, we would have to give up many or most of these folk beliefs as false. This has engendered various defenses of folk psychology arguing that we can't do without these beliefs and that any possible conceptual revision that was adequate to the phenomena would have to preserve them in some form.

By contrast, *this* account of innateness is an eliminativist account that better and more richly accounts for and anchors the intuitive phenomena that innateness was invoked to explain than innateness itself. Because of this unusual collection of features, it may or may not technically be a better analysis of innateness as currently conceived that Ariew's (though the criteria offered for it shouldn't be irrelevant to this judgment!). In any case, GE seems both a better concept for organizing this domain of phenomena and better adapted to the future of theory in these areas. If any of the recent eliminativist theories of mind had even nearly the promise of this analysis of saving so many of the phenomena, they would never have been so roundly attacked.

A Closer Look at Canalization

Even if I urge a different account than Ariew's (or Waddington's), canalization is an extremely important concept, and it remains so both in developmental biology and in thoughts about the evolution of development. There is growing interest in this latter area concerning the experimental assessment of and conditions favoring the evolution of genetic and environmental canalization, and how they interact with each other, with stabilizing selection of different intensities, with inbreeding, and with the evolution of modularity. (See, e.g., Stearns et al. 1995; Wagner et al. 1997; Rice 1998; Schank and Wimsatt forthcoming.) These kinds of studies and of the relation between canalization and generative entrenchment should be a natural outgrowth of growing interest in relations between evolution and development. I want to consider briefly (1) canalization's relation to generative entrenchment, and (2) its character as a phenomenological concept and its consequent plasticity for fitting a variety of diverse cases. The first emphasizes its importance for thinking about the evolution of development, and it helps to explain why canalization and generative entrenchment so often go hand in hand. The second suggests a dangerous looseness that is better avoided.

Relations between Canalization and Generative Entrenchment. Because canalization is a kind of regulatory phenomenon and because GE induces stabilizing selection for that feature, both can be expected to share sets of criteria in many respects. Indeed, one can wonder how many things quickly glossed as canalization instead reflect significant GE with at most modest canalization—as might be reflected in significant mortality of early deviant embryos, followed by modest regulation of smaller and later deviations. Nonetheless, GE and canalization are distinct—both as concepts, and also in the manner in which they are described. Waddington (and Lerner) characterize canalization phenomenologically, in terms of regulation of an outcome across environmental and genetic variation. This does not mean that one cannot do productive experiments or theory involving these concepts, but it does mean that there may be more play in how they are realized in any situation. GE is characterized more mechanistically, in teams of causal dependency relations in the production of phenotypic traits. This means that as we learn more about the mechanisms, the sophistication of our GE accounts, and what we can infer from them, will automatically rise.

It would usually be adaptive for entrenched things to be canalized: if it is strongly deleterious that an organism deviates from a state (whether static or developmental), it is advantageous to regulate its production as tightly as necessary to avoid deleterious effects, if that is possible (or to abort without expending further resources if not).[24] Thus one should expect selection for such regulation. In different circumstances, this might include some or all of environmental canalization, genetic canalization, and regulation of developmental trajectories (Waddington's 1957 *homeorhesis*, later assimilated to canalization.) (See Rice 1998.)

In yielding a stable state, canalization invites accumulation over macroevolutionary time of features that depend on that state. (See Wimsatt and Schank 1988 for elaboration of this argument.) Thus canalization breeds generative entrenchment.

(One can't come to depend upon things that aren't reliably there.) Stable states may become generatively entrenched, whether we think of them as internal or external. In perhaps the most extreme demonstration of this, Morowitz (1992) argues that primary metabolism is the entrenched remains of the prebiotic and early biotic "organic soup" in which life first evolved. In subsequent evolution, the external environment was materially internalized, becoming the "milieu interior" to control and regulate diverse organisms' entrenched states and to permit reliable and efficient operation of the processes that depended upon it.

Given that either canalization or generative entrenchment will under a wide range of circumstances favor selective enhancement of the other, we need to be very careful in analyzing cases that purport to explain innateness in terms of one while denying the need for the other. If evolution naturally builds organisms which have both it may be all too easy to utilize the properties of the "silent partner" while making the argument for the preferred criterion. Canalizations and generative entrenchments have obviously been interleaved many times in constructing our layered architecture of kluged exaptations. In the next section we will see various cases where canalization and generative entrenchment are almost inextricably interdigitated.

On the plasticity of canalization.   Canalization is a phenomenological concept, suggesting a kind or kinds of developmental regulation without specifying any mechanisms for that regulation. Waddington (1957) explicates it with specific examples and the metaphor of the "epigenetic landscape"—in effect a kind of state-space representation for developmental trajectories, which I will return to below. Nothing in Waddington's characterization tells how wide or deep the regulatory channels ("chreodes") are, how they are determined, or how they are supposed to relate to intra- or interspecific differences. This then leaves lots of room for interpretation. One could imagine trimming the canalization "chreodes" quite narrowly, construing canalization as contextual, and avoiding "genetic canalization" by having the canalized states change with different genotypes within the species. We would then have different "innate" phenotypes for different genotypes, as Ariew apparently wants. Then innateness Mendelizes: one can speak of baldness and blue vs. brown eyes as innate traits, and criteria (E9) and (E10) are restored. This moves in the direction of merging the canalization account with the genetic.[25] But then it is not clear what function "innateness" serves that the term "genetic" doesn't already. It also then becomes only too easy to speak of innate intraspecific differences in all kinds of traits. But we've been there before, and it is dangerous and easily misused territory. Chomsky's scientific tastes for species-universals correspond with a good place to draw the distinction: keep innate differences at the species level or above. The GE account goes even further, since it would start with a strongly winnowed subset of these species-specific traits.[26]

Moreover, if we ignore genetic canalization and allow garden-variety intraspecific differences to be innate, it is hard to know how "phenotypic switching" could be easily included. With phenotypic switching, changed environments early in development yield characteristically and radically different adaptive phenotypes. Since the alternative phenotypes are significantly different, and since each appears to be relatively tightly regulated—one can't easily generate a continuum between them—we have a threshold-based switching structure. Maybe one could say that their disjunction is canalized, but this begins to be a slippery slope away from environmental invariance and toward admitting almost anything. (Waddington 1957 had images of "nudging a ball" into one trough-trajectory rather than another in the "epigenetic landscape" with a small stimulus early in development: this fits developmental switching all right, but rather metaphorically, and it is hard to know how to explicate it without violating canalization. The problem is that the physical analogy suggests that it is the magnitude of the nudge, together with the height of the walls of the chreode (together with other issues such as the timing of events) that matter. But it is not the magnitude of the nudge that matters, but the character of the stimulus. Movement in the visual field works for greylag hatchlings, and conspecific cries do for other species, but nudges or hot breaths—two other forms of energy transfer— don't work for either. Migratory vs. nonmigratory forms in locust species (Maynard-Smith 1975) and different morphs for caterpillars (Greene 1989) are mediated by phenotypic switching early in development in response to specific environmental stimuli that are good predictors of the environments in which the alternative phenotypes are better adapted. The bithorax response of developing *Drosophila* to ether in Waddington's (1957) account of genetic assimilation (though maladaptive) and the Baldwin effect are reflections of this kind of sensitivity. This better fits a characterization of the system as a control structure with complex consequences that are both "programmed" and regulated. Neither term by itself is adequate. But the more structure there is to the consequences, the better the GE account fits, and the less revealing it is to claim that the behavior is canalized.

Another problem with this case for a "narrow" construal of canalization is that phenotypic switching is a species-specific characteristic that demands substantial *genetic* canalization because of the large genetic variability found in almost all natural species. If alternative trajectories and their environmental releasers were not stably present across different genotypes, they would never be recognized in nature as a stable response, could never serve their functions, and would never have had sufficient heritability to have evolved.

Couldn't one be a narrow canalizationist in some environments (to capture Mendelizing and selectable traits in natural populations and steal the ground from genetic accounts), a broad one in others (to capture species-universality and other macroevolutionarily relevant criteria), and a structured one in still others (to capture phenotypic switching and all kinds of complex adaptive programmed interactions with the environment)? But how do we know when to be which? Without criteria to tell us we don't have an analysis that is of much scientific use. And notice that the second and third alternatives are each equivalent in different ways to sneaking generative entrenchment in the back door—for what else are the releasing parameters and the complex coordinated changes responding to them but rich programmatic structures whose architectures and characteristic responses are deeply generatively entrenched?

Part of the problem is that it is easier to start with a chosen "innate" trait and say how it is canalized or not, in what respects, and at what times in development, and how all of this may vary by genotype than it is to start with an idea of canalization and using just that criterion decide which traits are classified as innate. (All of the preceding qualifications then appear as gerrymandering, and the sense that one has captured the distinction seems to slip through your fingers.) To some extent, this can also be said for generative entrenchment, or perhaps for any category in a science with richly textured objects where a lot of the details matter—that is, at least for any evolutionary process. But I think that GE has more structure and seems much less prone to this problem than canalization.

A more diffuse but nonetheless serious problem for canalization is that it becomes less informative as an explanation the more complex and conditional the behavior becomes, whereas GE, with its invocation both of stabilizing selection and the accumulation of layered exaptations (which brings in seemingly arbitrary contingency) does successively better. The phenotypic switches discussed above, or the rigidly stereotyped mating rituals of various species, seem designed for a GE account. One might on a GE account expect evolution to produce a growing succession of initially arbitrary display, feeding, and appeasement behaviors which become added to differentiate rituals of closely related species. In these cases, it is obvious enough that the behavior is canalized, but also that it is generatively entrenched (both relative to the next part of the ritual and with respect to mating success). On the GE account, entrenched arbitrary contingencies (it doesn't matter what differentiates mating rituals in similar species that overlap, only that they be successfully differentiated) and the emphasis on dependency structure lead naturally to the complex interactions through which the behavior is realized—appropriately tuned to circumstance and the conventionalized offerings. Here the causal richness and environmental sensitivity of the adaptive design couples naturally with the definition of GE and its intentionally relevant causal dependencies.

Table 8.1
Standard criteria

Criterion for "innateness":	Import	Import for GE	Canalization G = genetic E = environmental	Genetic	Generative entrenchment	Comments:
1. Developmental canalization or fixity	••	••	X	not implied	X	*Defining* criterion for "canalization" analysis
2. Universality within species	••	••	X (Ariew: no) (G ⇒ yes)	x (if fixed)	X	(for "normal" members of that species)
3. Presence in related species	••	•	(G ⇒ more likely)	x	X	Treated as weaker "inductive" criterion
4. Appears early in development or in naive individuals prior to opportunity for learning	••	••	no	no	X (conditional)	Traditionally important, and common, but not necessary or sufficient on GE analysis
5. Simple stimuli → complex char. behavior	••	•	(problematic)	X	X	Presupposed by Chomsky's "poverty of stimulus" argument
6. Deprivation → major malfunctions	••	•••	(possible: Canaliz. is often GE'd)	X	X	Often ignored in analyses but crucial to Lorenz and pivotal to GE account
7. Releaser elicits activity			yes?		(x*, case of #5)	From Lorenz's early "hydraulic" model
8. Teaching mechanisms are innate					X	Nativist counter to empiricism
9. Stereotypy of behavior	•	?	X?	?	x*	Contrasted with flexibility of learning
10. Parallels between behavior and phylogeny		•	(G ⇒ more likely)	?	X	Crucial for Ethologist's focus on behavior
11. Unusually easy, rapid 1-shot learning	•	•	X		X	Contrast with S-R learning paradigm
12. Critical periods for learning	••	••	?		x*	Connected w. poverty of stim. and imprinting
13. Resistance to evolutionary change	••	••	X?	X?	X	Derived but important property of GE
14. Spontaneous prod. of complex behavior	•		homeorhesis (but problematic)		X	Not learned, thus innate (related to 5 above)

15. Imprinting is irreversible	●	X	Lorenz 1937
16. Imprinting is not repeatable		X	Lorenz 1937

How different analyses of "innateness" deal with criteria or characterizations found in the ethological literature.

Relations between criteria and the specified analysis of innateness:

X = true by definition; $\mathbf{X}$ = presupposed by; $\mathbf{x}$ = explained by; x = consistent with; $*$ = explicable with addition of Mayr's closed/open program distinction; blank = no relation claimed, but consistent with; "yes," "no" indicated explicit claimed connection or denial.

Importance of criteria: ●●● = ineliminably important ●● = centrally important; ● = important.

Table 8.2
Criteria demoted and new predicted criteria

Criterion for "innateness": denied or not central on ge analysis:	Import	Import for GE	Canalization	Genetic	Generative entrenchment	Comments:
17. I/A as phylogenetic/ontogenetic acquired	●●		no	X	explicable but not central on GE	alt. characterization of I/A; **Definitional?** Q-able whether an independent criterion
18. Mendelian or other simple inheritance	?		possible if NO genetic canalization	yes	no	derived from "innate = genetic" equation Mayr:segregating behavior in *species* hybrids
19. (currently) selectable ($\Rightarrow$ heritability — 0)	?		yes	yes	no?	derived from "innate = genetic" equation?
20. (physical) modularity of functional trait	?				no?	emphasized by Chomsky, linguists, and no one else. [a mistake!]
Predicted on GE analysis:						
21. reappearance of early simpler (reflexive) traits after (brain) damage in adult	●				x	noted by Teitelbaum for rooting and grasping reflexes
22. phylogenetic ancestral traits in hybrids	●	●	X?		X	Darwin's argument that all pigeon varieties are descended from ancestral rock pigeon; relates both to canalization and to generative entr.
23. early devel traits more evol conservative		●●			X	closely conn. w. GE and von Baer's "law."
24. deeply entrenched generative role	●	●●●			X	**defining** criterion for "GE" analysis; generativity noted as important by Chomsky
25. existence of supernormal stimuli					X	like poverty of stimulus, a consequence of simplicity of releasing stimuli
26. analogies in cognitive structures	●●	●●			X	centrally important for "innateness" in mental, and philosophical origin of ethological concept.
27. restricted modes of deep GE modification	●				X	plausible consequence of GE for innateness
28. association between "habitual" & "innate"		●	no	no	X	vernacular: plausible on GE account, by analogy; contradicts genetic account

How different analyses of "innateness" deal with criteria or characterizations found in the ethological literature.

Relations between criteria and the specified analysis of innateness:

X = true by definition; X = presupposed by; $\mathbf{X}$ = explained by; x = consistent with; $\mathbf{x}$ = explained by with addition of subsidiary hypothesis; * = explicable with addition of Mayr's closed/open program distinction; blank = no relation claimed, but consistent with; "yes," "no" indicated explicit claimed connection or denial.

Importance of criteria: ●●● = ineliminably important ●● = centrally important; ● = important.

Tables 8.1 and 8.2 include all criteria I have been able to find in the literature or in conversation with practicing scientists in the disciplines affected. Most are found widely in the literature, with 1, 2, 4, 5, 12 and 13 probably the most common. (I have not done a systematic census and tally). Most writers use multiple criteria, often drawing on more than they list explicitly. Thus Lorenz (1965) uses many of them seriatim in his discussions, either criterially or as comments about innateness, and has probably used or discussed almost all of the first 16 at one time or another. Mayr (1974) lists six explicitly, (10, 4, 9, 18, 19, 1), but I know no other sources listing nearly this many. The judgments of importance are not based on a systematic careful survey, but on lots of reading and talking with practitioners. Different individuals may weight different criteria differently—if I am right, indicating how much use they have made of them themselves and reflecting the role that generative entrenchment plays in their own thinking.

A final observation indicates the significant differences between the GE approach and either the genetic or canalization accounts. This is the natural association in common speech (no. 28 in table 8.2) between instinctive behavior and things that have become habitual—especially intentionally, through practice, and especially if done smoothly, or sometimes stereotypically. Thus, "on hearing the faint click, he *instinctively* went for his gun"—a plausible description in the middle of any "dime" Western. There are myriad varieties of this statement. It fits (at least roughly) criteria 5, 7, and 9 from table 8.1, but without satisfying either the genetic or the canalization accounts. It is complex behavior which must be learned, even trained for, which has become "chunked," "black-boxed" (Latour 1987), or modularized. It cannot any longer be executed piecemeal, but only as a unit, and it must be started "at the beginning," not in the middle—showing that its early pieces (and sometimes its learned "releasers") are generatively entrenched relative to the rest. This example midwifes the transference of the GE account from biology and developmental psychology to the unchallenged common minutia and practices—and the most deeply entrenched principles—of science and culture. But that is a story for another time.

Conclusion

Canalization seems an intuitive reading of much of what many past theorists (including Lorenz) have had in mind by innateness, though it fails to capture some crucially important criteria. GE provides an extremely fruitful reconstructive analysis which fits existing claims (including Lorenz's) better than any other, offers new fruitful connections and predictions, is consistent with modern accounts of the relation between genes and development, and provides an engine for evolutionary change for the developmental systems view. Particularly intriguing (but only hinted at here) is the ease with which the GE account captures or explains many of the traditional philosophical criteria (P1)–(P7) for innateness. This suggests a fruitful naturalistic analysis of the a priori–a posteriori and analytic-synthetic distinctions, which I provide elsewhere (Wimsatt unpublished, forthcoming-c). Accepting the GE account requires breaking some strong associations made in traditional accounts of innateness, but it allows making many others in a satisfying and unitary fashion. Since no analysis can consistently capture all of the criteria, the loss of these associations should not be taken as critical. Perhaps in the long run they will be regarded as important only for helping to understand why this alternative powerful reading of the phenomena known as "innate" has remained invisible for so long. Finally, the strongly relational character of this analysis can torpedo the basis for recurrent nativist claims that something is innate and therefore independent of environmental involvement or effects. It is time that we stopped basing faulty social analyses on mistaken conceptions of human nature anchored firmly in obsolete biology.

Acknowledgments

These acknowledgments should be appended to those of my earlier (1986). Those whose influence has lasted include Stuart Altmann (for his informed skeptical discussions—now extending for twenty-six years—of the usefulness or coherence of the distinction), Susan Oyama (for hers in the last dozen years), to my "colleagues in crime," Susan Goldin-Meadow and Martha McClintock, for twenty-one years of co-inspiration in our practice in the "Mind" course, and to Eric Lenneberg, who in the year before his death in his course and seminar at Cornell in 1974–75 expounded a relational and interactive view of development that was far ahead of his time. Continuing discussions with Jeff Schank on generative entrenchment and with Paul Nelson on GE and evolutionary contingency have been useful and thought-provoking. Finally, despite our disagreements, Andre Ariew's papers and supposed counterexamples have productively forced me to rethink, clarify, and to make some aspects of my analysis more explicit. Some of this material was presented at the Max Planck Institute for Adaptive Cognition in Munich, November, 1994, and at the ISHPSSB meetings in Seattle, WA, in July, 1997.

Notes

1. I avoid talk of maximization of fitness or even constrained optimization of design here: satisficing accounts better fit available mechanisms (Simon 1982, 1996.) Analysis of population genetic models show that fitness is maximized only in highly idealized and notoriously limited circumstances—rarely if ever found in nature. Satisficers can use maximization accounts as heuristics to qualitatively identify possible attractors and to structure and conceptualize many problems, but this has no deeper significance. For a useful classification of such optimization methods in the related context of rational choice, see the introductory chapter in Gigerenzer, T. et al., 1999. See also Wimsatt forthcoming-b, ch. 1.

2. Evolutionary biologists use "contingency" differently than modern metaphysicians: the opposition is not between "necessary" or "analytic" vs. "contingent" (as possibly false), but law-bound or probable vs. unlikely or arbitrary. The distinction is not unlike Aristotle's distinction between necessity vs. chance or accident.

3. Gould and Vrba (1982) define an exaptation as a feature of the phenotype that is not itself an adaptation (i.e., a product of selection for that feature), but which provides a base for the evolution of a new adaptation or function. After selection has elaborated this feature it has become an adaptation or a part of one. In emphasizing this aspect of evolution, Gould and Vrba wanted to emphasize the "fortuitous opportunism" and sometimes circuitous paths taken in the creation of adaptations.

4. This applies not only to the contingencies, but to their detection. If you find a fossil, even if it is the only known instance of its type, the chance that it *represented* a rare type in its time and place is small.

5. I do not suggest that elements cannot change earlier in development (they may be early without being entrenched, and some changes in them may be selectively neutral). Nor is it impossible to change deeply entrenched elements: it is more improbable, and changes must meet more constraints. Various mechanisms, both in biology and in culture, can make deeply entrenched change significantly less unlikely. Some such mechanisms *must* exist: Raff (1996) documents relatively common changes early in development (in pre-Bauplan stages) in some phyla and urges an "hourglass model" with variation necked at the Bauplan stage. And for scientific theories (which show many of the same patterns favoring conservation of deeper theoretical structures), many of the deepest structures of current theory date only to the last scientific revolution—so called for just that reason (Wimsatt 1987, forthcoming-c). Closer study of the nature and evolution of mechanisms permitting deep change (such as modularity—see Schank and Wimsatt 1998) is an important elaboration of this theory.

6. Interest in exploiting a GE-like perspective and in problems with similar characteristics is growing. These related ideas have been invented independently at least four times. (See Riedl 1978; Wimsatt 1981b; W. Arthur 1982; Glassman and Wimsatt 1984; W. Arthur 1984; Wimsatt 1986, unpublished; Rasmussen 1987; W. Arthur 1988; Schank and Wimsatt 1988; Wimsatt and Schank 1988; Turner 1991; Griffiths 1996; B. Arthur 1994; W. Arthur 1997; Schank and Wimsatt forthcoming; and Griesemer forthcoming-a,b.) In his rich review of approaches, Nelson (1998) documents an increasing number of attempts to move in this direction within paleontology and developmental biology over the last decade. Riedl (1978) is the classic first development of a theory of this kind. W. Arthur (1997) currently has the fullest account of generative entrenchment for evolution and development. B. Arthur's (1994) work is having growing impact in economics and other social sciences. My further work on it (published and unpublished, in biology, cognitive development, cultural evolution, and scientific change) will be included in a book now in process (Wimsatt forthcoming-c).

7. This can actually happen in genetics, if a virus caught through normal epidemiological means leaves through its action a change in the germ line of its host. Some now think that this can be a significant cause of informational change over macroevolutionary time. Nonetheless, the relative frequency of such biological events is many orders of magnitude smaller than the corresponding cultural events would be. I ask my students to imagine what genetics would be like if we caught, in each generation, (an average of) half of our genome from viruses, and which viruses we had been exposed to affected our sensitivity to selected other families of viruses and modulated our behavior toward placing ourselves in situations where we could be infected (again by perhaps different specific families) of other viruses. This begins to suggest but does not exhaust the complexity of cultural transmission and evolution.

8. Many of these crucial differences are discussed—from a somewhat different perspective—in Boyd and Richerson's pivotal review of the characteristics of cultural transmission in chapter 2 of their (1985). Their influence in this area is deservedly immense.

9. We *do* sometimes have to live in the house while it is being rebuilt. But this only works because the conceptual organization of science, and of engineering practice, is usually robust, modular, and local, each of which reduces GE. Shaking (local) foundations usually doesn't bring the house down, and we still have a place to stand (on neighboring timbers) while we do it. For a sense of life at the critical edge (!) read Rhodes (1986) on revising theory and practice in the design, construction, and testing of the first nuclear reactor and atom bomb, or Feynman's view (Gleick 1992) of the groping development of theory and computation at Los Alamos when they *had* to have accurate results without direct experiments.

10. Though his ultimate conclusion favors the account in terms of GE given here, Hull's careful review, analysis, and discussion (1988, pp. 379–382) of "Planck's Platitude" (that older scientists are slower to acccept new theories) shows how dangerous easy generalizations are in this area.

11. On functional differentiation, aggregativity, and emergence, see Wimsatt (1997b), (1997a), and forthcoming-b, respectively.

12. There are more complexities to Griesemer's story than I can address here, and it is much richer than my (1981b) proposal to define genes in terms of their heterocatalytic function, as is Naumann-Held's. Griesemer argues that material transmission from generation to generation is a crucial feature of

reproduction—which may differentiate it from cultural transmission, with further interesting consequences. He also has many things to say on the distorted representations of the biology induced by the "informational" or "replicator" conception of the gene. Griesemer's account is now getting increasing attention from well-known "gene-centered" biologists. Thus Szathmary and Maynard-Smith (1997) quote his account approvingly as delivering more clearly the conceptual revision they had sought.

13. From my notes of Boulding's lecture at Max Black's seminar in the "humanities, science, and technology" program at Cornell University, 1974–75. The following elaboration is mine, but Boulding got a widely appreciative laugh, indicating that his audience quickly drew just this interpretation.

14. "Lock in" is Brian Arthur's (1994) term for the same process—which he explores in economic models of the development of technology, the formation of cities, and other cases where increased adoptions of an action or technology reduces the relative cost of doing so to others, generating a positive feedback loop. Arthur rigorously develops relatively simple models (usually with just two alternatives, and often with analytical results), whereas the work on GE has focused with less rigor on larger adaptive structures with greater differences in depth of GE, using simulations and qualitative discussions.

15. My first model of GE—the developmental lock—goes back to 1972, but it is not discussed in print until 1981b. See also Glassman and Wimsatt 1984, and Callebaut 1993.

16. Ariew suggests that the appearance of pubic hair is both late and innate, so thus a counterexample to the GE analysis. But it is not terribly late (appearing during adrenarche, in middle childhood), and it is clearly part of a very pleiotropic complex of activities initiated by adrenal hormones (DHEA and its products) playing an important role in the developmental emergence of sexuality. McClintock and Herdt (1996) show that sexuality starts earlier (at adrenarche, around age 10) than formerly supposed (at gonadarche, around ages 12 for girls and 14 for boys). Furthermore, mammals have a variety of social activities mediated by pheromone secretions, many from the regions of the secondary sex characteristics. (Odor plays a critical role in mating in many species. Its role in other behaviors is not denied, but simply unknown.) Chemical compounds emitted from the axillary (underarm) region in humans have recently been shown to speed up or slow down ovulatory cycles in other women as a function of when in the cycle they are emitted (Stern and McClintock 1998). The glands in this area as well is in the pubic and areolar areas complete development and become active as pubic hair emerges, and the hair follicles may be foci of secretion and the oily and curly form of the hairs may aid in their dissemination. This area of research is still relatively new (see news and views commentary by Weller in the same issue of *Nature* as Stern and McClintock), but it is at least as plausible at this stage to claim that pubic hair and the consequences of its emergence then are deeply generatively entrenched as it is to claim that they are not. Indeed, McClintock notes (personal conversation) that because many of these structures are already morphologically well developed in the fetus and in the neonate, they may well be deeply generatively entrenched in other respects.

17. Or it may be a relatively ineliminable consequence of such a deeply generatively entrenched trait. This qualification is required to deal with cases where an entrenched and adaptive trait may have maladaptive or nonfunctional side effects. This suggests that there can be such things as "innate" or "intrinsic" design flaws or limitations. And of course, in the real world, every design must make its compromises among conflicting design constraints and desiderata.

18. Conceptual analyses can make predictions just as theories can. A good analysis of a phenomenon will generate criteria in terms of some deeper understanding of its character, just as a theory of a mechanism can be used to generate good indicators of its presence. Test the analysis by asking knowledgeable observers what they think of the criteria. My new criteria are successful in this sense. See my (1986), (forthcoming-c).

19. (P4) is arguably false for an important subclass of cases in ethology: imprinting may be to an individual (parent or offspring) or a class (species-typical mating song). But (P4) and (P7) *are* captured at least qualitatively in applying a GE analysis to the analytic-synthetic distinction via a dynamical foundationalist analysis of scientific knowledge and change (Wimsatt unpublished, forthcoming-c; Griffiths 1996). The strong *cognitive* associations of the innate-acquired distinction persist: curiously, it seems never to have been applied to the behavior of plants—something neither predictable nor explicable on *either* Ariew's analysis or mine!

20. This numbering follows Wimsatt 1986. (E5) and (E7) from that list are not discussed here, but appear as criteria 9 and 10 in table 8.1.

21. Among genes, maternal-effect genes fit this account best, acting before conception in the formation of the egg cytoplasm, or in placental mammals, in the embryo's environment. Maternal effects should be paradoxical on traditional "genetic" accounts of innateness because they are examples of extranuclear (and even extra-organismal) inheritance, and the genes are acting not in the offspring, but in the mother.

22. We have exploited the biases of the gene-centered perspective since early in this century. Its enormous inertia (generative entrenchment!) informs how we conceptualize all sorts of fundamental relationships in biology. We must stand outside it to see its limits (as developmental-systems theory does) to better assess its true strength. Many urge that the "innate-acquired" distinction simply be trashed. Productive thought in the new paradigm—or effectively avoiding the missteps of the old distinction—may require it.

GE combines elements of genetic and selectionist theories in developmental form. It is essential to a developmental-systems theory, as the closest availablle thing to a motor that could drive evolutionary change. GE actually explains trajectories more in terms of highway construction than propulsion technology, but things that look like constraints or pathways on shorter time scales may act as generative motors on longer ones. But developmental-systems rhetoric is sometimes overstated: we should not hasten to reject genetic explanations, especially newer perspectives from developmental genetics: new theories must remain in contact with well-developed tools already in the discipline or be dismissed as obvious heresy or pointless unprincipled worry. But one can use these tools without regarding them as foundational. (Griesemer 1999a, 1999b is forging this as yet poorly marked path.)

23. Capturing or making clear why it should *appear* to capture—*pace* Hume on explaining our vulgar notions.

24. Like all "adaptive design" arguments, this requires detailed qualifications, which won't be provided here.

25. Mendel (1866/1902) made clear in choosing the characteristics for his experiment that he sought traits that were relatively insensitive to environmental conditions, so as to get clear ratios and avoid possible confounding effects of environmental variation. So he too made use of canalization. This was indeed an elegant aspect of his experimental design.

26. This is at least partially a tactical decision of how to draw the boundary, rather than simply of who is right and who is wrong. The issuer of where to draw boundaries is an unavoidable source of argument with a degree property. But as should be obvious, I favor the GE account over the canalization account for reasons stronger than tactics.

References

Ariew, A. (1996) "Innateness and Canalization," *Philosophy of Science* 63:S19–S27.

Ariew, A. (1999) "Innateness Is Canalization: In Defense of a Developmental Account of Innateness," this volume.

Arthur, B. (1994) *Increasing Returns and Path-Dependence in the Economy.* Mighigan University Press.

Arthur, W. (1982) "A developmental approach to the problem of variation in evolutionary rates," *Biological Journal of the Linnaean Society* 18:243–261.

Arthur, W. (1984) *Mechanisms of Morphological Evolution: A Combined Genetic, Developmental, and Ecological Approach.* Chichester: Wiley.

Arthur, W. (1988) *A Theory of the Evolution of Development.* New York: Wiley.

Arthur, W. (1997) *The Origin of Animal Body Plans: A Study in Evolutionary Developmental Biology.* Cambridge: Cambridge University Press.

Boyd, R. and P. Richerson. (1985) *Culture and the Evolutionary Process.* Chicago: University of Chicago Press.

Callebaut, W. (1993) *Taking the Naturalistic Turn: How to Do Real Philosophy of Science.* Chicago: University of Chicago Press.

Campbell, D. T. (1974) "Evolutionary Epistemology," in P. A. Schilpp (ed.), *The Philosophy of Karl Popper*, v. II:413–463. LaSalle, IL: Open Court.

Chomsky, N. (1967) "Recent Contributions to the Theory of Innate Ideas," *Synthese* 17:2–11.

Churchland, P. S. (1986) *Neurophilosophy: Toward a Unified Science of the Mind/Brain.* Cambridge, MA: The MIT Press.

Darwin, C. (1859) *The Origin of Species.* London: John Murray (reprinted by Harvard University Press, 1959).

Diamond, J. (1997) *Guns, Germs, and Steel: The Fates of Human Societies.* New York: Norton.

Garrod, A. E. (1909) *Inborn Errors of Metabolism.* London: Frowed, Hodder, and Staughton.

Gigerenzer, G., Todd, P. and the ABC Group. (1999) *Simple Rules That Make Us Smart.* Oxford: Oxford University Press.

Glassmann, R. B. and Wimsatt, W. C. (1984) "Evolutionary Advantages and Limitations of Early Plasticity, in R. Almli and S. Finger, eds., *Early Brain Damage*, v. I, pp. 35–58. Academic Press.

Gleick, J. (1992) *Genius: A Biography of Richard Feynman.* New York: Pantheon.

Gould, S. J., and Vrba, E. (1982) "Exaptation—A Missing Term in the Science of Form," *Paleobiology* 8:4–15.

Greene, E. (1989) "A Diet-Induced Developmental Polymorphism in a Caterpillar," *Science* 243:643–646.

Griesemer, J. R. (forthcoming-a) "Reproduction and the Reduction of Genetics," forthcoming in H.-J. Rheinberger, P. Beurton, and R. Falk (eds.), *The Gene Concept in Development and Evolution* (Workshop II, Max Plank Institute for History of Science, Berlin, October 17–19, 1996).

Griesemer, J. R. (forthcoming-b) *Reproduction in the Evolutionary Process.*

Griesemer, J. R. and Wimsatt, W. C. (1988) "Picturing Weismannism: A Case Study in Conceptual Evolution," in M. Ruse, ed., *What Philosophy of Biology Is*, pp. 75–137. Dordrecht: Martinus-Nijhoff.

Griffiths, P. (1996) "Darwinism, Process Structuralism, and Natural Kinds," *Philosophy of Science* 63(3) Supplement: PSA-1996, v.I:1–9.

Griffiths, P. and Gray, R. (1994) "Developmental Systems and Evolutionary Explanation," *Journal of Philosophy* 91:227–304.

Hull, D. L. (1980) "Individuality and Selection," *Annual Review of Ecology and Systematics* 11:311–332.

Hull, D. L. (1988) *Science as a Process.* Chicago: University of Chicago Press.

Latour, B. (1987) *Science in Action.* Cambridge: Harvard University Press.

Lehrman, D. S. (1970) "Semantic and Conceptual Issues in the Nature-Nurture Controversy," in L. R. Aronson et al., eds., *Evolution and Development of Behavior*, pp. 17–52. San Francisco: Freeman.

Lerner, I. M. (1954) *Genetic Homeostasis* (reprint by Dover Books, New York).

Lewontin, R. C. (1966) "Is Nature Probable or Capricious?" *Bioscience* 16:25–27.

Lewontin, R. C. (1970) "The Units of Selection," *Annual Review of Ecology and Systematics* 1:1–18.

Lorenz, K. Z. (1965) *Evolution and Modification of Behavior.* Chicago: University of Chicago Press.

Maynard-Smith, J. (1975) *The Theory of Evolution*, 3rd ed., London: Pelican.

Mayr, E. (1974) "Behavior Programs and Evolutionary Strategies," *American Scientist* 62:650–659.

McClintock, M. K. and Herdt, G. (1996) "Rethinking Puberty: The Development of Sexual Attraction," *Current Directions in Psychological Science* 5:178–183.

Mendel, G. (1866/1902) *Experiments in Plant-Hybridization* (reprint of Bateson 1902 translation). Cambridge: Harvard University Press.

Morowitz, H. (1992) *The Origins of Cellular Life*. New Haven: Yale University Press.

Moss, L., (1992) "A Kernel of Truth? On the Reality of the Genetic Program," in D. Hull, A. Fine, and M. Forbes, eds., *PSA-1992, vol. 1*: 335–348. East Lansing: The Philosophy of Science Association.

Naumann-Held, E. M. (1998) "The Gene is Dead—Long Live the Gene! Conceptualizing Genes the Constructionist Way," in P. Koslowski (ed.), *Sociobiology and Bioeconomics. The Theory of Evolution in Biological and Economic Theory*, pp. 105–137. New York: Springer.

Nelson, P. (1998) "Common Descent, Generative Entrenchment, and the Epistemology of Evolutionary Inference." Ph.D. dissertation, Department of Philosophy, The University of Chicago.

Oyama, S. (1985) *The Ontogeny of Information*. Cambridge: Cambridge University Press.

Putnam, H. (1967) "The 'Innateness Hypothesis' and Explanatory Models in Linguistics," *Synthese* 17:12–22.

Raff, R. (1996) *The Shape of Life: Genes, Development, and the Evolution of Animal Form*. Chicago: University of Chicago Press.

Rasmussen, N. (1987) "A New Model of Developmental Constraints as applied to the *Drosophila* System," *Journal of Theoretical Biology* 127(3):271–301.

Rhodes, R. (1986) *The Making of the Atomic Bomb*. New York: Simon and Schuster.

Rice, S. H. (1998) "The Evolution of Canalization and the Breaking of von Baer's Laws: Modeling the Evolution of Development with Epistasis," *Evolution* 52:647–657.

Riedl, R. (1978) *Order in Living Organisms: A Systems Analysis of Evolution*. New York: Wiley.

Schank, J. C. and Wimsatt, W. C. (1988) "Generative Entrenchment and Evolution," in A. Fine and P. K. Machamer, eds., *PSA-1986, volume II*: 33–60. East Lansing, MI: The Philosophy of Science Association.

Schank, J. C. and Wimsatt, W. C. (forthcoming) *Evolvability: Adaptation, Construction, and Modularity, festschrift for Richard Lewontin*, volume 2, eds. J. Singh, C. Krimbas, S. Paul, and J. Beatty. Cambridge: Cambridge University Press.

Simon, H. A. (1982) *Rationality and Human Affairs*. Palo Alto, CA: Stanford University Press.

Simon, H. A. (1996) *The Sciences of the Artificial*, 3rd. ed. Cambridge, MA: MIT Press.

Sperber, D. (1995) *Explaining Culture*. London: Blackwell.

Stearns, S. C., Kaiser, M., and Kawecki, T. J. (1995) "The Differential Canalization of Fitness Components against Environmental Perturbations in *Drosophila melanogaster*." *Journal of Evolutionary Biology* 8:539–557.

Stern, K., and McClintock, M. K. (1998) "Regulation of Ovulation by Human Pheromones," *Nature* 392:177–179.

Szathmary, E. and Maynard-Smith, J. (1997) "From Replicators to Reproducers: The First Major Transitions Leading to Life," *Journal of Theoretical Biology* 187:555–571.

Todd, P. and Miller, G. (1998) "Biodiversity through Sexual Selection," in C. Langton, ed., *Artificial Life V*. Cambridge, MA: MIT Press/Bradford Books.

Turner, M. (1991) *Reading Minds: The Study of English in the Age of Cognitive Science*. Princeton, NJ: Princeton University Press.

Van Valen, L. (1973) "A New Evolutionary Law," *Evolutionary Theory* I:1–30.

Waddington, C. H. (1957) *The Strategy of the Genes*. London: Routledge.

Wagner, G. P., Booth, B. and Bagheri-Chaichian, H. (1997) "A Population Genetic Theory of Canalization," *Evolution* 51:329–347.

Weller, A. (1998) "Communication through Body Odor," *Nature* 392:126–127.

Williams, G. C. (1966) *Adaptation and Natural Selection*. Princeton, NJ: Princeton University Press.

Wimsatt, W. C. (1972) "Teleology and the Logical Structure of Function Statements," *Studies in History and Philosophy of Science* 3:1–80.

Wimsatt, W. C. (1981a) "Robustness, Reliability, and Overdetermination," in M. Brewer and B. Collins, eds., *Scientific Inquiry and the Social Sciences*, pp. 124–163. San Francisco: Jossey-Bass.

Wimsatt, W. C. (1981b) "Units of Selection and the Structure of the Multi-level Genome," in P. D. Asquith and R. N. Giere, eds., *PSA-1980, vol. 2*: 122–183. Lansing, MI: The Philosophy of Science Association.

Wimsatt, W. C. (1986) "Developmental Constraints, Generative Entrenchment, and the Innate-Acquired Distinction," in W. Bechtel, ed., *Integrating Scientific Disciplines*, pp. 185–208. Martinus-Nijhoff.

Wimsatt, W. C. (1994) "The Ontology of Complex Systems: Levels, Perspectives, and Causal Thickets," *Canadian Journal of Philosophy*, supplementary volume 20:207–274.

Wimsatt, W. C. (1997a) "Aggregativity as a Heuristic for Finding Emergence," in L. Darden, ed., *PSA-1996, v. 2* (*Philosophy of Science*, supplement to v.4, #4), S372–S384.

Wimsatt, W. C. (1997b) "Functional Organization, Functional Analogy, and Functional Inference," *Evolution and Cognition* 3 (2):102–132.

Wimsatt, W. C. (forthcoming-a) *The Analytic Geometry of Genetics: The Evolution of Punnett Squares.*

Wimsatt, W. C. (forthcoming-b) *Re-Engineering Philosophy for Limited Beings: Piecewise Approximations to Reality*, Cambridge: Harvard University Press.

Wimsatt, W. C. (forthcoming-c) *The Evolution of Generative Structures: Dynamical Foundationalism in Biology and Culture.*

Wimsatt, W. C. (unpublished) "Generative Entrenchment, Scientific Change, and the Analytic-Synthetic Distinction," invited address at 1987 Western (now Central) Division APA meetings.

Wimsatt, W. C. and Schank, J. C. (1988) "Two Constraints on the Evolution of Complex Adaptations and the Means for Their Avoidance," in M. Nitecki, ed., *Evolutionary Progress*, pp. 231–273. Chicago: University of Chicago Press.

9 Feelings as the Proximate Cause of Behavior

Daniel W. McShea

Male chimpanzees spend considerable time patrolling their territories. Sometimes this behavior involves long scouting excursions around the periphery of a territory. At other times, it might involve quiet, watchful sitting. At still others, patrolling includes noisy displays, perhaps to frighten potential intruders.

It would be hard to believe that patrolling behavior, or any major part of it, is rigidly preprogrammed, or "hardwired." Scouting a territory's periphery, for example, can be undertaken in an almost limitless number of specific ways, varying from episode to episode in the precise path the chimp takes, in its body postures and the sounds it makes at each moment, and so on. The behavior seems too complex, too flexible, and too variable to be orchestrated by a biologically plausible program.

What is the alternative? What other types of proximate causes are available, in theory, to account for such complex and flexible behaviors? An obvious alternative, and perhaps the only one, is that such behaviors are "motivated." We might say that the chimpanzee experiences a kind of "patrolling feeling," a desire or inclination to behave in some way that satisfies some particular feeling, or group of feelings, associated with patrolling. In putting it this way, we are in effect inserting a new variable, an extra degree of freedom, in the (mostly unknown) chain of causation that leads to the behavior, in order to account for the observed flexibility. For this to work, I will argue, the relationship between feeling and behavior must be hierarchical in a special sense: it must be that of the general to the specific. More precisely, a feeling is a demand made upon an animal that some specific behavior be chosen that meets a set of general criteria prescribed by the feeling.

In psychology, feelings are generally understood to be one of three aspects of emotion. Emotion has a physiological aspect (e.g., an increase in heart rate during moments of intense fear). It also has outward manifestations (e.g., facial expressions). Finally, it has a subjective aspect, namely the feelings themselves. The first two can be studied directly, and for both, plausible and satisfactory suggestions have been offered regarding their function; for example, physiological changes might help prepare the organism for rapid action, and facial expressions might function as means of communicating internal states. The third aspect, arguably the least understood in functional terms, is the focus here.

The discussion will begin with a rough conceptual scheme in which I argue that feelings are (in organisms that have them) the proximate cause of conscious, deliberate behavior, with "cause" understood in a hierarchical sense. Then I offer a model from developmental and evolutionary biology to help explain this sense of causality. Applied to feelings and behavior, the model makes a number of testable predictions.

The scheme also provides an in-principle resolution of certain aspects of the nature-nurture controversy, a resolution that accommodates many of the most extreme claims both of universalists, who argue that human nature is relatively invariant across cultures, and culturalists, who argue that it is quite plastic. Ultimately, my claim will be that by introducing a variable that is causally "upstream" of behavior (the feelings), we can have it both ways. Specifically, to put it somewhat crudely, behavior is plastic; it is feelings that are universal. A longer explanation is necessary to see what sense this claim might make, and a number of caveats are required for it to be defensible.

The notion of universal feelings is not new. Indeed, it has a long and distinguished history, mainly in moral philosophy (McShea 1990; see also McShea and McShea in press). But it has not appeared, to my knowledge, in the most recent incarnation of the nature-nurture debate in biology and psychology, which spans the last two decades since E. O. Wilson's book on sociobiology (Wilson 1975). Both the proposed resolution of this debate and the conceptual scheme on which it is based are somewhat sketchy; much needs to be filled in, and significant changes may be necessary as thinking about it progresses. My motivation in presenting them here, perhaps prematurely, flows from the conviction that the core concept—that feelings are the proximate cause of behavior—will survive the necessary revisions, and that it has the potential now to change the terms of that debate in a productive way.

I am an evolutionary biologist and an outsider in the serious discussion underway in theoretical psychology and philosophy about the mechanisms and functions of emotions, and therefore, at least indirectly, of feelings. Thus it is difficult for me to determine to what degree the suggestion offered here is consistent with the various views that have been developed in that literature (Buck 1985; Damasio 1994; Frijda 1993; Griffiths 1997; Izard 1993; Johnson-Laird and Oatley 1992; Lazarus 1991; LeDoux 1996; Plutchik 1980; Sloman 1987; Solomon 1977; Young 1973; Zajonc and McIntosh 1992). Based on the sources just listed, at least, the present suggestion seems neither to have been proposed nor to have been ruled out. But for the present, I leave such judgments to scholars more appropriately positioned to make them.

Also, as an outsider, I am doubtless prone to certain kinds of errors, such as misusing terms with established technical meanings and mistakenly indicating by choice of words either affiliation or disaffiliation with this or that school of thought. I apologize for these gaffes and hope they will not draw critical attention from my main point.

1 Preliminaries

An example of feelings in action will help to illustrate the proposed scheme: a dozing mother cat awakens as a dog approaches the hiding place of her kittens. She sits up, her ears straighten, her tail swishes, and her gaze becomes fixed on the dog. As the situation develops, she might well relax and even begin to feel frisky and playful. On the other hand, she might experience a combination of fear for her own safety and a kind of "brood-defensive feeling." Which package of feelings is evoked will depend on her cognitive analysis of the situation, which might include an assessment of the dog's present mood and intentions, his position relative to the kittens and to herself, the likelihood that he will notice the kittens, and her own present condition and ability to defend herself. Also relevant are her memories of past encounters with dogs generally, and with this one in particular.

Suppose that she concludes the dog is hostile, and that she experiences both fear for her own safety and a brood-defensive feeling. Both feelings demand satisfaction, and a struggle for dominance ensues. One feeling eventually triumphs—suppose it is the brood-defensive feeling—and results in some behavior. What behavior, specifically? She might attack, but she might also run, not in fear but in an attempt to distract the dog's attention from the kittens. Her past experience and present cognitive analysis of the situation will have much to do with the behavioral choice she makes. But her goal is nothing other than the satisfaction of the brood-defensive feeling, and her chosen behavior is her best guess at which behavior will achieve that end. We will suppose that she decides to attack the dog.

The example is speculative, of course, in that I may not have correctly identified the specific feelings that would typically be evoked in a house cat in an encounter of this sort. But this is beside the point; its purpose is mainly heuristic.

Two Distinctions

Two preliminary distinctions are necessary. The first is between, on the one hand, mental activities such as feeling and cognition, and on the other, behavior. Behavior here refers only to motor activity. The cat's swishing of her tail is behavior. Her fixed gaze is also behavior, because it involves muscle activity, even though no net movement occurs. Any vocalizations she makes are behavior; in humans, speech is behavior. But the cat's fear and her calculations and evaluations are not behavior. In what follows, I will use the term "behavior" in a slightly more restricted sense, to refer to motor activity that is both conscious and deliberate. For the cat, this would likely include her attack on the dog, but would probably exclude her tail-swishing.

The second distinction is between two categories of mental activity, cognition and feeling. A feeling is a conscious state of dissatisfaction or disequilibrium, more specifically, of desiring, wanting, or preferring. Notice that, in contrast with colloquial usage, in the present scheme a feeling is not a mood (i.e., not a feeling in the sense of "feeling good" or "feeling depressed"). Also, a feeling is an affective experience, not a form of knowledge or an awareness of knowledge, such as an intuition or insight—in other words, not a feeling in the sense of "having a feeling *about* something." The connection with emotion will be discussed later.

Cognition here is a wastebasket category to hold all of the nonfeeling (i.e., preference-neutral), conscious mental functions, such as perception, memory, reason, the various learning and language mechanisms, and so on. Thus the cat's brood-defensiveness was a feeling, but her recollections, spatial calculations, and evaluation of the dog's disposition were all parts of cognition.

Peripheral Issues

Feelings Are Physical. My decision to focus on the subjective is not intended to have any dualist implications. I do not intend to suggest that the subjective aspect of emotion is not physical or that it is not translatable into physical terms. Indeed, it must be so translatable—at least in principle—if feelings are to be efficacious, if they are to cause behavior. However, the discussion here is agnostic about the nature of the physical basis of feelings—for example, about whether they occur in discrete structures within the brain or arise from activity more globally distributed (although some evidence on this matter exists; see LeDoux 1996).

Feelings in Nonhuman Species. At least some nonhuman animals have cognitive and emotional mechanisms sufficiently similar to ours that we can meaningfully speak of them as thinking and feeling in the same sense that we do. Humans have direct introspective evidence of their own feelings, of course. The claim that other species have them also is somewhat speculative but not entirely baseless. From an evolutionary standpoint, it would be surprising if our close relatives had utterly different brain mechanisms mediating their behavior. But it is an open question whether this closely related group includes only the great apes or extends to the rest of the mammals or perhaps to all vertebrates and other phyla as well. It is also possible that the existence of feelings is a matter of degree, that some species have more of them than other species, or that the degree of flexibility in behavior that they offer varies from species to species. Here, for the sake of discussion, I assume that feelings occur in and cause behavior in all mammal species; but this restriction is somewhat arbitrary, and the reasoning that follows does not depend on it.

2   A Rough Conceptual Scheme

Three Basic Claims

The scheme is constituted by three basic claims; each is discussed further in the sections that follow.

(1)   Mammals use their cognitive abilities to sense and interpret the world and to anticipate future events. (In the above example, the cat "reads the situation" as the dog approaches.) These interpretations and anticipations in turn evoke feelings. (The cat experiences either playfulness or brood-defensiveness together with fear.)

(2)   The range of possible feelings that individuals routinely experience is large, and in any given situation, more than one felling may be evoked, orienting the animal to a number of different purposes at once. (Again, in the story, brood-defensiveness and fear are evoked simultaneously.)

(3)   A struggle for supremacy occurs among the evoked feelings. Eventually one feeling, or coalition of feelings, triumphs over all others and causes some behavior. (Brood-defensiveness wins, and the cat attacks.)

The most important claim, for present purposes, is (3), the notion that feeling causes behavior. The evocation of the feelings and the range of feeling are side issues and are discussed briefly in the following subsections; I present them mainly to fill out the larger conceptual picture within which the central claim is to be understood.

How Feelings are Evoked.   The various mechanisms by which emotions are evoked have been studied aggressively, especially recently (LeDoux 1996), but they are still only imperfectly known. The evocation of their subjective aspect, the feelings, is even less well understood, but in any case, a deep understanding is unnecessary for present purposes. Here, the only requirement is that feelings are reactions to interpretations of situations, and that therefore variations in how situations are interpreted may account for differences in which feelings are experienced. Thus two individuals of the same species with identical feeling repertoires (what I later call "feeling profiles") and who are placed in the same situation nevertheless may not experience the same feelings. For example, one cat may not have met this particular dog before, and may judge it dangerous based on its exuberant behavior, whereas a second cat may know the dog quite well and understand that, contrary to appearances, it is friendly and safe. Both cats experience the same *external* world, but as a result of their differing histories, their cognitive processes analyze that world differently, and thus their feeling profiles are presented with different *interpreted* worlds. It is the interpreted world to which feelings react.

The suggestion that differences in which feelings are evoked (and thus ultimately in which behavior is caused) are the result of differences not in the underlying feeling profile but in the cognitive processes involved in interpretation will be a central element in the treatment offered later of the human nature-nurture problem.

The Feeling Profile.   The repertoire of feelings that individuals in a given species normally experience, each weighted according to the degree to which it is experienced in the full range of problematic situations that individuals of the species normally encounter, constitutes the species' "feeling profile" (McShea and McShea in press). Thus the feeling profile for a house cat would consist of a set of feelings that cats typically experience, including the feelings mentioned in the example— playfulness, fear for one's own safety, and brood-defensiveness—and many more, along with the situations in which each is normally evoked. In addition, the profile would specify the situation-specific intensity with which each feeling is normally experienced.

I am inclined to believe that the feeling profile in most mammal species is complex —in other words, that the number of different feelings is large, and that in addition to some very general types, such as fear and envy, numerous variants also exist. Fear, for example, would seem to have a number of different subtypes; the fear produced in us by the rumble of thunder seems to be qualitatively different from the fear produced by a prolonged stare of a rival. Of course, the possibility remains that complex profiles are constructed or built up, neurologically, from a small set of elemental mental states. Conceivably, some group of basic states—perhaps closely paralleling the set of "basic emotions" that has been suggested—might combine in various proportions to produce the great variety of qualitatively different feelings we recognize subjectively. But this would not contradict the claim that the profile is complex, because it is the subjective variety of feelings that constitutes the profile.

Given this complexity, a complete description of the feeling profile of even a single species is out of the question. Attempting a partial description would be a useful exercise and an important part of a more fully developed theory of feeling, but I will not attempt it here. First, feelings are secondary qualities, like colors, and therefore they are extremely difficult to describe in words. My use of awkward expressions like "brood-defensive feeling" is a symptom of this inadequacy of language. And second, the flaws in my attempt would only introduce unnecessary controversy and distract from the main point, which is the causal relation between feeling and behavior.

Later I will argue that feeling profiles are likely to be very similar among individuals in a species. And finally, feeling profiles are presumably adaptive to some extent. However, the degree to which they are functional, the issue of whether profiles are optimal in some sense or merely minimally functional, is beside the point, at least at present. (Presumably, this varies from species to species anyway.)

Feelings as Causes of Behavior

Hierarchical Causation.   The precise mechanism by which the brood-defensive feeling causes the cat to attack the dog is unknown, but we can nevertheless characterize it in two ways. First, the feeling causes the behavior in the sense that it galvanizes the animal or energizes it to act. Second, the causation is hierarchical in the same sense that the flow of orders in a military chain of command is hierarchical. A vague or general order given at a high level (e.g., "Prepare the base for inspection!") evokes a range of specific commands and ultimately specific behaviors at lower levels (policing the lawn, painting the flagpole, etc.). Likewise, feelings orient or direct the animal toward vague or general goals, which can be understood as the causes of the behavior they precede.

Another way to understand hierarchical causation is to think of higher-level causes as boundary conditions on lower-level behavior (Salthe 1985). A balloon is the cause of, or determines, the spatial positions of the gas molecules it contains, but only to the extent of limiting their movements to within a small region. Within the balloon, their movements are free. If the balloon is falling, it causes the gas molecules to move downward (on average), but movement within the balloon is still unconstrained. Likewise, in a military chain of command, high-level orders are boundary conditions. Many different specific behaviors are consistent with the vague order to prepare the base for inspection. Indeed, the high-ranking officer may not have had any specific sequence in mind and may not care about the details (what procedure is used to police the lawn, who paints the flagpole, etc.), provided that the base is prepared.

Feelings are causes of behavior in just this sense; they are demands (or at least emphatic requests) that the range of behavioral choices be limited to those that achieve some end. Feelings limit behavior, but within the bounded range they specify no particular behaviors, no precise motor sequences. The brood-defensive feeling is a demand that the cat limit her behavioral choices—from the vast range of possible motor sequences—to those that are likely to result in the safety of her kittens.

Another analogy: Feeling causes behavior in much the same sense that the object and rules of a sport cause the activities of the players, not by specifying those activities in detail, but by constraining them. Indeed, feelings might rightly be called the "object" of behavior in just this sense. That is, the object of behavior is the satisfaction of the feelings.

Proximate vs. Distal Causes.   The notion that feelings are the proximate cause of certain behaviors does not rule out the possibility that those same behaviors have more distal causes as well. Ultimately, a tiger hunts because historically natural selection acting on tiger ancestors favored individuals who were predisposed to hunt and who did it effectively. More proximally, a tiger hunts because she is hungry. But even hunger may not be the most proximal cause, if—as seems to be the case with humans—tigers need not actually experience hunger during the act of acquiring food. Instead, a tiger hunts because she wants to, because she experiences a hunting feeling or an inclination to hunt. Such a feeling might be triggered initially by hunger, but the feeling itself and not hunger likely occupies the mental foreground while a hunt is in progress.

A Developmental Model

Physical development in organisms is at least partly hierarchical in the following sense. Early in development, an embryo consists of a relatively small number of structures. These interact to give rise to more structures. Which in turn give rise to yet more, and so on, in a widening cascade. One result of this organization is that early structures have more structures developmentally downstream from them than later ones do. Thus variation occurring in early structures tends to produce long cascades of changes throughout the downstream portion of the developmental trajectory. Most of these changes are likely to be deleterious. For example, a variation early in the development of a human embryo in which the anterior neuropore of the neural tube fails to close produces widespread and catastrophic consequences, notably anencephaly, a condition in which the brain as a whole fails to develop properly. Finally, because the consequences of variation in early structures are often so widespread and deleterious, variation in them will on average be more strongly opposed by natural selection and thus will occur less often than variation in later structures.

   This hierarchical outlook, or model, has been developed independently a number of times in biology in recent years, notably by Riedl (1977), Arthur (1984, 1988), Wimsatt (1986), Kauffman (1993), and Salthe (1993). In Wimsatt's terms, structures with many downstream developmental consequences are "generatively entrenched." Riedl describes such structures as "burdened." (See Raff 1996 for an alternative view.)

   Based on the model, it should be easy to see that generative entrenchment—in combination with natural selection—might account for the conservativeness of certain features of organisms, both in their development and in their evolution. In particular, it might account for the basic or foundational similarities among groups of related species, for what are known as "bodyplans." For example, a two-part, six-legged body characterizes almost all of the many millions of insect species and constitutes a feature of the insect bodyplan. For the vertebrates, a dorsal, segmented, axial column—the vertebral column—is a bodyplan feature.

In standard usage, the term "bodyplan" is restricted to features that characterize a number of related species at a high taxonomic level, such as vertebral columns, but in principle it could be applied at lower levels. For example, in humans, a large brain (i.e., relative to body size) might be described as a bodyplan feature at the species level. The prediction of the entrenchment model would be that the structural features involved in a large brain arise later in development than the vertebral column, but earlier than features that are more variable within our species (e.g., eye color). More generally, the prediction is that both conservativeness of traits in development and how widespread they are among species in higher taxa, will be strongly associated with degree of entrenchment, with the more entrenched structures expected to be both more conservative and more widespread.

Entrenchment, Vagueness, and Feelings.   For present purposes, the key property of entrenched structure is that they limit or constrain the form of less entrenched structures downstream—constrain them, that is, without precisely determining them. The limitation is similar to that of a foundation, which constrains but does not precisely determine the design of the house that will be built upon it. Likewise, I argue, feeling constrains behavior and is foundational for behavior in the same sense. A feeling is an attempt by the animal to impose a boundary condition, or a constraint, on its own behavior, an attempt to limit the behaviors actually performed to those that will satisfy it. Bodyplans are also foundational and constraining in this sense, and thus the feeling profile as a whole might be understood as a kind of "behavioral bodyplan." (Wimsatt 1986 has already described the sense in which the entrenchment model might apply to behavior. My proposal modifies his treatment to incorporate what I argue are the causes of behavior, viz. the feelings.)

Salthe (1993) provides a different language in which to make the same point. In his view, early embryonic structures have fewer features—or more precisely, fewer features that are regular in their appearance—that are reliably present from moment to moment. Thus, in a sense, their composition is less specific, less well defined. They are, in Salthe's terms, "vaguer" than later structures. As development proceeds and these early structures become elaborated, they also become more regular, that is, they become better defined, more "specified." In his language, a feeling is behavior that is still vague, or behavior-not-yet-completely-specified.

We can make the connection between the developmental model and the feeling-behavior relationship more explicit. Imagine each feeling-behavior sequence in a human individual—each instance in which a feeling occurs and causes some behavior—as a miniature ontogeny, a short developmental subroutine. As such a sequence begins, some strong feeling has been evoked, by some mechanism unknown, and will cause some behavior, chosen from among the many that would satisfy it. The feeling acts as a constraint, limiting the range of acceptable behaviors to those that would satisfy it, but it does so without specifying precisely the behavior to be performed, just as the earlier-arising structure limits the range of later-arising structures without specifying them. The main difference is that in structural development, all of the terminal downstream structures are (eventually) expressed simultaneously in adult form, whereas in the brief feeling behavior developmental subroutine, only one behavior is expressed at a time; the entire range of behaviors satisfying a given feeling can only be expressed over time and exists at any one moment only as a set of potentialities.

We can now describe a sense in which feelings are entrenched with respect to behavior. Behaviors, like later-arising structures, may vary with few consequences; one behavior may be substituted for another in a behavioral repertoire, so long as it satisfies the constraints imposed by the feeling with which it is associated. However, a variation occurring in a feeling, one that even slightly changes the goal or purpose it represents, may have enormous consequences, perhaps rendering inappropriate all or most of the existing repertoire of behaviors associated with it. Thus feelings are entrenched in the sense that variations in them will tend to have more consequences and therefore will have more dire consequences than would variations in behavior.

Speculations and Predictions

The Specification of Feeling in Development.   The suggestion so far is quite modest: Causality has the same sort of hierarchical organization in both structural development and in the feeling-behavior relationship. More speculatively, we can draw a closer connection between development and feeling/behavior. Again, imagine each feeling-behavior sequence as a miniature ontogeny, a brief developmental subroutine. In the course of a lifetime, each subroutine will be run many times, and in each run, some behavior will be chosen or devised, in order to satisfy the feeling, from among the many possible behaviors of which the animal is capable. Eventually a set of fairly specific behavioral options will come to be associated with each feeling.

Presumably, the individual learns that one (or a small number of) behaviors are more effective than others in satisfying a given feeling, and eventually he will acquire the habit of behaving that way when that feeling is experienced. As an individual matures, habitual behavioral selections become more finely tuned. Also, behaviors become more elaborated, more routinized, and more fixed. The result is that, in effect, the upstream feeling becomes more entrenched, more specified. We might say that a feeling begins in childhood as a vague desire and becomes transformed into a specific desire, a desire to do one or a small number of particular things or to behave in a number of specific ways. Notice that a result of this progressive entrenchment is that the many-to-one relationship between feelings and behavior is gradually eroded; therefore, to the extent that my speculation here is true, the earlier discussion about the source of behavioral flexibility and variability is more applicable to *young* organisms and less so to older ones.

Another result of this progressive entrenchment is that as individuals age, change in the feelings becomes more and more unlikely, because the consequences of change become more and more severe. For example, consider an older but still pre-adolescent child, in whom the feeling profile has become somewhat entrenched. At this point, any change occurring in an entrenched feeling would be highly disruptive, rendering ineffectual all of the highly elaborated and specified behavioral options that have become fixed downstream of it. An entirely new behavioral repertoire would have to be developed to satisfy the modified feeling. Something like this may occur in human adolescence, at least for some portion of the feeling profile. Thus, given this admittedly speculative understanding of the ontogeny of feeling and behavior, the entrenchment model enables us to make a prediction: to the very limited extent that feeling profiles are susceptible to change, or even modifiable in some deliberate way, they should be more modifiable in younger mammals than in adults. And because behavior lies causally downstream of the feelings, behavior should be more modifiable than feelings at any age. In clinical terms, behavior modification should be easier than feeling-profile restructuring.

For these speculations, I have no evidence of the sort that would convince a skeptic. I present them mainly because they enable us to make some predictions that are potentially testable. (Obviously such variables as the modifiability of behavior and of the feeling profile would first have to be operationalized.) Also, the second speculation—that feelings are developmentally entrenched relative to behavior—will have a key role in my discussion of the nature-nurture problem. Importantly, the hierarchical model of behavioral causation does not depend on these speculations and would stand even if both were shown to be false.

Disparity among Feeling Profiles.   Speculating even further, suppose that the physical brain structures that produce feelings, whatever they are, arise earlier than and give rise to the brain structures that produce motor activity. In other words, suppose that causal pathways by which feelings produce behavior, moment to moment, mirror the pathways by the which the corresponding neurological structures are produced in embryology. The suggestion is that feeling structures are entrenched relative to behavior structures—entrenched in the original structural sense. Again a prediction follows, namely that, among individuals and across higher taxa, feeling profiles should vary less than behavioral repertoires. For example, feelings should be more similar than behaviors among members of the same species. All chimpanzees are expected to have similar feeling profiles,but the behavior of individual chimps will vary widely. And although differences in feeling profile are expected to increase at higher taxonomic levels—a mouse species profile will be very different from that of an elephant—the prediction is that disparity in behavior will grow much faster.

The Role of Cognition

The mechanism by which cognition evokes feelings is unknown and will not concern us further here (see above). But more can be said about some of the many functions or roles of cognition. The discussion of these roles will necessarily be cursory, and a number of claims will be made with little supporting argument. My purpose is not to prove these claims, but rather to fill out the conceptual picture within which the view of feelings as causal was developed. For a more thorough treatment, see McShea (1990).

Cognition as the Slave of the Feelings.   A major function of cognition is the perception of and interpretation of situations to which the feelings react. In humans (at least), cognition has the further mission of constructing hypothetical scenarios to which the feelings also respond. Consider the following somewhat outlandish example of cognition and feeling in action:

Walking along a city street, I encounter an armored truck parked at a local bank. The street is deserted, the back of the truck is open, the bags of cash are lying in full view, and the guards have mysteriously fallen asleep. In my imagination, I see myself grabbing the money, flying to some thieves' haven, safe from extradition and basking in the sun (and in the envy of my friends) on the deck of my yacht, while well-paid servants cater to my every whim. Some feeling or coalition of feelings reacts with approval to this imagined scenario.

But before any behavior results, my imagination races to prepare another scenario. I see the guards waking up, shots are fired, sirens are wailing, and I am rushed to the hospital, and later—if I live—to jail. Some bundle of feelings reacts in horror to this imagined sequence of events. I walk right past the armored truck without pausing.

The point is that, among their many other functions, cognitive (i.e., nonfeeling) mental processes are involved in projecting possible futures and also in devising various behavioral options and anticipating their consequence. The difference between humans and cats is that our greater cognitive powers enable us, among other things, to pursue longer and more detailed imaginative sequences, to hold a number of such scenarios in mind at almost the same time, and to pause long enough for a larger proportion of the feeling profile to react, coalitions of feelings to form, and well-considered behaviors to be chosen or devised. But equally in humans and cats, cognition is nothing more than a tool deployed in the service of the feelings. The argument is originally Hume's, who wrote: "Reason is, and ought only to be the slave of the passions, and can never pretend to any other office than to serve and obey them" (1978, p. 415).

This view of reason as the slave of the passions raises the possibility (not to be explored in this essay) that an argument precisely parallel to the one about could be devised, the conclusion of which would be that feeling is the proximate cause of all conscious cognition, as well as of all conscious behavior.

Cognition Alone Cannot Cause Behavior. In humans, we commonly speak of thinking as causing behavior. For example, we might speak of an act resulting from and therefore being caused by an observation or a chain of reasoning or both. To see why the suggestion that cognition causes behavior is mistaken (if taken literally), consider a modification of the earlier example, substituting a human mother for the cat: The human mother sees the dog bounding toward her baby and responds by picking the baby up and carrying it indoors.

It might seem at first that her observation of the dog together with her reasoning about what it might do are possible causes of her behavior. But observations and reasoning alone are preference-neutral; they provide no motivating force. Like the cat, the human mother might use her cognitive powers to project the dog's probable path, assess the likelihood that the baby will be noticed, and judge the dog's probable behavior if it does notice. But no behavior follows directly from these calculations, Her behavior can be seen to have a cause only by interposing an intermediate variable between preference-neutral cognition and action, only by interposing a feeling, in this case, a feeling of protectiveness for the child. The same argument applies to all other aspects of cognition—memory, language capacity, and so on—which likewise deliver no motive, no impulse to act, at least by themselves.

To see this more clearly yet, imagine a creature with great cognitive powers who is devoid of feeling, a pure intellect. Without feeling, such a creature would have no preferences, and therefore for it, no possible or imagined state of affairs in the world would be preferable to any actual state. Thus it would never have any motivation to *do* anything, not even to move about. Crucially, to only the subjective aspects of these processes are relevant here. Pure intellects (e.g., modern computers) can be physically constructed so that they do act, and it is possible in principle that our brains are so wired that, contrary to the evidence of introspection, behavior is caused by purely cognitive processes. And feeling might be epiphenomenal. The point here is that the connection cannot be made subjectively: logically, it is impossible to see how any behavior, any action, follows directly from pure cognition.

Other Issues.   With the nonfeeling mental processes removed as sufficient causes, feelings become essential to conscious behavior, in species that have them. Consistent with this view, we can see decision making, choices among behavioral options, as essentially contests of feelings. Situations reported and interpreted by cognition, or hypothetical scenarios devised by cognition, evoke a number of feelings. In some cases, a single feeling instantly dominates: I see the bus bearing down on me and some variety of fear instantly overwhelms me. In others, no clear winner emerges right away. My house needs painting to appease my neighbors and to prevent the shingles from rotting, but paint is expensive and I am tired at the moment. Thoughts of peevish neighbors, leaking walls, a dwindling bank account, my aching muscles, and so on, evoke a flood of feelings, which then vie among themselves for dominance. A temporary coalition eventually forms, say, among the various fears associated with the neighbors and the leaks, and then I paint the house. A decision has been made. (I have simplified the process considerably, of course, omitting the role of memory, the consequences of further trains of thought inspired by the feelings themselves, which in turn evoke other feelings, and many other complications.)

These arguments run against the grain of the usual view of feelings in one respect. We ordinarily think of feelings as a source of bias in cognition and in behavior. Ordinarily, feelings are thought to cause only rash or "emotional" behavior, and to do so against our better judgment, against reason. Feelings are often considered the more "primitive" mental processes, which drive emotional behavior—for example, rage, which can cause violent behavior—whereas reason is said to underlie more temperate behavior. In the present scheme, these claims are seen to be non sequiturs. Feelings cannot bias judgment, because feelings are judgment itself in action. Feelings do not cause only emotional behavior while cognition causes temperate behavior; rather, feelings are essential in all conscious behavior. An appropriate alternative distinction might be one between passionate feelings and calm feelings, or between the short-term feelings, which can only be satisfied by immediate action (e.g., getting out of the way of the bus), and the long-term feelings, which can only be satisfied by consistent patterns of behavior over long periods of time (e.g., care of children). Again, for further discussion of this issue, see McShea (1990).

3   The Nature-Nurture Problem

The understanding of feelings as distinct from behavior and cognition opens up some conceptual space in the nature-nurture controversy, creating the possibility of a new theoretical position, which I only outline here. Before turning to it, I should make clear that the controversy has a number of different aspects, such as the possible genetic basis of certain diseases, the modifiability of certain behavioral tendencies, and the efficacy of education in overcoming various innate predispositions. All are potentially relevant here, but I am presently concerned only with the debate between culturalists and universalists. Culturalists argue that because of the plasticity of the mind to environmental influences, human nature varies enormously from one culture to the next. The universalists contend that these differences are superficial, that beneath an outer layer of cultural variation lies an inner core of common characteristics, a human nature that is everywhere the same.

The new theoretical position follows from the conceptual scheme above. Cognition produces interpretations of life situations, which in turn evoke the feelings. In any given situation, multiple feelings may be evoked, but eventually one feeling or coalition of feelings triumphs and causes some behavior, chosen from the various options that have been learned or devised. The new position begins with a recognition that, in the realm of interpretation and behavior, culture is virtually omnipotent. We associate flag with country, and country with a strong, familial authority figure, thereby evoking—in certain contexts—reverence and awe in its presence. The associations are implanted by culture. Further, any of a large range of behaviors would satisfy these feelings, but culture has narrowed the range options to a context-specific few: on the right occasions, we might stand, perhaps at attention, and maybe we even sing. (Or we might choose *not* to stand or sing, perhaps in protest, but if so, it will be because other stronger feelings have also been evoked, also via associations implanted by the culture; cultures do not speak with one voice.)

Thus the basic culturalist insight is supported. By manipulating associations and circumscribing behavioral options, a culture can train people to say and do almost anything. Further, in moving from one culture to the next, we shift from one culturally structured system of interpretation to a different one. Different cultures serve up entirely different symbolic worlds to the feelings. The visitor from a distant culture looks up at our proud symbol of national unity and sees only a bit of colored cloth, curiously mounted on a stick. Indifferent, he walks away. Both we and the visitor see the same flag, but on account of our cultural differences, our cognitive processes analyze that object differently. Our feeling profiles are presented with different *interpreted* worlds, different feelings are evoked, and different behaviors result.

Notice that none of this in any way contradicts the basic universalist insight. All that this shows is that differences in interpretative system and in behavior—the standard evidence against the universalist view—do not necessarily point to differences in feeling. It shows that individuals with identical feeling profiles may nevertheless be expected to behave very differently if they have different interpretative schemes. The conclusion is that cultural differences in behavior, in what people say and do, can be largely, if not wholly, accounted for *without* positing differences in feeling profile.

As the debate is usually framed, the culturalist and universalist views seem to contradict each other. For example, the universalist wants to claim that warfare and the care of children are human universals. But the culturalist points out that so-called warfare and child care differ so substantially among cultures in their meaning and in the practices they involve that there is hardly any commonality. Seemingly, both cannot be right.

However, if we make the appropriate distinctions, both *can* be right. Warfare involves simultaneously a set of feelings (motivations), a set of behaviors (practices), and a cognitive interpretative scheme (a set of culturally assigned meanings). In view of these distinctions, the universalist might revise her claim and argue not that warfare per se is universal, but rather that there is a common set of feelings that all or most cultures mobilize and which tends to produce intergroup violence of some kind. And the culturalist, in turn, might revise his claim, arguing that only interpretative schemes (the meaning of warfare to its participants) and practices (actual behaviors) vary significantly among cultures. Whether either or both is right is a purely empirical matter; the point here is that in principle they both could be, that they are not contradicting each other.

Given the existence of a common feeling profile, two consequences emerge. The first is that communication across cultures becomes possible. If we did not share a set of common interests arising from a common feeling profile, communication across cultures would be impossible, and not merely difficult as we observe it to be. The second consequence of a shared feeling profile is a common basis for evaluating alternative cultural interpretative schemes and sets of behavioral options. We ought, in principle, to be able to agree on which sorts of understandings of the world and which courses of action tend to satisfy the feelings better in this or that situation. In other words, a common feeling profile gives us a common foundation for morality, for judgments about the better and the worse (McShea 1990; McShea and McShea in press).

Clarifications

Universality. For convenience, I have presented the universalist argument as though it requires all individuals in a species to have identical feeling profiles. In fact, it does not. All traits are expected to vary (at least somewhat) in all species. Consistent with the developmental model, the expectation is that cognitive systems of interpretation and ranges of behavioral options will be shallowly entrenched and therefore highly variable both among individuals and among cultures. Feeling profiles are expected to be more deeply entrenched and thus less variable, but no trait is truly universal.

Also, some systematic differences in feeling profile are expected among cultures. The reason is that the development of the feeling profile in an individual is, like the development of any physical structure, a dynamic, interactive process, one undoubtedly requiring input from the environment, which includes the culture. The reason for not emphasizing these differences earlier is that the developmental model partly obviates them. That is, to the degree that feelings are entrenched, environmental variation is expected to have less effect, and feelings are expected to be less variable. (Genetic variation is also expected to have less effect; see below.) More importantly, given the much greater variability that the developmental model predicts for cognition and behavior, large cultural differences in feelings are unnecessary to account for the tremendous differences in behavior observed among cultures.

Finally, nonnegligible differences may exist in feeling profiles among age groups and between the sexes, but these matters require a much longer treatment that I can given them here.

Genes vs. Environment. In its most recent incarnation, the nature-nurture problem has taken the form of a debate over the relative importance of genes and environment to the determination of human nature. Arguably, the problem has been misconstrued. What really concerns us is not genes versus environment but the degree of flexibility in human nature and the extent to which a common nature is shared by all humans. For these issues, the relevant factors are the degree of hierarchical organization in behavioral development and the depth of entrenchment of the various behaviors in that hierarchy, not whether the source of variation happens to be genetic or environmental (Wimsatt 1986).

The present scheme embraces this critique. Both feelings and behavior (understood as motor activity) have essential genetic and environmental (i.e., cultural) components, and differences in their contributions are not significant in this context. For example, I have argued that the feeling profile is likely to be somewhat entrenched and therefore less variable than behavior within a species. But proper development of the profile could well require substantial information from the environment, information that is "expected," so to speak, in development (for example, in "critical periods") and that is perhaps just as essential as any genetic information. "Entrenched" does not mean the same thing as "genetic." Thus one of the main virtues of the developmental model is that it ignores the often misleading opposition between genes and environment (Oyama 1985).

4 Summary

I have proposed a conceptual scheme in which feelings—the subjective aspect of emotion—are the proximate cause of all conscious behavior in species that have them, here assumed (for the sake of argument) to be limited to mammals. To see the sense in which this might be true, two preliminary distinctions are necessary: first, mental processes must be distinguished from behavior, which refers only to motor activity. Second, two types of mental process must be distinguished: feeling, which refers to wanting, preferring, desiring, etc., and cognition, which refers to all of the nonfeeling, or preference-neutral, conscious mental processes, including those described colloquially as memory, reason, language capacity, and so on.

In the scheme, the function of the cognitive processes includes perceiving and interpreting the world and anticipating future events. In species with highly developed cognitive powers, such as humans, these anticipations may involve long narratives and elaborate scenarios. But in all such species, cognitive processes are understood to be perfectly passive, completely powerless to cause any behavior, at least directly.

However, cognition evokes feelings, which do cause behavior. The list of the entire range of feelings that an individual is capable of experiencing, together with a specification of the situation-specific intensity with which each is experienced, is called the feeling profile. Feeling profiles in most species are complex, and in various life situations many different feelings may be evoked, orienting the animal to a number of different purposes at once. When this occurs, a struggle for dominance among the feelings ensues, and eventually one feeling or a coalition of feelings triumphs over all others.

The victorious feeling or coalition then causes some behavior. The causal relationship between feeling and behavior is hierarchical, which in the present context means that the range of behavioral options is constrained (to those that will satisfy the feeling), but no particular behavior, no particular motor activity, is specified. A model from developmental biology helps to explain this sort of hierarchical causation further, and with the addition of certain speculative assumptions, the model enables us to make some (at least in principle) testable predictions: first, as individuals age and behavioral sequences become more elaborated and less easily modified, the consequences of feeling modification become more severe. Second, at any age, modification of behavior should be easier to achieve than modification of the feeling profile. Further, it can be argued that if the developmental relationship between the physical brain structures involved in feeling and those involved in behavior mirrors their relationship subjectively, then feeling profiles should vary less than behavioral repertoires at all taxonomic levels.

Assuming this is true at the species level in humans, the feeling profile is expected to be less variable, more nearly universal, among individuals and across cultures than is behavior. Finally, the possibility of a nearly species-universal feeling profile suggests an in-principle resolution to the debate between culturalists and universalists over the degree to which human nature varies among cultures. The suggestion is that cognitive interpretative schemes and ranges of behavioral options vary among cultures, whereas the feeling profile is more nearly universal. The main virtue of this proposal is that it accommodates the widely shared intuition that both the culturalists and the universalists are right, that tremendous differences exist among cultures in mode of understanding and in behavior, and also that despite these differences, and underlying them, there is a human nature that is everywhere the same.

Acknowledgments

The ideas presented in this essay arose from discussions over a period of about twenty years with my father, Robert J. McShea, who was a political philosopher. The notion of the feelings as causal was a secondary theme in his 1990 book, *Morality and Human Nature* (where the focus was on feelings as a foundation for ethics), and was developed further in a more recent collaborative effort (McShea and McShea in press). I am also grateful for recent useful discussions with D. Alchin, P. Griffiths, S. Salthe, and the participants in a 1996 Behavior Discussion Group at the Santa Fe Institute. I thank D. Ritchie, K. Sterelny, and N. McShea for thoughtful reviews of the manuscript.

References

Arthur, W. (1984) *Mechanisms of Morphological Evolution*. New York: Wiley.

Arthur, W. (1988) *A Theory of the Evolution of Development*. New York: Wiley.

Buck, R. (1985) "Prime Theory: An Integrated View of Motivation and Emotion," *Psychological Review* 92:389–413.

Damasio, A. R. (1994) *Descartes's Error*. New York: G. P. Putnam's Sons.

Frijda, N. H. (1993) "The Place of Appraisal in Emotion," *Cognition and Emotion* 7:357–387.

Griffiths, P. E. (1997) *What Emotions Really Are*. Chicago: University of Chicago Press.

Hall, B. K. (1996) "Baupläne, Phylotypic Stages, and Constraint," *Evolutionary Biology* 29:215–261.

Hume, D. (1978) *A Treatise of Human Nature*. L. A. Selby-Bigge, ed. Oxford: Clarendon Press.

Izard, C. E. (1993) "Four Systems for Emotion Activation: Cognitive and Noncognitive Processes," *Psychological Review* 100: 69–90

Johnson-Laird, P. N. and Oatley, K. (1992) "Basic Emotions, Rationality, and Folk Theory," *Cognition and Emotion* 6:201–223.

Kauffman, S. A. (1993) *The Origins of Order*. New York: Oxford University Press.

Lazarus, R. S. (1991) *Emotion and Adaptation*. New York: Oxford University Press.

LeDoux, J. (1996) *The Emotional Brain*. New York: Simon & Schuster.

McShea, R. J. (1990) *Morality and Human Nature*. Philadelphia: Temple University Press.

McShea, R. J. and McShea, D. W. (in press) "Biology and Value Theory," in J. Maienschein and M. Ruse (eds.), *Biology and the Foundations of Ethics*. Cambridge: Cambridge University Press.

Oyama, S. (1985) *The Ontogeny of Information*. Cambridge: Cambridge University Press.

Plutchik, R. (1980) *Emotion*. New York: Harper & Row.

Raff, R. A. (1996) *The Shape of Life*. Chicago: University of Chicago Press.

Riedl, R. (1977) "A Systems-analytical Approach to Macroevolutionary Phenomena," *Quarterly Review of Biology* 52:351–370.

Salthe, S. N. (1985) *Evolving Hierarchical Systems*. New York: Columbia University Press.

Salthe, S. N. (1993) *Development and Evolution*. Cambridge: The MIT Press.

Sloman, A. (1987) "Motives, Mechanisms, and Emotions," *Cognition and Emotion* 1:217–233.

Solomon, R. C. (1977) *The Passions*. Garden City, New York: Anchor Press.

Wilson, E. O. (1975) *Sociobiology*. Cambridge: Harvard University Press.

Wimsatt, W. C. (1986) "Developmental Constraints, Generative Entrenchment, and the Innate-acquired Distinction," in W. Bechtel, ed., *Integrating Scientific Disciplines*, pp. 185–208. Dordrecht, Holland: Martinus Nijhoff.

Young, P. T. (1973) *Emotion in Man and Animal*, second edition. New York: R. E. Krieger.

Zajonc, R. B. and McIntosh, D. N. (1992) "Emotions Research: Some Promising Questions and Some Questionable Promises," *Psychological Science* 3:70–74.

IV PHILOSOPHY OF MIND

10 Situated Agency and the Descent of Desire

Kim Sterelny

1 Situated Agents

This paper focuses on the transition from *detection* to *representation*. Virtually all living creatures have mechanisms that are adapted to, and that direct response to, specific features of their environment. No living thing is simple, but the simplest of such creatures, bacteria, have such adaptations. So, famously, *E. coli* have a gene that switches on in the presence of lactose, and whose product enables them to use that food source. So what distinguishes the capacity to detect and respond from the capacity to represent and act, and under what circumstances will the capacity to represent and act evolve from the capacity to detect and respond?

Godfrey-Smith defends the idea that environmental complexity selects for behavioral plasticity. If environments vary in ways relevant to the organism, then responding differently in different circumstances will pay better than a single fixed behavioral pattern. This is true, of course, only if the organism has some means of tracking variation in the environment and matching its behavior appropriately to that variation. The world must be complex, but in ways the organism can map, if minds are to pay their way. Minds are not free. They require resources to build and maintain them. The adult human brain is only about two percent of total body mass, but consumes around twenty percent of total energy absorbed (Dunbar forthcoming). Further, operating one has the cost of error, too. Trying to do the right thing at the right time risks doing it at the wrong time. In general, what ain't there can't break, malfunction, or be misused (Godfrey-Smith 1996). Godfrey-Smith spends most of his time defending this idea for very simple forms of plastic response. One of his key examples is inducible defense by sea moss—animals not widely regarded as intellectual giants. But I shall suggest some ways of extending his idea from the simple detection mechanisms on which he focuses.

Mental representation has been the focal problem in philosophy of mind for the last twenty years. The orthodox view is that human behavior in particular, and intelligent behavior in general, is explained by explaining the capacity of the mind to generate, transform, and use representations of the world. This theory, the representational theory of mind, is haunted by the notorious "frame problem." If we consider the complexity of real environments and the menu of actions available to agents, each of which has different potential effects on the environment, the problem of maintaining and using an accurate and relevant representation of the world seems intractable (Dennett 1984). So perhaps this is a misconceived approach to intelligence. Developmental-systems theorists deny that the genotype is preformed information that guides development (Oyama 1985; Griffiths and Gray 1994). In a similar vein, there has been the development in the artificial life (AL) literature of antirepresentational models of intelligent action. These models deny that intelligent behavior must be guided by preexistent beliefs and goal structures within the organism. Sometimes this line of thought suggests that the information to guide behavior is constructed in behavioral interaction with the world, by organisms actively searching for specific, relevant cues. On these views, intelligence is the result of the interaction between organism and environment. Sometimes the idea is that information is stored in the environment, not just the brain, of the organism. Many spider mating rituals fit the interactionist idea rather well, for the action of each participant is contingent on signals from the other at each step. There is no obvious need to suppose that either participant has a stored template of the whole procedure if, instead, each stage completes the precondition for the next. Ant pheromone trails, on the other hand, obviously fit the picture of organisms' storing information in the world rather than in themselves (Brooks 1991; Hendriks-Jansen 1996; for a critique see Kirsh 1996).

This conception of intelligent action without an internal representation of the agent's world is often called a theory of *situated agency*, to emphasise the importance of context in the control of action. Agents that act intelligently in the world without benefit of a prior representation are *situated agents*. This theory of agency makes four central claims:

1. Behavior can be partitioned into task-oriented skills. These are behavioral modules, each with distinct sensing/control requirements.

2. The behavioral repertoire of even complex creatures can be built from these modules, adding increments to a base of simple skills.

3. "Classical AI" has underestimated the information available in the environment. In the view of those sympathetic to antirepresentational views, that extra information suffices to control these basic skills. These skills can be appropriately triggered and guided by local cues. "World models" are almost always unnecessary.

4. Organisms coordinate their behavior through a built-in motivational structure using information that the environment provides.

So as the organism moves through its environment, it interacts with it in ways that generate a variety of appropriate behaviors without it having to have and update a model of its world. At its most ambitious, the program suggests that the frame problem is a pseudo-problem. It depends on mistaken presuppositions about intelligent action. But for the idea to be generally applicable, an organism must typically find itself in an environment that provides it with reliable cues. Moreover, these cues must be *local* cues. Situated agents escape the frame problem by avoiding having any *overall model* of their world, even any overall representation of their immediate environment. For that is how they escape the problem of update. The behavior of cue-driven organisms does not derive from the execution of stored templates nor is it otherwise controlled by information held in the organism.

Here we have the classic AL theme of emergence. Complex behavior—the intelligent, adaptive behavior of organisms in their environment—emerges from locally governed interactions between relatively simple components. A complexly behaving system need have neither complex parts nor central direction. A standard methodological moral emerges as well, in the idea that the *interaction* of the components determine system-level behavior. Hence we do not get much of a handle on what the system will be like by studying the components in isolation. The emergence of complex behavior can be understood only through new models of scientific explanation (Burian and Richardson 1996; Clark 1996; Hendriks-Jansen 1996; Clark 1997; Sterelny 1997b).

This general methodological thesis links naturally with the tradition within ethology of skepticism about lab-based, experimental studies of behavior. That tradition is still alive; for example, Russon, in writing about the social expertise of primates, grumbles that "What expertise they actually do exploit this way is an empirical question. The relevant evidence comes from field observations, not laboratory experiments—the concern is how primates themselves use social learning, not how clever experimenters can induce them to use it" (Russon 1997, p. 183). If organisms are situated agents, storing some of their operational information in the environment or developing it in interaction with very specific cues in their natural environment, caution about the experimental, manipulative approach make sense. We would need to be very cautious in extrapolating from laboratory behavior to natural behavior and back again. In shifting an organism from its natural environment we may be extracting part of the cognitive system, not the whole cognitive system. Furthermore, interactions between the components, rather than the intrinsic features of the components themselves, are of most importance in generating system-level behavior. So studying a component in isolation will have limited utility for understanding the system. Equally, this picture vindicates skepticism about comparative psychology based on a few model organisms. The basic neural equipment—the basic internal mechanisms of control and learning—may well be fairly similar from organism to organism. But if the explanation of behavior depends critically on the interaction between these intrinsic mechanisms and their environments, then neural similarity is no reason to expect overall similarity.

2 Thought in a Hostile World

What are the scope and limits of situated agency? When should we expect acting on local cues to be good enough? Just about all organisms, by design, track some features of their environments. I have previously argued that a significant change occurs when creatures become able to track their environment via more than one kind of proximal stimulus (Sterelny 1995). Organisms that can track features of their world in several ways represent rather than merely detect features of their environment. This capacity to track functionally relevant features of the environment in more than one way is required for behavioral capacities to be robust.[1] Arthropods often have beautifully ingenious ways of detecting relevant features of their environment, but they are often dependent on a single proximal cue. Thus the hygienic behavior of ants and bees—their disposal of dead nestmates—depends on a single cue, the oleic acid produced by decay. They have nothing equivalent to perceptual constancy mechanisms, mechanisms that would enable them to track the liveliness of their nestmates in other ways. Since ants recognize and react to one another by specific chemical and mechanical cues, parasites bearing no physical or other kind of resemblance to the ants can invade and exploit their nests by mimicking the right specific signals. Thus Holldobler and Wilson (1990) describe a number of beetle species that live in ant nests, persuade their hosts to feed them and even to tolerate their feeding upon the ants' larvae by mimicking the ants' chemical signature and food-begging gestures (pp. 498–505). So though efficient in the right circumstances, their capacities are fragile.[2]

An organism that can track its environment only through a single specific cue is very limited in its ability to use feedback to control and modulate its behavior, for it is restricted to reliance on variation over time in that single cue. If the cue can be misleading, or if there are time lags in response (as there will be with chemical cues and other physical signals whose transmission speed is low) feedback will at best be crude. Moreover, cue-bound organisms are unlikely to have behavioral abilities that are robust over a range of different environments. Environmental shifts are likely to disrupt situated cues. If animals have several channels to features of the world that are important to them (some of which may be mediated by memory), their capacities to act will be less subject to disruption through environmental instability. For example, an organism with a mental map of its territory will be less apt to get lost than one that simply follows a list of procedures ("right at the big tree, left at the wombat hole') cued by specific stimuli.

So cue-driven organisms will often struggle if their action plans depend on rapid feedback. More generally, they will struggle if ecologically relevant features of their environment—their functional world—map in complex, one-many ways onto their transducible world. Such organisms live in *informationally translucent environments.* If food, shelter, predators, mates, friends, and foe map in complex ways onto physical signals they can detect, cue-driven organisms' behavior will often misfire. Such organisms face the problem that many different sensory registrations form a single functional category, and similar physical signals may derive from distinct functional sources.

But why and when would organisms find themselves in such environments? What evolutionary or selective histories might result in an environment's being translucent? For translucence is a feature determined in part by the evolutionary history of the lineage, not just fundamental physical processes. "Nestmate" is a transparent property of the environment for many social insects. One reason is that organisms are often distributed through many different environments. The greater the variety of environments in which an organism might find itself, the less it will be able to rely on constant physical signals to identify food, shelter, and its other needs. When a population is spread across many different niches, selection cannot predict the informational specifics of the organism's world. Koalas and echidnas may well be able to rely on cues—may well be situated agents—in recognizing food. I doubt whether impala are. The relationship between the physical signals generalists can detect and the functional features they need to know about is likely to be complex. Hence the organism will need to represent its world. It will need multiple routes to the environmental features of interest to it. Organisms can be generalists not just through inhabiting a wide range of different environments but also through opportunistically exploiting a wide variety of resources in a single environment.

So ecologically more generalized organisms will be under some pressure to escape cue-driven behavior. I think hostility also matters. Situated agency is a plausible picture of adaptive action only with respect to indifferent features of the environment. It is no accident that in this program, AL models are of robots interacting with their physical surrounds; succeeding, for instance, in navigating their way around the walls of a room without having any overall representation of the room and its layout. Cue-driven behavior offers a plausible picture of interaction with an indifferent physical environment. Thus we can conceive of, say, a beaver's ability to repair a dam as a sequence of skills driven by local cues. The sound of running water could initiate a random search along the inside of the dam walls; the feel of the current near a break could then induce a local search, and then simple behavioral rules could guide action in dam repair. For this task, the beaver need not have any overall model of the dam and its state. Equally, cues can be sufficient to drive behavior in cooperative interaction, where each organism is trying to make its intentions as explicit as possible. Ants communicate with one another by producing local cues and mammals often have a single distinctive way of communicating the desire to play. But much animal behavior takes place in a hostile world of predation and competition. Predation is not just a danger to life and limb; predation results in epistemic pollution. Prey, too, pollute the epistemic environment of their predators. Hiding, camouflage, and mimicry all complicate an animal's epistemic problems.

For cue-driven behavior to be adaptive, the cue itself must be detectable and discriminable, though detecting a cue may require the organism to probe its environment. Further, there must be a stable cue-world relationship; that is, the organism's local environments must be homogeneous with respect to the cue-world relationship. What the cue tells the organism about its world needs to be fairly independent of what else is in the local scene. That is why the organism does not need to represent the local scene as a whole. Hostility imposes a cost of probing. It imposes a cost on animal action taken to disambiguate a cue, or to locate one ("If called by a panther, don't anther"). Second, it makes local environments heterogenous with respect to easily discriminated cues and the functional properties they signal. Deceptive fireflies mimicking the female's signal to the male decrease the overall reliability of the signal-mate relationship, as the firefly environment becomes heterogenous with respect to the "species-specific" signal. Further, they impose a cost on probing.

Let's pull all this together. Godfrey-Smith discusses the evolution of simple detection mechanisms, arguing that they evolve in environments that are heterogenous. Environments vary in ways relevant to the organism and to which the organism can respond. Godfrey-Smith models the evolution of signal pick-up, suggesting that it's a function of the reliability of the signal, the benefit of acting on it, and the cost of error. So he considers varied environments, but ones in which the variation is signaled to the organism. These environments are *transparently heterogenous*.[3]

I have used the idea of situated agency to stalk the shift from these simple systems to more complex ones in which organisms can "see through" variant or misleading proximal stimuli to the relevant features of their environment, through having the capacity to track particular features of their environment in more than one way. In *translucent worlds* there is a complex relationship between incoming stimuli that the organism can detect and the features it needs to know about. In these worlds there is selection for representation, not just for detection. In turn, I have suggested that generalist organisms will typically live in partially translucent worlds, and that hostility, too, has the effect of making critical aspects of an organism's environment translucent.

No one would suggest that translucence is sufficient for representation. As the parasites and predators that crack the ant's recognition and response system show, plenty of arthropods and other organisms are cue-bound with respect to their ene-mies. Nonetheless, if representation is within the range of evolutionary possibilities for the lineage, translucence will select for it. We need representation rather than detection in certain kinds of complex environment, for example, in those where the mapping between biological significance and proximate cue is complex. This complex of ideas is not easy to test, but I think it's a very plausible model of the evolution of belief-like states. In is easy to see how the capacity to represent key aspects of your environments might be a "fuel for success." One point of Godfrey-Smith's project is to synthesize the pragmatist idea that the point of cognition is the appropriate con-trol of action with the reliabilist idea that the point of cognition is to accurately track the world. Accurate tracking is a means to appropriate behavior, and helps explain the success of that behavior. In turn, that pattern explains the evolution of the cog-nitive mechanisms that construct representations that track the environment.

3 Nike-Organisms

In thinking about the evolution of cognitive capacities, Godfrey-Smith is typical in focusing on our capacity to represent our world; that is, on the evolution of belief. But what of desire? What explains the evolution of preferences? We digest, breathe, and vary our heartbeat rate without any cognitive representation of the metabolic needs these activities service. Over longer time periods, organisms partition their metabolic resources between growth, somatic maintenance, and reproduction, quite often in somewhat flexible ways, again without any internal representation of the needs these trade-offs serve. The gonad weight of birds changes dramatically with the season, as does, in migrants, their body weight and composition as they build up their resources for migration. So it is not always necessary to represent your needs in order to act on your needs. Hence many organisms do not have preferences.

In McFarland's useful terminology, the "cost function" of an organism is its real trade-offs between dangers and resources (McFarland 1996). He points out that many organisms do not represent their own cost function. Instead of representing their needs they simply have a built-in motivational hierarchy. These are *Nike-organisms*; they "just do it." For them, motivation derives directly from the value of internal variables that are keyed to internal metabolic conditions of the organism. Their behavior is a response to a mix of external and internal signals. They act on hunger, thirst and environmental states. They forage when their internal food reserves drop and the value of a specific internal variable is cranked up. They drink upon internal signals of dehydration. Kirsh calls this the "capacitor" model of moti-vation (Kirsh 1996), and it is perhaps most vividly illustrated by the "hydraulic"

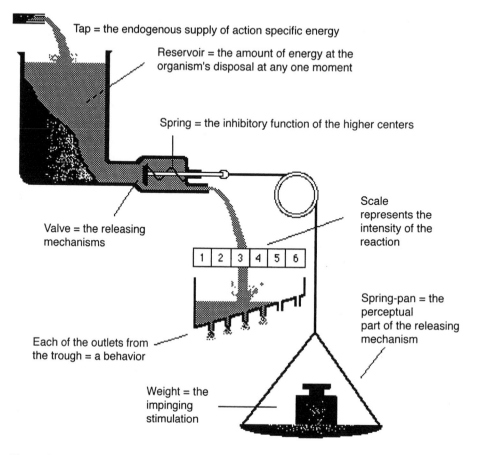

Tap = the endogenous supply of action specific energy

Reservoir = the amount of energy at the organism's disposal at any one moment

Spring = the inhibitory function of the higher centers

Valve = the releasing mechanisms

Scale represents the intensity of the reaction

Spring-pan = the perceptual part of the releasing mechanism

Each of the outlets from the trough = a behavior

Weight = the impinging stimulation

Figure 10.1

models of motivation in early ethology, illustrated above (Lorenz 1950). Motivation impulses of various kinds ("action-specific energies") are released by specific stimuli. The greater the energy build up, the lower the threshold on the releasing stimuli.

The existence of Nike organisms shows that desires are not mandatory. What then selects for the evolution of desire? Under what circumstances is it important for the organism to have a preference structure? For preferences have a cost. Any representation entails the possibility of misrepresentation, so representing the cost function brings with it the cost of misrepresenting it. Humans, obviously, have embraced the opportunity to misrepresent their cost function many times. We cannot assume that there has been selection for the capacity to represent cost functions. Desire might be a side effect. Perhaps representing motivational states of yourself is a side effect of representing the motivational states of others: prey, predators, conspecifics. Perhaps it's a side effect of a fancy capacity to represent the world. But if a preference structure is an adaptation, it seems most likely to be an adaptation to the complexity of *choices* facing the organism. That is, the descent of desire is a response to complexity, but complexity in the output side: complexity of responses to the environment. So once we shift focus from the evolution of belief-like states to the evolution of desire-like states, the environmental complexity hypothesis faces a different test. How is the complexity of choice and action related to the complexity of the environment?

Let's first consider the nature of the difference between Nike-organisms and organisms with preferences. "Hydraulic" models of motivation are just models; their implementation details are not meant to be taken seriously. But even as models, they are no longer widely accepted. But homeostatic models, which fit equally easily into Kirsh's general conception, remain quite popular. On these models a behavior—for example, drinking—is generated as an organism responds to a variance in its internal economy between an ideal value of some state and its actual value. At its simplest, an animal detects, say, the partial dehydration of its cells, and drinks until the signal of dehydration ceases. Animals with motivational systems of this general kind differ from animals that act on their preferences in at least two and possibly three ways. Animals whose motivation systems are well captured by homeostatic models are stimulus-bound. Their behavior is a response to stimuli. It is just that some of these stimuli are signals of important internal states of the organism while others are signals of important external states. So (first) one transition relevant to the evolution of preferences is the change from *internal detection* to *internal representation*. However, an internally represented motivating state is still not a preference, for preferences are typically representations of how an animal's environment might be, not of how it is or is taken to be. So (second) the evolution of preference from capacitor-systems of motivation involves not just the emergence of representation, but also shifting its content outside the organism.

How and why might this happen? The internal environment of an organism is not hostile, so hostility does not drive organisms to represent rather than merely detect their metabolic states. But in discussing the shift from detection to representation, I noted that one problem with single channel tracking is feedback. If a signal changes slowly in response to changes at its source, it will not provide good feedback. Manning and Dawkins point out that cell dehydration has this feature. Though cell dehydration is important in initiating a rat's drinking, it is not a good mechanism for telling the rat when to stop drinking. For rehydration takes 10–15 minutes (Manning and Dawkins 1992, pp. 87–88). When an animal needs a halting criterion as

well as an initiating criterion, *the mere cessation* of the initiating signal often will not
do. We can see here a circumstance that might select for a change from internal detection to internal representation. Start signals often will not be good as stop signals
and vice versa. The rat can control its drinking adequately by tracking an internal
condition, though using two different cues. But for some behaviors, the halting criterion might be external not internal. To know when to stop, the animal needs to
track *a change* in the environment. Antipredator behavior, I suggest, will often have
its "off" switch tuned to an external environmental change.

So organisms with preferences contrast with organisms with capacitor-style motivations by representing rather than merely detecting internal states that produce
their behavior, and perhaps at the most primitive level of preferences by representing
environmental changes as halting criteria in the control of behavior.[4] Finally, (third)
we often think of preferences as forming a *system*, a preference order that enables an
agent to choose between competing means and competing ends. On this view,
organisms have preferences only if their motivations satisfy the conditions of commensurability, transitivity, and so on that enable utility functions to be calculated.
Moreover, motivations can themselves be the object of motivation: we have preferences about our preferences. Since it is not obvious that humans fully satisfy these
systemic aspects of preference structure, I will assume here that they are evolutionarily recent and inessential to the evolution of preferences from simpler motivation
systems.

I have assumed that the complexity and control of behavior must be central to the
evolution of preferences. But behavior can be complex in at least two ways. Behavior
can stand in a more or less complex relation to its initiating stimulus. A cock that
gives false food-calls only when a female is present and unable to see that the calls
are misleading behaves more complexly than a cock that cannot tune its behavior to
such a subtle feature of its environment. More obviously, the behaviors themselves
can vary in complexity. Thus Povinell argues that arboreal locomotion increases
significantly in complexity as organisms become heavier. In part, this increase in
complexity is in the stimulus-behavior relation. As animals become heavier, branches
deform under the animals weight, and so line-of-sight routes cease to be possible.
The relationship between what the animal sees and the route through which it moves
becomes more complex. But actual movement itself becomes more complex: it is less
stereotyped and involves multiple support points (Povinelli and Cant 1995). To a
first approximation, complexity of the first kind—of stimulus-behavior relations—is
more tied to the evolution of belief-like representation, complexity of the second type
to preference-like representation. For the second kind of complexity poses problems
of control.

We can draw the same distinction between the different ways behavior can be complex by imagining two different ways an animal's behavior can become stimulus-independent. One is via belief: organisms become stimulus-independent as their behavior becomes increasingly affected not just by their motivating states and their registration of their current environment but by their belief-like states. Their stored information becomes increasingly relevant to their behavior. But there is another potential route, running through the evolution of a preference structure. In the same circumstances different preferences will generate different actions.

I have suggested that motivations shift from internal detection to internal representation through the importance of feedback in controlling behavior; in the simplest cases, when switching the behavior off depends on a different cue than switching it on. Here the problem is to control a single element (drinking) in the rat's behavioral repertoire. But there is a second way in which control becomes complex, and one which may be even more important. Animals behave complexly in this second sense to the extent that we are unable to model their behavior as consisting of a repertoire of discrete capacities, each with its own initiating feedback and halting requirements. Motivating conditions can be relevant to *more than one element* in an animal's behavioral repertoire. As motivational states interact with one another and with other internal states, the integration of behavior becomes more complex. As a consequence, the system as a whole will register the existence of the motivational state in its behavioral dispositions in more than one way. For internal states will affect behavior through more than one causal route. In virtue of complex control the same motivational state can (help) produce different behaviors, for its impact will be modified by other internal states. The relations between environmental input, motivational states, and behavior will become less direct. Contrast this situation of complex control with the simple homeostatic models. These are classic instances of capacitor models of motivation, and they are models in which the relation between motivational state and behavior is simple and single tracked. We shift beyond mere motivations as a given motivational state—say, a given value on the lust capacitor—can interact in different ways with other states to produce different behaviors. The state contributes to the shape of different behaviors rather than succeeding or failing to get its hands on the rudder. "Fight or flight" decisions may well be motivation without preference, for one motivation can simply be trumped and is silent in behavior. But if the "flight" motivation affects the tactics of the fight without causing flight, we are shifting toward more complex control. Ethologists since Tinbergen and Lorenz have noted many behaviors apparently under complex control of this kind. The "tactical deception" literature provides many examples. One clear example is the suppression of copulation cries by chimps who rightly suspect intervention if their activity is discovered. The "mate" routine is not running independently of the animal's other motivation and action systems. Control has become more complex.

Hence I think more complex behavior evolves at least in three ways. (1) Increasing information about the world, including information not given in immediate perception, liberates behavior from stimulus control, making behavior less predictable given only information about a creature's immediate environment. (2) The need for independent initiating and halting criteria, and more generally the demands of feedback, require animals to multitrack both their own internal states and changes in those states and their environment. (3) Behavior becomes more complex as metabolic motivators become relevant to more than one element of their behavioral repertoires. Under such conditions, control becomes less discrete, less modular. My hypothesis is that (2) and (3) are central to the evolution of desire-like states from brute causal motivational states. In the rest of this chapter, I propose to consider (2) in a little more detail.

I suggest that preference structures evolve as an animal's behavior becomes more intricate and less stereotyped. A squirrel monkey scurrying along a branch may not need preferences; arguably, an orangutan does. Some animals need to represent instrumental goals, or subgoals, because they are faced with a variety of means to one of their ultimate ends. You need to represent your goals when you need to be able to represent subgoals and choose between them. Organisms that *plan* need to know what they want. The most developed case for planning in nonhuman animals has been constructed by Richard Byrne through his idea of a "behavioral program" (Byrne 1997, p. 68). He introduces this idea in discussing the vexed issue of primate imitative capacities. Though there is plenty of anecdotal evidence for great ape imitation, experimental evidence for imitation is surprisingly thin (Byrne 1997; Russon 1997). He points out that behavioral routines may have an overall functional organization, rather than consisting of a mere chain of independent behavioral atoms. Gorilla food-preparation illustrates this idea of a program. Since they often eat thistles and other rather awkward plants, gorillas often need to do a good deal of manual processing of their food before they can eat it. This processing is quite complex, and involves a division of labor between the hands that changes through different stages of the processing.

Byrne's idea is that if skills depend on behavioral programs, imitation can involve copying that program rather than a specific motor pattern. If young gorillas acquire this ability through imitation, they may be copying the program rather than the motor sequences. Indeed, if an animal has the representational capacity to represent another's behavior as a program, we might expect it to use this representation to direct its own attempts at the same behavior. For direct parent-child behavioral copying will often not be a suitable vehicle for social learning. Adults differ from their young in size and strength. But they also differ in coordination, for many adult motor subroutines are already assembled and automatized. So it would often be very difficult for a juvenile to use adult motor routines as templates for her own behavior even if she could manage the required transformation in point of view.[5] True imitation may be rare because adult motor templates are not useful, and few animals have the representational power to develop an abstract functional representation of a skill.

Byrne argues that the primate lineage has seen three major episodes of cognitive evolution. One distinguishes the haplorhine clade (apes and monkeys) from the strepsirhine clade (lemurs and lorises); a second distinguishes the great ape clade from the monkeys; and a third singles out the hominid lineage. He thinks that the first of these episodes is driven by social complexity, but he argues that there is little reason to think that the evolution of the distinctive capacities of great apes are so explained. Great ape social groups are themselves of very varied size and apparent complexity. Their social groups show no obvious overall increase in complexity over those typical of haplorhines.

So Byrne argues that the distinctive feature of great ape cognition is the capacity to plan. In his view, this is manifest in their complex processing of food (as in the gorilla example above, but it's typical of orangutans as well), in tool construction and use (chimps), in bed construction (most great apes), and in unstereotyped and complex motion through an arboreal environment (orangutans). The same seems to be true of the use by one animal of another as a social tool, as they form and exploit alliances. When the top-ranking chimp tolerates the number three chimp having sex, because he needs support in a coalition against the number two, the first is preparing to use the third as a social tool, and that requires goal representation as well as world representation. The alpha chimp must represent what he himself wants to do. These activities all involve coordinated actions unfolding over time. They require an ape, "in other words, to *plan*" (Byrne 1997 p. 21).[6] Byrne goes on to suggest that the representational capacities required here are then redirected onto physical objects and other individuals. But we do not have to accept this extension of Byrne's idea to accept that flexible plan-building requires the evolution of preference structures.

How does this view of the evolution of preference structure bear on the environmental complexity hypothesis and on the extent to which the evolution of beliefs and preferences are linked? On this picture, we certainly would not expect the complete decoupling of the evolution of belief-like structure from desire-like structure. If Byrne's picture of great ape cognitive evolution is right, environmental complexity is part of the story. Gorillas and orangutans have evolved behavior programs because of hostility: their food is defended by "spines, stings, hard casing, tiny clinging hooks" (Byrne 1997, p. 19). Tool manufacture and use, and perhaps use of social tools too, strongly suggests that they categorize some aspects of their environment in functional terms rather than just in terms of sensory similarity. There is, for instance, quite wide physical variation in termite fishing twigs, but the complexity of the environment is not the whole story. The pressure for planning derives as well from apes' large body size and unspecialized digestive systems. These factors intensify the nutritional challenge facing the apes. These are internal features of the lineage. So too are features of great ape social life. The life-history patterns of great apes may also be relevant: they have long periods as juveniles in which they have the opportunity to learn from adults.

So it seems unlikely that anything like preferences could evolve in the absence of belief-like structures, for environmental complexity and its representation do seem part of the story. Moreover, the stimulus-independent behavior that the evolution of preferences would produce would in itself make the environment translucent for any animal that needs to predict what its fellows are likely to do. It is much less obvious that the converse is true: animals might have belief-like representational capacities without having anything much like preferences. If, for example, the social intelligence hypothesis is right about the haplorhine cognitive burst, they will need to represent their social world. For such monkeys form, maintain (often through grooming), and track long-term social relations, since gains and losses in social encounters often depend on these relations. As Byrne notes, in this clade triadic interaction is often critical. It does not follow that such monkeys need complex control mechanisms. Once you have successfully tracked your environment, perhaps there is no problem deciding what to do and how to do it. Hauser reports that rhesus monkeys that discover food-catches scan their environment before either calling to advertise the food or eating silently. Those caught eating silently are likely to face attack, so Hauser suspects that prior to food consumption, discoverers are searching for both friends and foes[7] (Hauser 1997). Here, plausibly, we have a belief-mediated route to a looser link between stimulus (sighting food) and behavior (food calling). If the rhesus monkey can determine what situation he is in, the decision problem is solved.

I have been discussing a specific hypothesis about the evolutionary origins of planning, and hence of preferences, in the great ape lineage. This might suggest that the evolution of preference-like representations is very rare, restricted to a single tiny twig in the mammalian clade. There are possibilities that suggest a broader distribution of the capacity. Change over time might select for the capacity to represent your needs by selecting for the capacity to act now to satisfy urges that you do not now have. Many animals store food rather than lose interest in food when satiated:

leopards and some canids conceal unconsumed parts of kills. Squirrels and a good number of bird species cache food too. So, does future-oriented behavior show representation of goals? Not by itself. But perhaps it does when it's flexible. Do acorn woodpeckers store acorns when they live in less seasonal environments with a steady supply of food all year round? Do leopards and dogs continue to cache food when food is freely and predictably available? If not, then it's likely that they represent their own needs. If the capacity to act for the future is flexible and is sensitive to what the organism is really likely to need in the future, the case for the representation of goals looks strong. Environmental variability in the form of boom and bust food cycles generates the need to represent intermediate goals.[8]

This essay is a first pass at the problem of the evolution of preferences. At best, it makes the problem explicit and sketches a few hypotheses about that evolutionary process. But I have tentatively endorsed a version of the environmental complexity hypothesis about the evolution of belief-like states, and I have argued that a certain species of complexity—informational translucence—is important in the shift from detection to representation. So how does preference—if my speculations are on the right track—fit into the picture? I do not think the answer is obvious. For we need to understand the relationship between behavioral planning and environmental complexity. Is planning necessary, or desirable, only in *certain kinds* of complex environments? At the very least, Byrne's examples make clear that environmental complexity is one factor in demanding complex control. It may also be important in explaining the process through which behavioral repertoires become less modular, though that process remains murky. It is much less clear that environmental complexity is the whole story. The complexity of the environment seems to play no obvious role in the problem of knowing when to stop; when to stop drinking, when to stop burrowing and make the nest chamber, when to stop laying and start brooding, and so on. Some control problems, in other words, seem to be generated by features internal to the lineage. So while the environmental complexity hypothesis offers a plausible model of the evolution not just of simple behavioral plasticity but also of more complex genuinely representational states that are ancestral to belief, it is much less clear that it is the key to the evolution of preference structures.

Acknowledgments

Thanks to Russell Brown, Fiona Cowie, and Peter Godfrey-Smith for their comments on an earlier draft of this essay. I thank James Maclaurin for his reconstruction of Lorenz's hydraulic model.

Notes

1. How do I count channels between organism and an environmental feature? Though I have no formal definition to hand, I should emphasize that channels are not sensory modalities. A baboon that monitors a rival by visually reacting to facial expression, body posture and relative location is tracking intent through three cues, not one.

2. Godfrey-Smith (personal communication) has urged upon me Dretske's idea that the multiple channel condition should be used to solve the distality problem rather than to draw the distinction between perception and detection (Dretkse 1988). Thus perceptual constancy mechanisms show that the representational content of perceptual states are features of the environment, not features of the proximal stimulus. My response to Godfrey-Smith here is programmatic. In a series of papers (including this one), I have tried to show the theoretical productivity of drawing the distinction this way; see also my forthcoming.

3. More exactly, Godfrey-Smith does not explicitly distinguish between transparent and translucent environments, but his examples are of transparent environments, for he sets himself to explain the most basic behavioral flexibility.

4. I would assume that tracking changes in the external world will be important in other aspects of feedback, too.

5. It is this transformation in point of view that has lead some to argue that true imitation requires metarepresentational capacities, and that is why it's so rarely found except in humans. I argue against this view in Sterelny forthcoming, while agreeing that imitation is cognitively sophisticated because it does require a functional rather than a sensory representation of the skill imitated.

6. Prima facie tactical deception also involves planning. Yet Byrne does not regard it as confined to the great apes. So tactical deception outside the great apes looks somewhat anomalous on Byrne's view of the evolution of planning.

7. It is not clear how Hauser excludes the possibility of predator surveillance. He does note that solitary individuals who find food neither call nor are attacked if discovered with food, though they are displaced. But he does not note whether they scan before eating.

8. At least for K-selected organisms. Obviously, there are many other evolutionary responses to environmental uncertainty of this kind. Many Australian animals are nomadic, or breed only in response to clear environmental signals of a good season, and this is presumably a response to the uncertainties of the Australian environment. Obviously features of the lineage itself are of great significance in determining the character of the lineage's evolutionary response to uncertainty. But cognitive sophistication seems to be one possibility.

References

Brooks, R. A. (1991) "Intelligence without Representation," *Artificial Intelligence* 47:139–159.

Burian, R. M. and Richardson, R. C. (1996) "Form and Order in Evolutionary Biology," in M. Boden (ed.), *The Philosophy of Artificial Life*, pp. 146–172. Oxford: Oxford University Press.

Byrne, R. W. (1997) "The Technical Intelligence Hypothesis: An Additional Evolutionary Stimulus to Intelligence?" in R. Byrne and A. Whiten (eds.), *Machiavellian Intelligence II: Extensions and Evaluations*, pp. 289–311. Cambridge: Cambridge University Press.

Clark, A. (1996) "Happy Couplings: Emergence and Explanatory Interlock," in M. Boden (ed.), *The Philosophy of Artificial Life*, pp. 262–281. Oxford: Oxford University Press.

Clark, A. (1997) *Being There: Putting Brain, Body, and World Together Again*. Cambridge, MA: The MIT Press.

Dennett, D. C. (1984) "Cognitive Wheels: The Frame Problem of AI," in C. Hookway (ed.), *Minds, Machines, and Evolution*, pp. 129–152. Cambridge: Cambridge University Press.

Dretkse, F. (1988) *Explaining Behavior: Reasons in a World of Causes.* Cambridge, MA: The MIT Press.

Dunbar, R. I. (forthcoming) "The Social Brain Hypothesis," *Evolutionary Anthropology.*

Godfrey-Smith, P. (1996) *Complexity and the Function of Mind in Nature.* Cambridge: Cambridge University Press.

Griffiths, P. E. and Gray, R. (1994) "Developmental Systems and Evolutionary Explanation," *Journal of Philosophy* XCI6:277–304.

Hauser, M. (1997) "Minding the Behavior of Deception," in R. Byrne and A. Whiten (eds.), *Machiavellian Intelligence II: Extensions and Evaluations*, pp. 112–143. Cambridge: Cambridge University Press.

Hendriks-Jansen, H. (1996) "In Praise of Interactive Emergence, or Why Explanation Doesn't Have to Wait for Implementations," in M. Boden (ed.), *The Philosophy of Artificial Life*, pp. 282–302. Oxford: Oxford University Press.

Holldobler, B. and Wilson, E. O. (1990) *The Ants.* Cambridge, MA: Harvard University Press.

Kirsh, D. 1996. "Today the Earwig, Tomorrow Man?" in M. Boden (ed.), *The Philosophy of Artificial Life*, pp. 237–261. Oxford: Oxford University Press.

Lorenz, K. (1950) "The Comparative Method in Studying Innate Behaviour Patterns,"*Symposium of the Society for Experimental Biology* 4:221–268.

Manning, A. and Dawkins, M. S, (1992) *An Introduction to Animal Behaviour.* Cambridge: Cambridge University Press.

McFarland, D. J. (1996) "Animals as Cost-Based Robots," in M. Boden (ed.), *The Philosophy of Artificial Life*, pp. 179–208. Oxford: Oxford University Press.

Oyama, S. (1985) *The Ontogeny of Information.* Cambridge: Cambridge University Press.

Povinelli, D. and Cant, J. G. H. (1995) "Arboreal Clambering and the Evolution of Self-Conception," *Quarterly Review of Biology* 704:393–421.

Russon, A. E. (1997) "Exploiting the Expertise of Others," in R. Byrne and A. Whiten (eds.), *Machiavellian Intelligence II: Extensions and Evaluations.* Oxford: Oxford University Press.

Sterelny, K. (1995) "Basic Minds," *Philosophical Perspectives* 9:251–270.

Sterelny, K. (1997a) "Universal Biology," *British Journal for the Philosophy of Science* 484:587–601.

Sterelny, K. (1997b) "Where Does Thinking Come From? A Commentary on Perter Godfrey Smith's *Complexity and the Function of Mind In Nature*," *Biology and Philosophy* 12:551–566.

Sterelny, K. (1998) "International Agency and the Metarepresentational Hypothesis," *Mind and Language* 13:11–28.

Sterelny, K. (forthcoming) "Primate Worlds," in K. Sterelny (ed.), *Evolution, Replication and Cognition.* Cambridge: Cambridge University Press.

11 Natural Answers to Natural Questions

Thomas Polger and Owen Flanagan

In this chapter we consolidate and elaborate on our earlier work to show how the natural method, the method of seeking reflective equilibrium from psychology, neuroscience, and phenomenology, can lead to progress on the central questions about consciousness (Polger and Flanagan in press; Flanagan and Polger 1995; Flanagan 1992, 1995, 1996, forthcoming).

1   What Naturalists Believe

An important strand of naturalism in philosophy is metaphysical naturalism about mental phenomena in general, and about consciousness in particular. Metaphysical naturalism is the view that the mind-body relation is a natural one—not a nonnatural one, as between a human body and an incorporeal soul or between a human body and a supernatural being.

Although adherents to views that are metaphysically naturalistic make up a majority of philosophers and scientists concerned with questions about consciousness, the precise way to formulate the doctrine is controversial. Some quotes from true believers are revealing:

Sensations are nothing over and above brain processes.   (Smart 1962, p. 103)

It seems increasingly likely that the body and the brain of man are constituted and work according to exactly the same principles as those physical principles that govern other, nonorganic, matter.... [T]here is rather strong evidence that it is the state of our brain that completely determines the state of our consciousness and our mental state generally.   (Armstrong 1980, p. 19)

I think we know enough about the universe to know that consciousness did not arise by miracle, by a sudden infusion from a supernatural realm. It arose by natural processes from natural materials—ultimately from the expanding matter that formed into clumps early on in the history of the cosmos.   (McGinn 1991, pp. 87–88)

[V]arious phenomena that compose what we call consciousness ... are all physical effects of the brain's activities.   (Dennett 1991, p. 16)

Mental phenomena are caused by neurophysiological processes in the brain and are themselves features of the brain.... Mental events and processes are as much part of our biological natural history as digestion, mitosis, meiosis, or enzyme secretion.   (Searle 1992, p. 1)

Philosophers who would assent to statements like those above and thereby be counted as naturalists in our sense include many of the best-known philosophers of mind over the past thirty years. But by no means all. Thomas Nagel (1974) and Frank Jackson (1982) are perhaps the most familiar critics of naturalism about mind and consciousness.

Even among metaphysical naturalists there is vigorous disagreement about how metaphysical naturalism can be true, and about how precisely to fill out the doctrine. J. J. C. Smart famously held that the relation between mind and brain is one of physical type-identity, that the sensation of pain is identical with C-fibers firing (1962). Daniel Dennett (1991) and William Lycan (1987, 1996) hold that the relation is one of functional identity—although each token mental event is a brain event, brain event types are to be individuated functionally rather than physically. Ned Block (1980), Paul Churchland (1988), and Sydney Shoemaker (1982) have discussed a two-part theory, functionalism for intentional states and type-identity for qualitative states.[1] Other philosophers, notably Colin McGinn, claim to be devout naturalists but also claim that the exact nature of the mind-brain relation is forever beyond our understanding: Naturalism is true, but how it is true is an eternal mystery (1989, 1991).

Given this lack of consensus about how to formulate a naturalistic metaphysics of mind, one might wonder what sense there is to discussing a view so broad that it fails to distinguish the major positions in the debate over consciousness. Compare this to the debate over just what sort of creature platypuses are. No one in the debate seriously held a view that might be called "animalism," the view that platypuses are animals of some sort. Such a view would miss the point: it would obscure the serious question of which kind of animal (bird, mammal, other) platypuses are. One might think that metaphysical naturalism about consciousness suffers an analogous problem. Why treat metaphysical naturalism as a coherent view at all?

One reason is that versions of naturalism do share a common core: mind is part of nature. This belief is easily distinguished from the dualistic and spiritual views that have dominated the history of thinking about minds. It is too easy for those of us working in the mind sciences to get mired in the differences between naturalistic views and lose our appreciation for the progress that has been made. Second, the naturalistic view is a powerful one in any of its versions, and it supports attention to the mind sciences. Third, despite being the dominant view in philosophy of mind, naturalism is by no means without its critics. These critics treat naturalism as a unified doctrine, but one which is false, vague, or empty. Thomas Nagel (1974) and Alvin Plantinga (1993, unpublished) spring to mind. Other critics of naturalism who may nevertheless themselves be metaphysical naturalists include Barry Stroud (1996) and Michael Friedman (1996).

For all that it lacks—being but a schema for a theory—the naturalistic standpoint can address some traditional objections to materialist views of consciousness. Naturalism does not make the project of understanding consciousness easy, but it does make it tractable, pace McGinn. Once one adopts metaphysical naturalism about consciousness certain methodological avenues are opened. The one that we will discuss is the natural method (Flanagan 1992).

2 The Natural Method

From the single insight of metaphysical naturalism, a methodological strategy for approaching questions about consciousness suggests itself, to wit, the natural method. The natural method begins by considering all available lines of analysis. Consider phenomenological reports, consider theories of cognitive science, consider neurophysiological data: "The object of the natural method is to see whether and to what extent the three stories can be rendered coherent, meshed, and brought into reflective equilibrium" (Flanagan 1992, p. 11).

The idea behind the natural method can be neatly framed in terms of phenomenology, psychology, and neuroscience (see figure 11.1); but the spirit of the proposal is to allow all possible sources of information about consciousness to be introduced without privileging any ahead of time. The natural method aims to reflect three facts: Mind has phenomenal structure, mind has neural structure, and mind processes information. While phenomenology, psychology and neuroscience are the keystones to the natural method, contributions from evolutionary biology, anthropology, and linguistics may also be important.

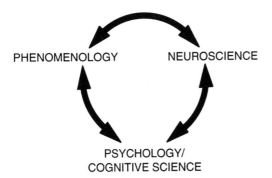

Figure 11.1
The natural method (after Flanagan 1995).

The idea behind the natural method is not to allow a simple democracy of data, but rather to discourage privileging one kind of evidence over others at the outset. Inevitably some kinds of information will be weighed more heavily than others. Some sources of information are simply more fruitful than others. But giving one discipline more weight than another can be done only after attempting to determine which ones in fact help accomplish our explanatory goals.

2.1   The Natural Method at Work

In this chapter we discuss two examples of the natural method at work: Alan Hobson on dreaming (1988, 1989; Hobson and Stickgold 1994) and C. L. Hardin on color vision (1988).

Alan Hobson's dream research provides a clear example of what we have in mind. Consider figure 11.2. Hobson and his colleagues illustrate how different sources of information regarding a phenomenon, dreaming, can be combined to yield a fuller understanding of the phenomenon. In this example, behavioral data (e.g., tossing

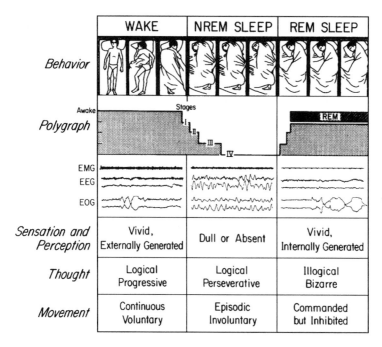

Figure 11.2
Sleep stages. (Reprinted from Hobson and Stickgold 1994, with permission of the authors and *Consciousness and Cognition.*)

and turning, immobility) are correlated with phenomenal reports (e.g., imageless, vivid), psychological data (e.g., memory), and physiological data (e.g., polygraph, EMG, EEG, EOG.)

Consider the project of extending the interdisciplinary approach by including a philosophical perspective along with considerations from evolutionary biology to yield a coherent hypothesis about dreaming that respects each source of information. Dreams are known in the first instance only phenomenologically. Viewed neuro-biologically, dreams differ in kind depending on whether they occur during NREM sleep or during REM sleep. Though we know that NREM dreams are boring and perseverative whereas REM dreams are wild and crazy, this we know by our developed theory. Without such a theory, which sleep stage the brain is in is not revealed on the phenomenological surface. Even with a theory, brain states are not directly revealed; they can be inferred, however, from phenomenological evidence.

It is not merely that the three key sources of information can be brought into reflective equilibrium in the case of dreams. If one considers only brain activity and neurochemistry, there simply is no special question about dreams over and above questions about sleep. Dream questions arise only when the phenomenology is taken into consideration. Without dream experiences as reported by dreamers there is no "problem" of dreaming, for there is no interesting phenomenon of dreaming. If we were to operate from purely information-processing and neurobiological standpoints, we might entirely omit any explanation of dreams. The phenomenon to be explained is not characterizable without reference to the phenomenology.

This simple observation—that which phenomena one considers depends upon the standpoint from which one is working—lies behind some important critiques of naturalistic views of mind (Nagel 1974; Chalmers 1995, 1996). It also inspires the natural method. The natural method agrees with Nagel, Chalmers, and others that consciousness cannot be understood if the phenomenological viewpoint is ignored or trivialized—indeed many (perhaps all) of its problems, such as that of dreaming, cannot even be framed. But whereas Nagel and Chalmers try to ensure that consciousness is not left out of our explanations by requiring us to formulate theories that explain from the subjective point of view (thereby ruling out any theory that adopts an alternative point of view), the natural method includes both the first-person point of view and other available points of view.

2.2   Historical and Ahistorical Questions about Consciousness

Questions about consciousness are numerous. They range from the broad ("What is consciousness?") to the particular ("Why should my visual sensation have the specific qualitative character that it does?''). Some questions about consciousness that have interested philosophers and scientists alike are these:

(a) How is it that consciousness arises from brain processes?

(b) What sort of thing is consciousness?

(c) What accounts for the structure of conscious experience?

(d) Is consciousness one phenomenon, or many?

(e) How did it come to be that we (and presumably some other animals) are conscious?

(f) What if anything does consciousness allow us to do that we could not do otherwise?

These questions can be divided into two groups. Questions like (a)–(c) can be understood as ahistorical questions. That is, they are questions about consciousness as it exists and is experienced in any of us at any moment. In contrast, (e) and (f) are most naturally understood as historical questions, questions about how it came to be that we are conscious, and about what evolutionary advantage (if any) consciousness might confer.

The distinction between historical and ahistorical questions is neither exclusive nor exhaustive. All of the questions above can be given historical or ahistorical interpretations, which may or may not be related to one another. There may be questions that do not belong in only one group or the other. "Is consciousness one phenomenon, or many?" is one such question; it can be considered in a phenomenological (ahistorical) or phylogenetic (historical) way, but it is best captured when both are taken into consideration. Still others are simply very different questions in their historical and ahistorical readings; the debate regarding what, if anything, consciousness allows us to do that we could not do otherwise is an example of how much confusion in the literature results from conflating the historical and ahistorical readings (see Polger and Flanagan in press). It may be that no interesting questions about consciousness can only be understood in one way, with no alternative readings.

The distinction between historical and ahistorical questions about consciousness is offered as a useful one, but not as the only way of dividing the issues. It is one way of sorting questions, answers, and objections. Let us examine one historical and one ahistorical question.

3 A Historical Question

The central historical question about consciousness is: How and why did consciousness come to be in human beings? This question has two forms. One reading is as a question about the development of consciousness in human beings from embryonic stages, through infancy, and into adulthood. The alternative reading is a phylogenetic question about the evolutionary history of consciousness in organisms of our type.

The two forms of this question illustrate the point, mentioned earlier, that the division between historical and ahistorical questions is not fixed. Both readings are diachronic. Relative to the evolutionary question, the developmental reading is ahistorical. Compared to questions about consciousness at a given moment, the developmental one can be treated as historical. Where one draws the line, and which questions one asks, will depend on what one is trying to explain.

3.1 What Is Consciousness For?

We concentrate here on the historical, evolutionary question. The question is: Is consciousness an adaptation, and if so what are the effects for which it was selected?

One possibility that has excited philosophers and scientists alike of late is that consciousness, or some varieties of it, is an adaptation.[2] Of course, providing adequate explanation of and evidence for such a claim is difficult—as it is for any claim that a trait is an adaptation. Still, it seems plausible that the presence of at least some varieties of consciousness will have adaptation explanations.

The alternative is that consciousness is not an adaptation, but that it is an evolutionary epiphenomenon (Polger and Flanagan in press). This could be the case for any of various reasons. Consciousness might be a spandrel (Gould and Lewontin 1978), just one of the things that happens when enough neurons are amassed in one place with a certain organization. On this story, consciousness is like the thumping noise that human hearts make.

Another reason could be that consciousness was produced by chance. Some creatures are conscious, and others are not—that it is so does not require any further explanation. If one feels the need for some sort of explanation, it will be a contingent historical story that will tell how, as a matter of fact, chance events led to the present state of things. Consider the following fable:

In a finite population of interbreeding organisms, random mutation caused a portion of the population to have some sort of conscious states (i.e., for those states, there is something that it is like for the organism to be in that state). In each case, the new phenotypic trait (speaking generally, consciousness) was heritable. Sadly, a nearby volcano erupted. By chance, the eruption killed all and only the nonconscious organisms. The conscious organisms, however, survived and reproduced successfully, passing on the trait—consciousness. Consciousness evolved. . . . Although evolution of consciousness occurred in this case, it was not evolution by natural selection but rather by random drift. Only by chance did the conscious organisms out-reproduce their nonconscious counterparts; it was not because they were conscious that they survived. (Polger and Flanagan in press)

Such a tale might explain how consciousness came to be in humans. But it is equiv-alent to denying that there is any explanation for why it came to be, if what we want to know when we ask "Why?" is for what purpose we are conscious—for according to this story there is no purpose.

3.2   Obstacles to Historical Questions, and Their Solutions

There are obstacles to giving any satisfactory answer to the historical questions—and especially to attempts to give adaptation explanations. Three obstacles are:

· No Fossil Evidence
· No Background Theory
· Heterogeneity of Consciousness

The No Fossil Evidence problem is simple enough: Consciousness, unlike other traits whose evolutionary history we wish to inquire about, leaves no fossil record. The metaphysical naturalist is in a position to claim that evidence about brains is ipso facto evidence about consciousness, but in this case that is little help—for fossilized brains are no more prevalent than fossilized experiences of orange leaves on a cool autumn day. So whatever claims we make about the evolutionary history of minds, we shall lack crucial evidence to make our explanations more than "just-so stories."

The response to this is that whereas it is indubitable that there is no direct fossil evidence of consciousness—no, say, well-preserved Pleistocene savorings of mam-moth à la mode—there is also no direct fossil evidence of many biological and eco-logical processes (e.g., whether dinosaurs were warm or cold blooded.) There is, however, a great deal of indirect evidence, including fossilized skulls and, what has seemed especially productive in other matters, artifactual evidence of lifestyle. That such evidence is bound up with general background theories, for example, about cultures today and their development over time, need not be a difficulty. It is far too late in the twentieth century to think that there is such thing as evidence that is not bound up with theory. If there is a special problem concerning evidence of con-sciousness, it cannnot be merely that the evidence is theory-laden.

But, and this is to raise the second obstacle for naturalistic answers to historical questions, it is not at all clear that we in fact have the background theories that could justify indirect evidential claims. This is the No Background Theory problem. That is, while it is true that background theories can support indirect evidence, we have no such theories.

But it is not the case that we have no background theories at all. Work on the relation between encephalization and intelligence, theory of mind, and increased capacity for language use is suggestive (Byrne 1995; Wills 1993; Nahmias unpublished). There is even one account that links metabolic factors to consciousness; if this is right, fossil evidence of certain respiratory structures (the "energy intake hub") would thereby be indirect evidence of consciousness (Fink unpublished). Whether the idea that consciousness is energy-hungry, therefore requiring larger and more efficient respiratory passages, will be borne out is an interesting question. The proposal, however speculative, is sufficient to make the point. Background theories exist that can add plausibility to evidential claims in the absence of well-fossilized qualia.

But even if there are no background theories now available, the present dearth of background theories, though frustrating, is not a fundamental problem for answering historical questions about consciousness. For there is no reason to think that this condition will persist—especially if it is recognized as an area of research that needs to be addressed.

A final obstacle to naturalized answers to the historical questions is the Heterogeneity of Consciousness problem. This is that consciousness is not a single, unified phenomena, and so to ask the question, "How did consciousness come to be?" is to fall victim to a confusion. Daniel Dennett has raised this objection against us:

> The question of adaptive advantage, however, is ill posed in the first place. If consciousness is ... not a single wonderful separable thing ("experiential sensitivity" [Flanagan 1991, 1992]) but a huge complex of many different information capacities that individually arise for a wide variety of reasons, there is no reason to suppose that "it" is something that stands in needs of its own separable status as fitness enhancing.   (1995, p. 324)

Here is a different way of making the point: Dirt is ubiquitous. Dirt comes in many forms. Dirt has many causes. If we want to know about dirt, we seek theories for different sorts of dirt, but we do not aim for a general Unified Theory of Dirt. We do not ask how and why dirt came to be, the objection goes, for dirt does not form a single kind that admits of such answers. And consciousness does not either. We should not expect a unified theory of consciousness for there is no unified phenomenon that answers to that name (P. S. Churchland 1983).

But perhaps consciousness is not so like dirt; perhaps it is more like jumping. Still, the ability to jump is only superficially a single trait; in fact it is a heterogeneous class involving different sorts of bones, muscles, joints, and so forth in the various creatures that jump. There is no one answer to "How and why did jumping come about?" that is common to grasshoppers, and dolphins, and dogs, and human beings. It is just a bad question.

The Heterogeneity of Consciousness is not a problem for answering the historical question unless one interprets the very question, "Is consciousness an adaptation, and if so what are the effects for which it was selected?" as begging the question of the unity of consciousness. But it is much more natural to understand the question as neutral on the question of whether consciousness is a unified or heterogeneous phenomenon: "Asking if consciousness is an adaptation does not depend on knowing ahead of time whether the phenomenal varieties of consciousness will turn out to be neurophysiologically homogeneous or heterogeneous phenomena" (Polger and Flanagan in press).

Once that clarification is made, there is no problem of heterogeneity. It just means that the varieties of consciousness, insofar as they are not a unified phenomenon, will not admit of a unified explanation of their origins. Each variety of consciousness will require its own answers to the historical (and ahistorical) questions. We should then expect that different explanations—adaptation, exaptation, spandrel, chance—will apply to different varieties of consciousness (Polger and Flanagan in press).

Some of these various explanations may even resemble each other; the phylogenetic and morphological diversity of eye structures in terrestrial animals does not disincline us generally from discussing the adaptive advantage of sight, though it may require us to loosen-up a bit on our human notion of what vision is.

4   An Ahistorical Question

Let us now consider how metaphysical naturalism and the natural method help us to understand one example of an ahistorical question about consciousness. Accepting that the mind-brain relation is a natural one constrains the possibilities, but still the question remains: What is the natural relationship between consciousness and the brain?

This mind-brain question is an ahistorical question when it is understood to ask for an explanation of how, for example, your mental states now (i.e., at each successive moment) are related to your brain states now. Of course that is simply the famous "mind-body problem" as it has existed for twentieth-century materialists. So has taking up metaphysical naturalism really gotten us anything? Yes; but to see this we need to examine some possible naturalistic answers to the ahistorical question, and the objections that are raised against them.

4.1 Answers to the Mind-Brain Question

Three styles of answer have been for the past twenty years the main contenders to explain the mind-brain relation. One answer is that conscious states are identical to physical states of the brain, where physical states are specified and picked out by their nonrelational properties. This is type-identity theory, often called simply identity theory (Smart 1962; Hill 1991; Macdonald 1989).[3]

The second option is one or another variety of functionalism. All versions of functionalism have in common that they hold that to be a conscious state is to occupy (instantiate, or realize) a specified functional role relative to some system. Versions of functionalism differ greatly in how systems are circumscribed, what sorts of functions are to be recognized, how functions are to be specified, and what metaphysical relations constitute instantiation. But they share a commitment to the view that the mind-brain relation can be understood as some kind of role-occupant relation, that what is important is not what minds are but what they do (see Putnam 1980; Dennett 1991; Lycan 1987; Van Gulick 1988; Hardcastle 1995). Like the identity theorist, the functionalist may identify conscious states with brain states—as long as brain states are themselves functional states, constituted by their relational properties.

A third possibility is that the relation between mind and brain is a supervenience relation. The idea behind supervenience, or emergence, is that conscious states or properties are higher-order states or properties of brain states and properties. On this view mental states depend on physical or functional states, but they are not identical to those physical or functional states (see, e.g., Kim 1993; Horgan 1987). Similarly, it might be said that the liquidity of water is not identical to a feature of its molecular structure, but supervenes on that structure.

As with the answers to historical question considered above, the purpose of describing these three views is not to decide among them. They are brought to the table to show what metaphysical naturalism about consciousness might look like.

There are important objections to metaphysically naturalistic theories of mind. But—and this is the point to be illustrated—the metaphysical naturalist has resources available to meet such concerns. Three objections are principled agnosticism, new mysterianism, and the explanatory gap.

4.2 Obstacles to Naturalistic Answers to the Mind-Brain Question

The principled agnostic view is best stated by Thomas Nagel in his well-known essay, "What Is It Like to Be a Bat?" (1974). Nagel writes, "Physicalism is a position that we cannot understand because we do not at present have any conception of how it might be true" (1974, p. 176). Nagel does not go so far as to argue that physicalism is not, or could not be, true. Still, we can agree that metaphysical naturalism would be a hollow doctrine indeed were it no more than an affirmation of faith in a metaphysics that was truly inconceivable to us. A "black-box" metaphysics—the conviction that there is some relation, we know not what, that explains how mind and brain are connected—would be powerless to address the questions about mind and consciousness that we would like to answer.

Colin McGinn's new mysterianism takes Nagel's concern one step further. Although he—unlike Nagel—is a card-carrying naturalist, he writes, "I do not believe that we can ever specify what it is about the brain that is responsible for consciousness, but I am sure whatever it is is not inherently miraculous" (1989, p. 349). Another may of putting McGinn's claim is that an understanding of the mind-brain relation is cognitively closed to us, as when he says, "there are properties of the brain that are necessarily closed to perception of the brain" (1989, p. 357, italics removed). Thus, according to McGinn, we are in a position relative to our own consciousness analogous to the position that platypuses are in relative to, say, relativity theory. It is not that relativity is inherently mysterious; it's just that platypus are not cognitively equipped to consider such matters. Likewise, McGinn thinks that some creatures might well have the appropriate cognitive apparatus to understand the mind-brain relation in humans; it's just that we are not such creatures (1991).

Joe Levine's explanatory gap concern is at once broader and yet less pessimistic that McGinn's new mysterianism:

What is left unexplained by the discovery of C-fiber firing is why pain should feel the way it does! For there seems to be nothing about C-fiber firing which makes it naturally "fit" the phenomenal properties of pain any more than it would fit some other set of phenomenal properties. . . . One might say that it makes the way pain feels into a brute fact.   (1983, p. 357)

Like McGinn, Levine finds inadequacy in our understanding of the mind-brain relation, but he locates it in the kinds of explanation that metaphysical naturalism permits rather than in our own cognitive limitations. It is not an insuperable problem of our cognitive makeup that makes consciousness a mystery; it is a problem with our explanations.

Principled agnosticism, new mysterianism, and the explanatory gap are skeptical worries that a metaphysical naturalist should take seriously. They do not dogmatically assert the falsity of the metaphysical view. Rather, they attack undeniable holes in our present understanding; they attempt to make the case that these are problems endemic to the naturalistic project. What is the metaphysical naturalist to say?

4.3 Solutions to Skeptical Obstacles

In fact there is a great deal that metaphysical naturalists can avail themselves of to overcome these obstacles. Four responses are (1) the History of Science reply, (2) the Limits of Explanation reply, (3) the Background Theory reply, and (4) the Structure of Consciousness reply.

The History of Science reply may be the first to come to mind, although it is weakest philosophically. This response simply points out that the history of science is filled with examples of problems that were considered unsolvable in their time, but that we now consider to have solved satisfactorily. Could Thales have known that water is H_2O? (If he had, would it have changed his theory?) Did it seem possible that Hesperus is Phosphorus, and that it is the planet Venus? Why should it be any less plausible that the relation between mind and brain will come to be understood by us? Our understanding of the temperature of a gas as its mean kinetic molecular energy does not raise questions about cognitive closure or explanatory gaps. It was not so long ago in the history of human knowledge that we had no concept of how it could be true that temperature be a microphysical relational fact. Does anyone now doubt that it is the case?

The History of Science reply is weakest when it is presented as merely an appeal to past successes, or as a promissory note on the future. With science and philosophy, as with mutual funds, past performance is not an unerring indicator of future performance. But there is a way of putting the point that lends it a modicum of credibility. What the History of Science reply does is shift the burden of proof to the skeptic. It is not enough to doubt that we can have a naturalistic understanding of consciousness; one needs to provide an argument to show why it is that questions about consciousness differ in kind from questions about, for example, water and temperature. New mysterians and explanatory gappists do not provide such arguments. Nagel's principled agnosticism, however, is based on an argument that distinguishes subjective phenomena from nonsubjective phenomena, so the History of Science reply is least effective in countering his concerns.

A second response is the Limits of Explanation reply. This reply agrees with Nagel, McGinn, and Levine that what they desire of explanations of consciousness is not deliverable by our present theories of mind, but it denies that this alleged lack is a problem for our explanations, for it is a requirement that we do not put on other natural theories. Why should we expect any more of explanations of consciousness than we do of explanations of water or temperature?

Levine is keenly aware of this response, and he acknowledges that his view requires a theory of intelligibility to account for what is required of explanations that is met in the cases of water and temperature but not, for example, in the case of pain (1983, pp. 358–359). But Levine thinks that he can motivate his claims even lacking such a theory of intelligibility, and that no technical formulation of the notion of intelligibility is needed to see the difference between the consciousness-brain case and the water-H_2O case. If this works, it would also address the concerns raised by the History of Science reply.

But the Limits of Explanation respondent may be happy to concede that the identifications of water with H_2O and temperature with mean kinetic molecular energy also suffer from explanatory gaps. What explanatory gaps show is that complete intelligibility is not a qualification on satisfactory explanations, that philosophers like Nagel, Levine, McGinn, and Jackson (1982) are expecting too much of naturalistic theories. They expect theories to "capture" the phenomena that are their objects in some overly strong sense; they expect satisfactory theories to also be fully satisfying (Flanagan 1992, pp. 93–97, 119). But it is a familiar and not the least bit troubling truth that an explanation of a hurricane does not capture the force of the hurricane in the same way that an actual hurricane does, and that an explanation of digestion does not actually digest, much less enjoy, a gourmet meal. We may admit that what it is like to be in a certain position relative to hurricanes, or digestion, or feelings of gustatory satisfaction is not reproduced by explanations of those phenomena. But explanations are not supposed to do that. We must simply recognize that there may be some disparity between satisfactory explanations and satisfying explanations (Flanagan 1992, p. 119). Daniel Dennett quotes Sydney Shoemaker in asking, "If what I want when I drink fine wine is information about its chemical properties, why don't I just read the label?" (in Dennett 1991, p. 383). Reading the label on a bottle of wine (either the chemical description or the nice little advertising blurb about how "spicy" or "fruity" the wine tastes) does not provide a satisfying explanation of what the contents taste like, but it may well be a satisfactory explanation.

If one still has concerns about what makes some explanations satisfying, the third reply may be of some help. The Background Theory reply says that what makes some explanations seem better than others—indeed, what makes some explanations better than others—is that they are embedded in the context of larger background theories that we accept. The identification of water with H_2O is not a lone claim; it is but one part of the theory of atomic chemistry, a theory that has great resources and vast explanatory and predictive success. If this is right, then we may readily understand why explanations of consciousness fail to be satisfying at the present time—for few if any general theories of psychology are as established as the theories that ground, for example, chemistry and molecular dynamics. Matters are even worse on the brain side of the equation, for the neurosciences are unquestionably nascent. So it is clear that mind-brain relations presently lack the broad theoretical foundations that give us confidence in our understanding of other such relations. But there is no reason to think this condition will persist indefinitely.

All of this sounds rather promissory, and we did claim that the natural method could do better than that. There is at least one body of well-established theory, discussed above in section 3, that the metaphysical naturalist can ground a theory in: the theory of evolution by natural selection. The division between historical and ahistorical questions is less than sharp. Understanding the synchronic mind-brain relation may require understanding how brains came to be formed as they are; likewise, making sense of how selection has acted on brains may require some understanding of how brains are conscious.

Let us indicate briefly the sort of things we have in mind. The view of the mind-brain relation that has most directly hitched itself to evolutionary explanation is teleological functionalism, a variety of functionalism that grounds its claims about functions, their individuation, and their instantiation in a biological notion of function known as etiological function (Lycan 1987; Van Gulick 1988; Flanagan 1992; for more about etiological function, see Wright 1973; Millikan 1989; Neander 1991; Godfrey-Smith 1994). But one does not need to be a teleofunctionalist to take advantage of the explanatory support that evolutionary theory provides. Any metaphysical naturalist can appeal to the ways in which conscious states contribute to the capacities of biological organisms. That is, the body of evolutionary explanation and theory supports all versions of metaphysical naturalism by providing a context against which ahistorical claims about the mind-brain relation may be both grounded and tested; nothing about this explanatory grounding is unique to those versions of metaphysical naturalism that appeal directly to evolution by natural selection as a means of individuating mental states.

The History of Science reply addresses the worries of all three skeptical challenges. The Limits of Explanation reply and the Background Theory reply are aimed at Nagel and Levine. The fourth response is directed specifically against Levine's explanatory gap concerns. The Structure of Consciousness reply provides perhaps the most underappreciated resource that metaphysical naturalists can avail themselves of.

What one gets, automatically, as it were, by taking up metaphysical naturalism and the natural method, is an immense body of data. Of course the data are all, as it were, out there for any philosopher or scientist; and it is widely accepted these days (though it was not always accepted, to be sure) that a philosopher ought not say something that contradicts accepted data, or at least not too much of it. It is an altogether different matter to use these data not only as a benchmark for and constraint on theorizing, but actually to make it do positive work in a philosophical theory. This is exactly what the Structure of Consciousness reply, as described by Robert Van Gulick, aims to do:

> The more we can articulate structure within the phenomenal realm, the greater the chances for physical explanation; without structure we have no place to attach our explanatory "hooks." There is indeed residue that continues to escape explanation, but the more we can explain relationally about the phenomenal realm, the more the leftover residue shrinks toward zero. (Van Gulick 1993, in Block, Flanagan, and Güzeldere 1997, p. 565)

The mind-brain questions will surely seem intractable if we try to explain the relation of some simple or unanalyzed brain states and properties to similarly simple or unanalyzed conscious states or properties. But neither brain states and their properties nor conscious states and their properties are simple or unanalyzable. Brain sciences provide more and more information each day about the structure of brains and nervous systems. Psychology, not to mention anthropology and linguistics, provides details on the structure of mental states and properties. Increasing information on each side provides more and more hooks by which the brain explanations and phenomenological explanations hang together.

The strategy is not a daunting search for some single unifying principle that will allow our explanations to cross freely between brains and consciousness, as McGinn suggests (1991). Rather than attempting to bridge the explanatory gap, the Structure of Consciousness reply aspires to close it! As the two kinds of explanation are attached at more and more points, the epistemological chasm is drawn ever narrower. The moorings are not one or a few great scientific or philosophical principles. They are innumerable bits of information about the structure of brains and experiences that, when taken into account, show us just how the two fit together—and why they do not fit in other arbitrary relations.

4.4   Color Vision and the Structure of Consciousness

C. L. Hardin (1987; 1988, p. 135) deploys the Structure of Consciousness strategy against Levine's explanatory gap concerns. Hardin sets up the problems of color vision in the way urged by the natural method, and he argues that better understanding of the science of color vision suggests solutions to age-old puzzles such as the inverted spectrum problem. The information about color vision that must be brought into reflective equilibrium includes external physical data about the surface properties of objects and about wavelengths of light, internal physiological data

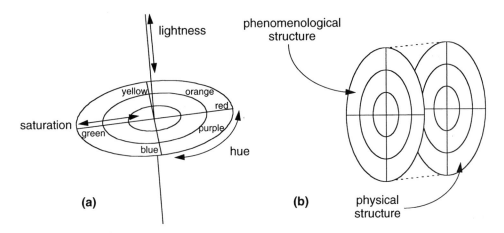

Figure 11.3
(a) Representation of phenomenal color space as a three dimensional solid. (b) The task of matching phenomenal color structure to physical structure.

about brains and retinas, cognitive cum behavioral data about our responses to various stimuli, and phenomenological data about our experience of color vision.

The project of understanding color vision begins with a detailed analysis of the structure of color experience as revealed not only by first-person phenomenological reports, but also, for example, cognitive testing and linguistic anthropology. One common way of representing the structure of color experience is as a spherical solid with axes that represent hue, saturation, and lightness or brightness (figure 11.3a). Such maps of color space are intended to describe the qualitative side of the explanatory gap. The task of understanding color vision as construed according to the natural method is to find some structure in the nonmental world that corresponds to the cognitive-phenomenal structure of color experience (figure 11.3b). The corresponding nonmental structure might be an "external" physical structure (e.g., surface properties of objects or the properties of light of varying wavelengths); it might also be an "internal" physiological property (e.g., patterns of activation in the primary visual cortex) or parts of the early visual system such as the retina or optic nerve.

Hardin's assessment is that there is no property of the external physical environment (surfaces of objects, wavelengths of light, ect.) that corresponds to the structure of all color experiences. He argues that the sought-after structure can be found instead in the functional organization of the early visual pathways (retinal ganglion cells play a crucial role) as accounted for by a particular empirical theory of color vision, the opponent process color theory. Thus, on Hardin's analysis, the physical

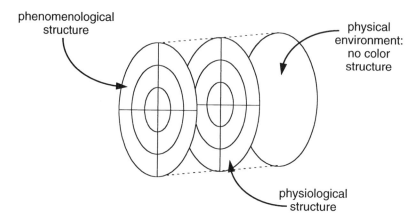

Figure 11.4
Phenomenal structure matches physiological structure, rather than structure in the external environment.

property identified with color is a physiological property of the visual system (figure 11.4). Since color is identified with an actual physiological property whose structure is open to empirical investigation, some previously recalcitrant questions can be answered. For example, for undetectable spectrum inversion—the sort that philosophers have often worried about—to be possible, color space must be symmetric. Hardin argues that opponent-process color theory does not yield a symmetric human color space (1988). If the asymmetry of human color space is tied to asymmetries in the visual system, then undetectable inverted spectra are empirically impossible for us. Moreover, if particular hues depend essentially on human physiology, then, pace Shoemaker (1982), the entire question of spectrum inversion for creatures that have a physiology that differs from our own may be undefined (similar to what Shoemaker calls the Frege-Schlick view).[4]

Hardin's treatment of color experience is controversial. Opponents charge, for instance, that he overestimates the evidence for the empirical theory (opponent-process color theory) on which he bases his philosophical analysis (e.g., Hardcastle 1995).

Suppose that Hardin's critics are correct that he is premature in his adoption of the opponent-process theory as the theory of color vision. That charge is not (and is not intended to be) a problem for either the natural method or the Structure of Consciousness strategy. Hardcastle, for example, does not fault Hardin's project; however, she finds that it comes up short within the bounds of the natural method:

If Hardin's story in fact details how color perception works, then he would be right and truly inverted spectra would not be possible. But the difficulty is that we don't know how color perception actually proceeds. (Hardcastle 1995, p. 30)

Hardcastle criticizes Hardin's results, but she does not reject his strategy. Hardin's project is an example of how one proceeds according to the natural method, and in particular of how the Structure of Consciousness strategy can be put to use. But Hardin's analysis is not supposed to go through if, as it were, the science on which it is based is wrong.

Another objection is raised by David Chalmers, who sets aside Hardin's theory on the grounds that it presupposes and so does not explain the subjective quality of color experience (1995, p. 206). Chalmers is right to note that knowing the intricate ways that neurophysiological structure is isomorphic with the structure of consciousness does not by itself tell us why that relationship obtains, or how it came to be so. Still, the Structure of Consciousness reply will suffice to explain why—given that brain states are associated with consciousness at all—a certain brain state is apt for correlation with one conscious state (sensation of red) rather than another (sensation of green).

Finally, Levine himself maintains that even granting Hardin's physiological theory, the explanatory gap remains as wide as ever (1991). To arbitrate this dispute is outside the scope of this essay.[5] An intermediate position might be that even if the Structure of Consciousness reply does not entirely close the explanatory gap, at least it narrows the margin.

5   Dreams: A Case Study

The case of dreams, introduced in section 2.1, is an example of an area in which progress has been made on both the historical and ahistorical questions.

If we are to understand the phenomenon of dreaming we must know what it is that happens in brains when we sleep, not just how it seems to us the next morning. We are all familiar with what it is like to dream. But knowing what it is like to dream reveals to the subject little about the hidden neurological structure of sleep or dreams, or about the function of dreams, if any.

On the other hand, it seems that we could account for brain activity during sleep without mention of the phenomenal character of the mentation, if any, that accompanies it. To put it in Levine's way: What is left unexplained by the discovery of certain neuronal activity during sleep is why dreaming should feel the way it does. For there seems to be nothing about any particular brain activity that makes it naturally "fit" the phenomenal properties of dreaming any more than it would fit some other set of phenomenal properties. The neuroscientific account of sleep leaves out what it is like to dream.

The Structure of Experience reply is aimed at just that assertion. What we need to do, it says, is to learn much more about the structure of experience, and about the structure of brain activity during sleep. One tactic for revealing the hidden structure of sleep and dreams is to study some animals, for example humans, who are awake and some who are asleep (according to our ordinary behavioral criteria for determining waking and sleeping). We can, for example, measure brain waves and see if there are differences between those that are awake and those that are asleep. And indeed there are.

When we are awake, brain waves occur very frequently, in quick succession. The rapidity of neuronal firing is reported in terms of frequency. Awake brains are characterized by high frequency waves—as many as fifteen per second. Despite the fact that the brain is firing rapidly while we are awake, its electrical activity has a shallow amplitude. (The greater the amplitude, the more energy emitted; in the case of brain waves, amplitude is a reliable measure of voltage.) Brain waves slow down as our heads hit the pillow, and within an hour or an hour and a half of going to sleep the brain has descended by steps into emitting slower and slower waves, with greater amplitudes than those of awake brain waves.

The waves that characterize wakefulness are called beta waves. As we doze off, beta waves are replaced by alpha waves. Alpha waves quickly give way to even lower-frequency theta waves: stage 1 NREM sleep (see figure 11.2, sec. 2.1 above.) Eventually, the peaceful, melodic string of music-like waves characterizing alpha and theta wave-orchestration is joined, in the wonderful metaphor of Peretz Lavie (1996), by occasional drumbeats and trumpet blares. These orchestral novelties (K-complexes and sleep spindles) mark a point in the sleep cycle when the sleeper can no longer be easily woken up, quickly clear her mind, and return to wakeful beta waving. This is stage 2 NREM-sleep. About ten to fifteen minutes after the drums and trumpet join the strings in stage 2 NREM-sleep, delta waves appear. Imagine a trombone or tuba joining the "cerebral symphony" (to borrow a phrase from William Calvin 1990). This is stage 3 NREM. The droning tuba drives the other players from the stage and the sleeper is in deep, delta sleep. This is stage 4 NREM-sleep and can last for as long as thirty or forty minutes.

Then comes the big surprise: REM-sleep. The brain's shift to REM-sleep—the type of sleep associated with bizarre mentation, delirium, and psychosis—begins with a rapid trip upward toward wakefulness through stages 4, 3, and 2 and then into REM-sleep.[6] On an EEG, REM looks very much like waking mentation: It is characterized by rapid firing (high frequency) and short amplitude (low voltage) waves.

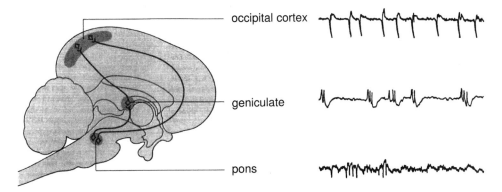

Figure 11.5
The visual brain stimulates itself in REM sleep via a mechanism reflected in EEG recordings as PGO waves. Originating in the pons (p) from the neurons that move the eye, these signals are conducted both to the lateral geniculate (G) body in the thalamus and to the occipital cortex (O) (from Hobson 1989).

The total sleep cycle (comprised of the descent through the four stages of NREM-sleep, the rapid ascent to REM-sleep, and a period of REM-ing lasting twenty to thirty minutes) usually takes ninety to a hundred minutes. An eight-hour sleep will typically divide into six hours of repeated descent into NREM; one to two hours will be spent in deep (delta) sleep and approximately one and a half to two hours in REM-sleep.

In REM-sleep, pulsing signals originate in the pontine brainstem (P, for short) and reach the lateral geniculate body (LGN) of the thalamus. When awake, that area (G, for short) is a relay between the retina—on certain views, part of the brain itself—and visual processing areas. Other pulses go to the occipital cortex (O, for short)—the main visual processing area of the brain. The general picture then is this: The pons (P) initiates waves that go to the lateral geniculate body of the thalamus (G) and to the occipital lobes (O). PGO waves are the prime movers of REM sleep (figure 11.5). Understanding PGO waves suggests an explanation for the salience of visual imagery in the dreams of sighted people. But the PGO noise is going to many different brain areas, and reverberating every which way. This helps explain why people who practice remembering their dreams will report auditory, olfactory, tactile-kinesthetic, and motor imagery in addition to visual imagery.

Recent studies have shown that the parts of the brain that reveal robust activity on PET, MRI, and magneto-encephalographs (MEG) indicate that "mentation during dreaming operates on the same anatomical substrate as does perception during the waking state."[7] PGO waves are dominant during REM-sleep and quiescent during

NREM-sleep. By inference to the best available explanation, this explains a good deal about why the mentation of REM-sleep involves vivid, bizarre, and multimodal imagery.[8]

Thus dreaming illustrates nicely the convergence of neuroscientific and phenomenological data according to the natural method. When we examine what is going on in the brain during sleep, we find that we do know why some brain activity is associated with vivid visual imagery while some other is not. It is just not the case that delta waves are equally well suited to give rise to bizarre dreams as PGO waves. There is no explanatory gap.

The cognitive function of dreaming and the psychological effects of sleep and dream deprivation are also subjects of serious study. Our all-things-considered view is that dreams are a by-product of the neurochemical processes of sleep (Polger and Flanagan in press; see also Flanagan 1995, 1996, forthcoming). Dreams are neither functional mechanisms in brain processes nor adapted features of our cognition:

The mechanisms required to turn off certain neurons and to turn on others cause waves that incidentally activate areas throughout the brain, especially in the visual areas. Some of these activations are experienced as "thoughts" and "sensations." Suppose the conscious brain is independently prone to try to make sense of thoughts it has. If so, there is no surprise that it tries—and in part succeeds—to supply a coherent story line to the noise it generates while the system as a whole is doing what it is does during sleep.   (Polger and Flanagan in press)

Understanding how dreams work thus suggests a plausible, if somewhat deflationary (in this case), account of their evolutionary origin.

That the explanatory gap can be shut in one small case does little by itself to close down other purported gaps. That we have some idea of what dreaming is not for fails to resolve the question of whether other kinds of consciousness are adaptations. Each case, as with dreaming, will depend on a detailed understanding of the structure of phenomenological and brain events, as well as other sources of information that did not figure in the brief sketch of an explanation of dreams just given. But this is enough to show the kinds of resources that the metaphysical naturalist has available to answer such tough challenges.

6   Reactions to the Natural Method

A familiar reaction to the natural method is that it is too permissive. We already know, one might think, that some sources of information are more valuable than others. Given a commitment to metaphysical naturalism, it is just obvious that neuroscience will be a more salient source of information than reports about the nature of consciousness from, say, folk medical practice.

Against this concern two things can be said. First, it seems likely that the natural method will bear out the intuition that neuroscience is, if not the most important, at least one extremely important source of information about consciousness. Indeed, it does not appear from our present state of knowledge that much or any progress can be made without significant contributions from the brain sciences. But it is equally clear that neuroscience cannot finish the project all by itself, for neuroscientists depend in practice and in principle on, for example, phenomenological information with which to correlate neuronal discoveries. Second, insofar as this first response is justified, it is justified by the natural method. That is, we know that neuroscience is valuable in the study of consciousness because we already implicitly engage, to some degree, in seeking reflective equilibrium in our scientific cum philosophical explanations.

This observation—that we sometimes engage in reflection much like that required by the natural method—highlights a common second reaction to the method: Isn't this what we already do? To this, the answer is straightforward: No, it is not. Almost everyone who has approached questions about consciousness has done so by privileging one source of information above the others. Years before advocating a version of the natural method that she calls the "co-evolutionary strategy," Patricia Smith Churchland privileged neuroscience and disparaged both phenomenology and psychology on the basis of a prior commitment to eliminativism, basically the view that those sources of information were not worth taking seriously (compare P. S. Churchland 1983 and 1986). Others, for example, functionalists like William Lycan and Daniel Dennett, although they acknowledge that neuroscientific and phenomenological information may be useful, favor cultivating alliances with work in psychology and cognitive science (Lycan 1987; Dennett 1991).

We do not mean to suggest that such preferences have been arbitrary or unjustified; for the most part they are not. But the justifications offered have not been via the reflective exercise described by the natural method. We do not claim that the natural method is the only method for studying consciousness, or the only way of justifying preference of one source of information about the mind over others. It is the method that we prefer, and the one that we discuss. But it should not be mistaken as the standard method of cognitive and neurosciences, much less of philosophical psychology.

6.1 The Bottom Line

There are multiple ways to study consciousness, multiple modes of access to information about qualitative experiences. We have shown that this is not an insurmountable obstacle to a natural theory of consciousness, as Nagel and others suggest.

Rather, that consciousness has many faces is but another part of its structure—another twist to be understood, but also another hook on which to hang our explanations.

Our goal has been to illustrate how the natural method can both take consciousness seriously and negotiate around well-known obstacles to metaphysical naturalism, yielding insights into the nature of consciousness. We have no illusions: Much work—most work—remains to be done. But even modest progress, on color or on dreams, for example, encourages the naturalistic line of inquiry.

Questions about consciousness are like any other questions about biological states, processes, and properties. If they seem different, perhaps this is because conscious experience is such a salient part of the human condition, and one about which we care deeply.

Acknowledgments

This chapter is based on our talk, "Biological Explanations of Subjectivity," presented at the International Society for the History, Philosophy, and Social Study of Biology (ISHPSSB), Seattle, Washington, USA, July 1997. Special thanks to Valerie Hardcastle, Eddy Nahmias, and Güven Güzeldere for their comments and assistance.

Notes

1. Block (1980, fn. 22) suggests that Hilary Putnam at a time advocated such a view.

2. The explication of the evolutionary question that follows depends on a particular view concerning the answers to question (d) from sec. 2.2 above, "Is consciousness one phenomenon, or many?" We noted when we raised the question that it was an example of one that has historical and ahistorical readings. The ahistorical reading of (d) asks whether the answers will be one or many to the main ahistorical question, "How is it that consciousness arises from brain processes?" The historical reading of (d) asks whether the answers will be one or many to the main historical question, which we have now phrased as "Is consciousness and adaptation, and if so what are the effects for which it was selected?"

Our view is that the answer to each of the readings of (d) is that we should expect multiple explanations; but the reasoning behind this conclusion is outside the scope of this chapter. The reader may treat the thesis that consciousness is a family phenomena united by the Nagel-property (that there is something that it is like for the subject of the state to be in that state) as an assumption (but to see why it is not an assumption for us, see Polger and Flanagan in press; Flanagan 1992, 1995, 1996, forthcoming.) We regard it as an empirical fact that consciousness is nonunitary. We do not believe the following discussion of methodological points in those terms hinges on the claim of heterogeneity.

3. Some, e.g., Hilary Putnam (1994), use the term "identity theory" more broadly.

4. Hardin puts the point somewhat differently, and not in terms of the Frege-Schlick view (1988, pp. 138, 142–154).

5. For more details see Levine "Cool Red" (1991) and Hardin's "Reply to Levine" (1991).

6. Hobson (1994). Hobson's view is not simply that REM-dreams are like psychoses, but that they are psychotic episodes.

7. R. R. Llinás and D. Paré (1991). This helps explain why prosopagnosiacs don't report dreaming of faces and why people with right parietal lobe lesions, who can't see the left side of the visual field, report related deficits in their dream imagery (p. 524).

8. Once such imagery is overrated, dreaming is equated with REM-ing and the sensorily dull, but thought-like, mentation of NREM sleep is overlooked. This then leads to the assumption that NREM sleep, especially stage 3 and 4 NREM sleep, is a period of unconsciousness.

References

Armstrong, D. M. (1968) *A Materialist Theory of Mind*. London: Routledge and Kegan Paul.

Armstrong, D. M. (1980) *The Nature of Mind and Other Essays*. Ithaca, NY: Cornell University Press.

Block, N. (1980) "Troubles with Functionalism," in N. Block (ed.), *Readings in Philosophy of Psychology*, Volume One, pp. 268–305. Cambridge, MA: Harvard University Press. Originally printed in C. W. Savage (ed.), *Perception and Cognition: Issues in the Foundations of Psychology*, (Minneapolis: University of Minnesota Press, 1978).

Block, N., Flanagan, O. and Güzeldere, G. (eds.) (1997) *The Nature of Consciousness: Philosophical Debates*. Cambridge, MA: The MIT Press.

Byrne, R. (1995) *The Thinking Ape: Evolutionary Origins of Intelligence*. New York: Oxford University Press.

Calvin, W. H. (1990) *The Cerebral Symphony: Seashore Reflections on the Structure of Consciousness*. New York: Bantam.

Chalmers, D. (1995) "Facing Up to the Hard Problem of Consciousness," *Journal of Consciousness Studies* 2(3):200–219.

Chalmers, D. (1996) *The Conscious Mind: In Search of a Fundamental Theory*. New York: Oxford University Press.

Churchland, P. M. (1988) *Matter and Consciousness*. Cambridge, MA: The MIT Press.

Churchland, P. M. (1995) *The Engine of Reason, the Seat of the Soul: A Philosophical Journey into the Brain*. Cambridge, MA: The MIT Press.

Churchland, P. S. (1983) "Consciousness: The Transmutation of a Concept," *Pacific Philosophical Quarterly* 64:80–95.

Churchland, P. S. (1986) *Neurophilosophy: Toward a Unified Science of the Mind-Brain*. Cambridge, MA: The MIT Press.

Dennett, D. C. (1989) "Quining Qualia," in Marcel and Bisiach (eds.), *Consciousness in Contemporary Science* pp. 42–77. New York: Oxford University Press.

Dennett, D. C. (1991) *Consciousness Explained*. Boston: Little, Brown, and Co.

Dennett, D. C. (1995) "The Unimagined Preposterousness of Zombies," *Journal of Consciousness Studies* 2(4):322–326.

Dretske, F. (1995) *Naturalizing the Mind*. Cambridge, MA: The MIT Press.

Fink, B. (unpublished). "The Evolution of Conscious Behavior: An Energy Analysis." Presentation at Towards a Science of Consciousness II, Tucson, Arizona, 1996.

Flanagan, O. (1991) *The Science of the Mind*, 2nd edition. Cambridge, MA: The MIT Press.

Flanagan, O. (1992) *Consciousness Reconsidered*. Cambridge, MA: The MIT Press.

Flanagan, O. (1995) "Neuroscience and Dreams: The Spandrels of Sleep," *Journal of Philosophy* XCII, 1:5–27. (Reprinted in Flanagan 1996.)

Flanagan, O. (1996) *Self Expressions: Mind, Morals, and the Meaning of Life*. New York: Oxford University Press.

Flanagan, O. (forthcoming) *Dreaming Souls: A Neurophilosophical Theory*. New York: Oxford University Press.

Flanagan, O. and Polger, T. (1995) "Zombies and the Function of Consciousness," *Journal of Consciousness Studies* 2(4):313–321.

Friedman, M. (1996) "Philosophical Naturalism." Presidential Address to the Ninety-Fifth Annual Central Division Meeting of The American Philosophical Association. *Proceedings and Addresses of the APA* 71:2 (Newark, DE: The American Philosophical Association, 1997).

Godfrey-Smith, P. (1994) "A Modern History Theory of Functions," *Noûs* 28(3):344–362.

Gould, S. J. and Lewontin, R. (1978) "The Spandrels of San Marco and the Panglossian Paradigm: A Critique of the Adaptationist Program," *Proceedings of the Royal Society, London* 205:581–598.

Hardcastle, V. G. (1995) *Locating Consciousness*. Amsterdam: John Benjamins.

Hardin, C. L. (1987) "Qualia and Materialism: Closing the Explanatory Gap," *Philosophy and Phenomenological Research* 48:281–298.

Hardin, C. L. (1988) *Color for Philosophers: Unweaving the Rainbow*. Indianapolis: Hackett.

Hardin, C. L. (1991) "Reply to Levine," *Philosophical Psychology* 4(1):41–50.

Hill, C. S. (1991) *Sensations: A Defense of Type Materialism*. Cambridge: Cambridge University Press.

Hobson, J. A. (1988) *The Dreaming Brain*. New York: Basic Books.

Hobson, J. A. (1989) *Sleep*. New York: Scientific American Library.

Hobson, J. A. (1994) *The Chemistry of Conscious States*. Boston: Little, Brown, and Company.

Hobson, J. A. and Stickgold, R. (1994) "Dreaming: A Neurocognitive Approach," *Consciousness and Cognition* 3:1–15.

Horgan, T. (1987) "Supervenient Qualia," *Philosophical Review* 96:491–520.

Jackson, F. (1982) "Epiphenomenal Qualia," *The Philosophical Quarterly* 32(127):127–136.

Kim, J. (1993) *Supervenience and Mind: Selected Philosophical Essays*. New York: Cambridge University Press.

Lackner, J. and Garrett, M. (1973) "Resolving Ambiguity: Effects of Biasing Context in the Unattended Ear," *Cognition* 1:359–372.

Lavie, P. (1996) *The Enchanted World of Sleep*. A. Berris, trans. New Haven, CT: Yale University Press.

Levine, J. (1983) "Materialism and Qualia: The Explanatory Gap," *Pacific Philosophical Quarterly* 64:354–361.

Levine, J. (1991) "Cool Red," *Philosophical Psychology* 4(1):27–40.

Llinás, R. R. and Paré, D. (1991) "Of Dreaming and Wakefulness," *Neuroscience* 44(3):521–535.

Logothestis, N. and Schall, D. (1989) "Neuronal Correlates of Subjective Visual Attention," *Science* 245:761–763.

Lycan, W. G. (1987) *Consciousness*. Cambridge, MA: The MIT Press.

Lycan, W. G. (1996) *Consciousness and Experience*. Cambridge, MA: The MIT Press.

Macdonald, C. (1989) *Mind-Body Identity Theories*. London: Routledge.

McGinn, C. (1989) "Can We Solve the Mind-Body Problem?" *Mind* 98:349–366. Reprinted in McGinn 1991.

McGinn, C. (1991) *The Problem of Consciousness*. Oxford: Basil Blackwell.

Millikan, R. (1989) "In Defense of Proper Functions," *Philosophy of Science* 56:288–302. Reprinted in Millikan 1993, pp. 13–29.

Millikan, R. (1993) *White Queen Psychology and Other Essays for Alice*. Cambridge, MA: The MIT Press.

Nagel, T. (1974) "What Is It Like to Be a Bat?" *Philosophical Review* LXXXIII (4):435–450.

Nahmias, E. (unpublished) "Why Our Brains Got So Big: Reciprocal Altruism, Deception, and Theory of Mind." Presented at the Southern Society for Philosophy and Psychology, Atlanta, Georgia, 1997.

Neander, K. (1991) "Functions as Selected Effects: The Conceptual Analyst's Defense," *Philosophy of Science* 58:168–184.

Place, U. T. (1956) "Is Consciousness a Brain Process?" *British Journal of Psychology* 47:44–50. Reprinted in V. C. Chappell (ed.), *The Philosophy of Mind* (Englewood Cliffs, NJ: Prentice Hall, 1962.)

Plantinga, A. (1993) *Warrant and Proper Function.* New York: Oxford University Press.

Plantinga, A. (unpublished) "Naturalism Defeated." Presented at the Chapel Hill Colloquium in Philosophy, Chapel Hill, North Carolina, 1997.

Polger, T. and Flanagan, O. (in press) "Consciousness, Adaptation, and Epiphenomenalism," in G. Mulhauser (ed.), *Evolving Consciousness.* Amsterdam: John Benjamins.

Putnam, H. (1980) "The Nature of Mental States," in N. Block (ed.), *Readings in Philosophy of Psychology*, Volume One, pp. 223–231 (Cambridge, MA: Harvard University Press). Originally published under the title "Psychological Predicates" in W. H. Capitan and D. D. Merrill (eds.), *Art, Mind, and Religion* (Pittsburgh: University of Pittsburgh Press, 1967).

Putnam, H. 1994. "Sense, Nonsense, and the Senses: An Inquiry into the Powers of the Human Mind," *Journal of Philosophy* XCI (9):445–517.

Rosenthal, D. (1986) "Two Concepts of Mind," *Philosophical Studies* 94(3):329–359. Reprinted in D. Rosenthal (ed.), *The Nature of Mind* (New York: Oxford University Press, 1991).

Searle, J. (1992) *The Rediscovery of the Mind.* Cambridge, MA: The MIT Press.

Shoemaker, S. (1982) "The Inverted Spectrum," *The Journal of Philosophy* LXXIX (7):357–381.

Smart, J. J. C. (1962) "Sensations and Brain Processes," in V. C. Chappell (ed.), *The Philosophy of Mind*, pp. 160–172. Englewood Cliffs, NJ: Prentice-Hall.

Squire, L. and Zola-Morgan, S. (1991) "The Medial Temporal Lobe Memory System," *Science* 253:1380–1386.

Stroud, B. (1996) "The Charm of Naturalism," *Proceedings and Addresses of the APA* 70:2–27.

Van Gulick, R. N. (1985) "Physicalism and the Subjectivity of the Mental," *Philosophical Topics* 13(3):51–70.

Van Gulick, R. N. (1988) "A Functionalist Plea for Self-Consciousness," *Philosophical Review* 97(2):149–188.

Van Gulick, R. N. (1993) "Understanding the Phenomenal Mind: Are We All Just Armadillos?" in M. Davies and G. Humphreys (eds.), *Consciousness: A Mind and Language Reader.* Oxford: Basil Blackwell. Reprinted in Block, Flanagan, and Güzeldere (eds.), 1997.

Wills, C. (1993) *The Runaway Brain.* New York: Basic Books.

Wright, L. (1973) "Functions," *Philosophical Review* 82:139–168.

V PHILOSOPHY OF SCIENCE

12 Mental Functions as Constraints on Neurophysiology: Biology and Psychology of Vision

Gary Hatfield

Scientific psychology and philosophy of psychology have made frequent use of concepts of function. In philosophy of psychology, two concepts have been prominent. The first is the input-output functionalism of the 1970s (as described by Block 1980). *Functions* relate inputs to outputs, and a *functional analysis* decomposes a system into contributing subprocesses, producing a flowchart of the division of causal labor in the processes mediating between inputs and outputs (or between inputs and internal system states). The causal-explanatory features of input-state-output (or "iso-") functionalism received their classical philosophical description from Cummins (1975; but see Shapiro 1994). The second notion of function is teleological and involves the attribution of a purpose to a structure or other trait. In Wright's (1973) classical analysis, functions are ascribed in biology (and psychology) in relation to the presumed evolutionary origin of a trait: A trait's ("teleo-") function is whatever the trait does that causes the unguided process of natural selection to select it. On the assumption that traits are selected because they bestow an advantage on their possessors, traits are said to be *for* whatever heritable thing they do that results in their selection.

These two concepts of function have been used in two areas of philosophy of psychology: in work on the mind-body problem and in the theory of mental representation. Putnam (1967) and Fodor (1974) produced antireductionist arguments using the first concept of function, contending that a functionally conceived system of (input-output) relations between internal states can admit of multiple physical realizations and so violates the classical reductionist requirement of bridge laws linking a given psychological type with a specific physical type. This multiple-realizability argument has also been applied to the teleofunctional conception (Sober 1990). In work on mental representation, Cummins-style isofunctionalism has been used in conceiving the mind as a symbolic engine. Fodor (1980) and Stich (1983) championed the view that psychological processes are decomposable into interactions of syntactically construed internal states. By contrast, the teleological concept of function has been employed in analyses of ascriptions of distally directed representational content. A sensory state might be ascribed a function to represent a distal state of affairs, on the assumption that the system of which it is a part evolved precisely to track such distal states. Dretske (1988), Matthen (1988), and Shapiro (1992), among others, have used or proposed teleofunctional analyses of representational content.

My purpose in this chapter is to examine a question at the intersection of the mind-body problem and the analysis of mental representation: the question of the directions of constraint between psychological fact and theory and neurophysiological or physical fact and theory. Does physiology constrain psychology?

Are physiological facts more basic than psychological facts? Or do psychological theories, including analyses of perceptual representational capacities, guide and constrain physiology? Despite the antireductionist bent of functionalist positions, it has generally been assumed that (physics or) physiology is more basic than, and hence constraining on, psychological fact and theory. In section 1 I respond to the intuitions the would lead one to adopt such a view. Section 2 examines whether a one-sided constraint from physiology has been and should be found in practice. Examination of developments in psychology shows that rigorous functional analyses can be carried out in advance of physiological knowledge and may even lead the way in relation to neuroscience. We will consider three cases from the study of visual perception (binocular single vision, stereopsis, and color vision) in which psychological results have provided a basis for physiological research.

   When psychology does guide physiology, it is tempting to think that it does so by discovering Wrightian teleofunctions that guide physiological investigations into the mechanisms for realizing those functions. Hence one might assume that psychologists and neuroscientist commonly take themselves to be investigating the evolved functions of neural structures. Neuroscientists surely investigate the eye and visual system under the assumption that these structures are for seeing, rather than (say) for detecting an electric discharge in the vicinity of the retina (which sighted humans can do, via the subjective light produced by such discharges), and this assumption guides the questions they pose about the eye, visual cortex, and intervening pathways. Accordingly, investigators view rods and cones in the eye as photoreceptors, even though these structures will produce "output" (will respond with hyperpolarization) to other forms of energy. Such assertions or tacit presuppositions of teleofunctions can be legitimate (Hatfield 1993; Proffitt 1993). Nonetheless, there may well be cases in which the functions that have guided research are simply precisely determined input-output functions whose teleofunctional role is absent or misdescribed. In sections 3 and 4 I argue that in the case of trichromatic color matches, the psychological function involved is such an isofunction, and that we must look elsewhere for an appropriate teleofunctional description of color vision as a psychological capacity.

1   Is Biology (Neurophysiology) "More Basic" than Psychology?

It seems natural to suppose that biology, or neurophysiology, is more basic than psychology. As one might reason, surely the structure of the brain limits the psychologies that can be realized by that brain, and surely psychological facts conform with physiological facts. Indeed, presumable they must. But then, if facts are facts, physiological facts must conform to psychological facts just as well. And if scientists find correlations between physiological states and psychological states, it should be possible, at least in principle, to read the correlation either way: from psychology to physiology, or from physiology to psychology.

The point of interest here is whether physiological facts and theories provide a special constraint on psychological theory, so that the range of plausible hypotheses in psychology is deeply beholden to findings in physiology. I would not dispute the general claim that there must be ways in which an organism's physiology constrains its psychology. An ant's brain does not enable the creature to discover and prove the fundamental theorem of calculus. There are limits. Size does matter.

But the claims for deep constraint go further. Those of a "narrow" content persuasion might argue that considerations of supervenience establish that psychological states are fully determined by microphysical states of the nervous system. Considered abstractly, it seems evident that two physically identical individuals would have to be in the same psychological state, and therefore that descriptions of physical states constrain the ascription of psychological states (Stich 1978). In opposition, the very notion of a description of the complete physical state of a human being can be questioned as utopian, as beyond the wildest dreams of current physical science. Or such a microphysical description might be considered irrelevant, since psychology individuates its states by considering organismic systems as they function in environmental settings (Shapiro 1993), and such descriptions typically collapse across physical kinds, let alone across organismic microphysical states. It is true that physiological psychology and psychophysics chart the relations between neural and psychological states, but it is hopeless to try to constrain such descriptions via unachievable "complete descriptions" of the microphysical states of one or more individuals.

Opposing intuitions may run strong on such matters, but my purpose is not to match intuition with intuition. Rather, I'd like to turn attention to the relations between psychology and neuroscience in actual scientific practice. Armchair speculation about supervenience may be useful in charting the boundaries of standard ways of conceptualizing mind-brain relations, but the new word in philosophy of science is attention to actual procedures of investigation, and reconstruction of the real course of thought. It is hard to know what use might be made of highly articulated conceptual analyses of concepts of mind-brain relations that have little or no relation to the current state of knowledge about mind and brain. By contrast, scientific psychology and neuroscience are flourishing enterprises, rich in data for philosophical reflection. Reflection on such data suggests that, in practice, psychology leads and constrains neuroscience more often than not.

Prior to considering actual cases, I will present an abstract argument—one that appeals to practice-related conceptual grounds—in support of the view that psychology can and even must guide neuroscience in the individuation and description of major functional units of the nervous system and brain. The argument's main point can be simply stated: the descriptions of perceptual and cognitive abilities found in psychology are the best, or the only, functional descriptions of molar neural structures that there are. The stereoscopic depth response as measured and described by perceptual psychologists is a function carried out by the visual system, comprising the eyes, visual pathways, and visual cortex. Binaurally guided localization of sounds is a function performed by the auditory system. Recalling the appearance of your favorite grade-school teacher is an instance of imagistic memory, another brain function.

Leaving aside eliminativist predictions about future but at present entirely unknown scientific vocabularies (Churchland 1986), there really is no alternative to the view that psychology provides the middle and large-scale functional vocabulary for describing neural anatomy and physiology. Neuroscience itself has a functional vocabulary of connectivity, transmission, graded potentials, spikes, chemical transmission, synchronous firing, and so on (Kandel, Schwartz, and Jessell 1991, Pts. II–III). This vocabulary represents (mainly) twentieth-century discoveries of the cellular and molecular structure of neurons and their local interactions. But no one has ever started from this vocabulary and built a full explanation of a psychological function, such as a motivational state like hunger or a cognitive learning mechanism, or produced an explanation of a molar behavior pattern described in nonpsychological terms. The vocabulary within which large-scale neural systems are described has been and continues to be a psychological vocabulary. As the neuroscientist Eric Kandel has put it, "what we commonly call mind is a range of functions carried out by the brain" (Kandel et al. 1991, p. 5). In describing the functional anatomy of the brain, Kandel and other neuroscientists use a psychological taxonomy: The brain is parsed into sensory, motivational, and motor systems (Kandel et al. 1991, Pt. IV, chap. 19), study of which is subdivided into such topics as sensation and perception, motor and motivational processes, language, and thought (Pts. V–IX). When neuroscientists turn to real-time measures of brain activity, they seek neurophysiological counterparts to psychologically described activities, such as reading, imagining, attending, and interpreting (Posner and Raichle 1995).

Neuroscience, when it looks beyond microanatomy, is a series of attempts to discover brain mechanisms for psychological capacities that are already known and perhaps measured. This is not to say that physiological results, once achieved, can't lead to new psychological questions. The so-called sense of touch was subdivided into various discriminatory systems—including special systems for pressure on the skin, stretching of the skin, warm and cold, and extreme conditions (resulting in pain)—as a result of psychophysical study of various modes of sensitivity and anatomical study of sensory end-organs in the skin. The physiological discovery of cone polymorphisms in the human retina led researchers to view "error" results in color psychophysics in a new light (Neitz and Jacobs 1990; Neitz, Neitz, and Jacobs 1993). Nonetheless, the general point stands that neuroscience has no other way to describe molar function than psychologically. The middle-scale functional vocabulary for describing the nervous system and brain is the long-standing (see Hatfield in press) and deeply entrenched vocabulary of perceptual modalities, imagination, attention, learning, and memory—which means that psychology provides the theoretical framework for the study of brain function.[1]

The claim that neuroscientific descriptions, or neuroscientific facts and theories, are "more basic" than psychological findings and descriptions may now seem difficult to interpret. It could be construed as an ontological claim, that the characteristics of individual cells and their activities constrain the molar functions that can be realized in an organism. Let us grant that no animal can possess psychological capacities its neurons can't realize. Granting this point so far has little relevance for the practice of neuroscience and psychology, since we might grant it independently of any general theory of the capacities neurons can and cannot realize. The claim is not without content, for it assumes that the fundamental unit of brain anatomy is the neuron, which is a basic assumption of twentieth-century brain science. Beyond this, the claim simply expresses faith that the properties of neurons limit possible psychologies. So let us look more closely at neurons considered as functional units of the brain.

When neurons are viewed as functional units, it seems fair to say that their ontology depends on their relation to the system they serve. A neuron isn't a neuron unless it has a relation to a suitable molar anatomical structure, that is, an organism, or, more particularly, a nervous system (large or small). And the framing functional descriptions of such molar anatomical structures are psychological, which means that the framing functional descriptions for individual neurons are also psychological. Furthermore, it is plausible that psychological function affects neural structure, inasmuch as the molar psychological functions that neurons perform affect the development of neuronal structure diachronically, through evolution. The capacities of neurons need not be seen as fixed for all time. New neural structures presumably become entrenched in populations because of the functions they realize. Thus a given microstructure may become prevalent in a population, and so come to exist in various descendant organisms, because of the psychological capacity it subserves.[2] That would be a case of psychological function affecting neural ontology.

The ontological interpretation of constraint could support considerable speculative discussion. But as it happens, this notion of constraint is of little practical consequence, given the present lack of knowledge of the detailed relations between micro-neural states and psychological functions. Consequently, one might proffer an epistemological interpretaion of the claim that neuroscience constrains psychology. It might be argued that knowledge of psychological functions requires knowledge of neural structures described in a "purely" neuroscientific vocabulary.

The view that knowledge of the mind generally relies on knowledge of the brain is implausible on the face of it. Not only have mental processes such as visual perception been studied with some success since antiquity, when Aristotle's conjecture that the brain serves for thermal regulation was still in play; but the study of mental phenomena became rigorous laboratory science in the mid–nineteenth century, decades prior to the rise of the neuron doctrine of brain function through the work of Ramón y Cajal and others (Finger 1994, pp. 43–48). There is of course room for the claim that physiological facts can permit inferences about psychological states. If brain-imaging techniques become more precise, it might become possible to determine whether someone is dreaming, paying attention, or forming a visual image, by examining patterns of brain activity. In these cases, discovery of a link between psychological activity and brain activity would allow one to read backward, from physiology to psychology. The more interesting question for present purposes is whether the investigation of brain physiology independently of psychological investigation could, under currently foreseeable circumstances, yield facts that strongly limit the ascription of psychological functions.

There is little basis for thinking that knowledge of neural properties gained independently of psychology can provide significant limits on psychological theory. Granted, no psychological theory would be accepted that posits, say, a reflex arc that exceeds the known speed of neural transmission. There are some properties of neurons, basic properties that could limit their functional capacities, that can be determined by direct measurement independently of specific psychological knowledge. Such might include the speed of transmission of impulses, the maximum number of impulses that can be carried per unit of time (which varies), the gain or loss of intensity of activation at various neural connections, the excitatory or inhibitory influences of one neuron on another, some small sample of the patterns of connectivity found in a brain, and so on.[3] But knowledge of these properties by themselves has not (and will not any time soon) permit inferences to the psychological functions carried out by neural structures.

More generally, the investigation of the functions served by neurons, such as transmission of information, depends on our seeing neurons as contributing to the activity of the brain considered as an organ. To see the brain as an organ is to consider the global functions it serves. But, as we shall see from cases in section 2, molar functions typically are known or conjectured prior to investigation of neuronal structure and activity. Researchers seek to understand the microactivities of neurons by asking how they contribute to one or another more global brain function, psychologically described. Typically, knowledge of psychological function has preceded and facilitated knowledge of neural functioning.

2   Psychological Fact and Theory Leads and Constrains Neurophysiological Investigation

I will provide three examples in which psychofunctional conjectures or theories have led the way in investigations of neural structure and function. One is from the period of early modern speculation, and the other two are from recent perceptual and physiological psychology.

2.1   Newton: Binocular Single Vision

Since ancient times visual theorists have wondered how binocular single vision is achieved, that is, how we can see a single world even though most objects are seen by two eyes at once (Ibn al-Haytham 1989, pp. 85–89; Ptolemy 1989, pp. 26–34). A few authors have denied that we actually achieve binocular single vision, arguing that we see with only one eye at a time (Priestley 1772, pp. 117, 666). But most people do see with two eyes at once and nonetheless see one world. From antiquity brain structures have been proposed by which a physiological unification of optical stimulation might be gained, to match the psychological fact of binocular single vision. Galen supposed that images from the two eyes are fused in the optic chiasma (not far behind the eyes), where the optic nerves meet (Siegel 1970, pp. 59–62). Ibn al-Haytham, Witelo, and others proposed that the stimulation from the two eyes is brought into point for point correspondence, perhaps in the chiasma (Hatfield and Epstein 1979; Ibn al-Haytham 1989, pp. 85–89). Anatomists had long observed that the optic nerves meet at the chiasma but continue into the brain, a fact portrayed in Descartes's prominent early modern treatise on optics (1637, pt. 5). So there remained a question about how post-chiasmally separate optic nerves mediate single vision.

In 1704 Newton speculatively advanced the basic anatomical scheme underlying binocular single vision. The remarkable passage in which he did so is worth quoting extensively. In the Fifteenth Query of the *Optics,* Newton asked:

Are not the Species of Objects seen with both Eyes united where the optick Nerves meet before they come into the Brain, the fibres on the right side of both Nerves uniting there, and after union going thence into the Brain in the Nerve which is on the right side of the Head, and the fibres on the left side of both Nerves uniting in the same place, and after union going into the Brain in the Nerve which is on the left side of the Head, and these two Nerves meeting in the Brain in such a manner that their fibres make but one entire Species or Picture, half of which on the right side of the Sensorium comes from the right side of both Eyes through the right side of both optick Nerves to the place where the Nerves meet, and from thence on the right side of the head into the Brain, and the other half on the left side of the Sensorium comes in like manner from the left side of both Eyes.   (Newton 1704, pp. 136–137)

Newton here was advancing the theory that nerve fibers from the two eyes "partially decussate," that is, partially cross, in the optic chiasma. He supported his conjecture with the claim that animals that use both eyes together, such as humans, dogs, sheep, and oxen, have an optic chiasma, whereas those that don't, such as fish and chameleons, don't. Although a problem of visual coordination between the eyes may be posed for animals whose visual fields do not overlap, no problem of achieving single vision from two images arises in those cases. So in (many) reptiles and fish, there is no need for stimulation from the two retinas to be brought together into a single visual image. Hence, the chiasma must serve to permit single vision, and the hypothesis of partial decussation shows how it could do so.

Although one might find Newton's reasoning plausible, in fact his comparative anatomy is incorrect. Reptiles and fish do have optic chiasmas. Nonetheless, his functional conjecture is partly right. There is no partial decussation in reptiles and fishes, many of which do not have binocular single vision. The nerves fully cross, so that the right eye projects into the left side of the brain and vice versa (Kuhlenbeck 1977, pp. 60–62).[4]

Remarkably, Newton did hit upon the anatomical structure by means of which stimulation from the two eyes is unified in mammals: The right hemifield of each eye projects into the left side of the brain, and the left hemifield into the right side. Newton had little data to go on, and some of it was erroneous. But his proposal was soon confirmed by clinical cases in which individuals lost vision in the same half of the visual field for both eyes, a finding that would be explained if the nerves projecting from each half of the retina are unified in the brain where they can be subject to a common pathology (Finger 1994, p. 83). More than two hundred years later, studies using single-cell recording techniques in cats and primates revealed the existence of binocularly driven neurons in the visual cortex, that is, of neurons that receive input from the two eyes (Barlow, Blakemore, and Pettigrew 1967; Poggio and Fischer 1977). Moreover, the primary projections from the two eyes are laid out retinotopically across the back of the brain (Barlow 1990). So Newton's speculative picture of half-images from each eye joining on respective sides of the brain has proven true.

Newton's speculations are an example of using a conception of psychological or mental function to guide hypotheses in neuroanatomy and neurophysiology. He started from the psychological fact that humans experience a unified visual world using two eyes. He assumed that there must be an anatomical solution to the "joining" of stimulation from the two eyes.[5] He had only rudimentary anatomical knowledge of the brain at his disposal. So it is clear that his understanding of the psychological function to be explained led the way to his neuroanatomical hypothesis.

2.2 Stereoscopic Depth Vision and Binocularly Driven Neurons

In the 1830s Charles Wheatstone discovered the powerful stereoscopic depth cue, through which we are able, from the slight differences between the images falling on the two eyes, to perceive the relative structural depth of objects. In the mid–nineteenth century, numerous authors examined the psychophysics of the stereoscopic depth response, including Peter Ludwig Panum, A. W. Volkmann, Wilhelm Wundt, Helmholtz, and Ewald Hering (Turner 1992, pp. 13–26). These investigators and their contemporaries studied the elicitation of the depth response under varying retinal disparities, including crossed disparities (image elements are reversed between the two eyes) and uncrossed disparities (image elements have the same relative order in the two eyes); the form of the geometrical and empirical horopters (positions yielding zero disparity between eyes, or yielding equal distance judgments); acuity for disparity; the temporal course of the depth response; the use of lines tilted relative to one another to achieve depth; and the relation of disparity, a relative depth cue, to convergence, which can indicate absolute distance. The discussion was framed by the anatomical question of whether single vision and depth perception rely on simultaneous stimulation of "corresponding" points on the two retinas or from stimulation of disparate points. Investigators debated whether stereoscopic depth perception results from innate anatomical mechanisms or is acquired through experience. These discussions of brain anatomy and physiology were carried out in total ignorance of the microphysiology of the brain. The discussion was driven by psychophysical data together with theoretical assumptions about what would count as evidence for an innate or learned basis for stereopsis (Hatfield 1990, pp. 180–188).

When binocularly driven neurons were discovered in the 1960s (see previous subsection), researchers found that some of them responded selectively to particular degrees of disparity (finely or coarsely tuned), and to crossed and uncrossed disparity. Investigators immediately posited that these disparity-detecting neurons serve the binocular depth response (Barlow et al. 1967). In this case, a newly discovered physiological mechanism was interpreted by appeal to a known psychological capacity. The neurons were discovered in the cat, and at the time there was no compelling laboratory evidence that cats have functional stereoscopic vision. Pettigrew, a codiscoverer of disparity sensitive neurons, later observed that in ascriptions of function to those neurons, "the final proof should come with the behavioral demonstration of stereoscopic abilities" (1986, p. 212). In the case of the cat, it was more than ten years before a convincing behavioral demonstration of stereoscopic ability was achieved (Mitchell, Kaye, and Timney 1979).

Subsequent research displays a pattern of interaction between neurophysiological work, computational simulations, and behavioral or psychophysical results. Bishop and Pettigrew (1986), leaders in the neurophysiological and psychological study of stereopsis, suggest that although the neurophysiological discovery of disparity-sensitive neurons has provided knowledge of cellular mechanisms that most likely contribute to stereopsis, and although progress has been made in constructing computational simulations of stereoscopic abilities, the interpretation of individual neurophysiological results is still based largely on psychophysics. Tychsen's more recent review of the field bears this out (1992). Tychsen found that the main conceptual shift in work on stereopsis from the past two decades was based on the neuroanatomical division of many visual functions into parvocellular and magnocellular layers in the lateral geniculate nucleus. He cited especially Tyler's (1990) application of these ideas to binocular vision. Tyler had in fact proposed three functionally distinct systems underlying stereopsis: (1) a "fine, global" system that is good for static targets, operating foveally at a fine grain but with a comparatively slow temporal response; (2) a "coarse, local" system operating about twice as fast, out to about ten degrees from the fovea, with greater sensitivity to moving targets; and (3) a central "protostereoscopic" mechanism for deriving tilt in depth from binocular stimulation without using local disparities. As Tyler observed, neuroanatomical studies provided "the inspiration for the psychophysical partition of sensory processing into categories of specialized analysis" (1990, p. 1877). But he also noted that previously (Tyler 1983) he had postulated similar multiple systems of stereopsis "on the basis of psychophysical evidence alone," drawing on the work of K. Ogle and B. Julesz. His aim in linking his psychological theories to neurophysiological findings was to permit mutual testing "of proposed associations between the separable processes in the two domains" (1990, p. 1894). In other cases, discovery of certain types of binocularly driven neurons naturally leads to the postulation of stereoscopic ability, thus stimulating further behavioral research (Pettigrew 1986). But there is no precedence of brain facts over psychological facts.

Stereoscopic vision thus provides an example of potentially useful interaction between psychological facts and theories and physiological facts and theories. At the same time, the functional vocabulary for describing binocular depth systems remains psychological: It is the vocabulary of spatial and temporal acuities, latency of response, retinotopic location (area of visual field), and color sensitivity (Tychsen 1992), all of which derive their significance in relation to the macrofunction of depth perception.

2.3 Color Vision and Opponent Pathways

The modern understanding of the physics of colored light stems from Newton's (1704) discovery that sunlight can be decomposed into lights of differing refrangibility, or, as we now say, into lights of various wavelengths. Knowledge of the basis for the human visual response to differing wavelengths and to surfaces with differing spectral reflectances (that is, to surfaces that differentially absorb or reflect lights of various wavelengths) has been much slower to achieve. In the second half of the nineteenth century it was established that (subject to minor qualifications) any given light (however composed) can always be matched by combining no more than three appropriately chosen monochromatic lights (pure lights of a single wavelength). In accordance with this finding, Helmholtz revived Thomas Young's proposal that there are only three distinct color sensitive elements in the retina, that is, three types of retinal element, each responding maximally to light of a different wavelength (Helmholtz 1866/1925, 2:134–146; Turner 1992, chap. 6). This proposal about retinal physiology was made solely on the basis of psychophysical results, that is, on the basis of observers' responses to various chromatic stimuli. In 1886 Arthur König and Conrad Dieterici estimated the sensitivity functions for the three types of color receptors with remarkable accruacy by renormalizing data taken from observers with various forms of color deficiency (see Turner 1992, 197–202). It was the second half of the twentieth century before the photic properties of visual pigment of the cones, the retinal elements subserving daytime vision, were directly measured (Kaiser and Boynton 1996, chap. 5).

In opposition to a simple trichromatic model of color vision, in the 1870s Hering argued that color vision results from three opponent processes in the central nervous system, one serving red-green perception, one blue-yellow, and one black-white (Hering 1875; Turner 1992, pp. 130–134). His arguments were based on the phenomenology of the color primaries: he contended that among the chromatic colors (that is, excluding white, black, and grey) there are four pure primaries, namely, red, yellow, blue, and green. He maintained that afterimages reveal adaptation effects among the four primaries: yellow produces a blue afterimage, red a green one, and so on. He also observed that there are no color-deficient individuals who are simply "red blind" or "green blind"; they are "red and green blind." Based on these observations, he speculated that the neural mechanisms of color vision, whether central or peripheral, show opponent organization. By this he meant to postulate a yoked physiological process underlying red-green perception, which produces a red sensation when it is driven one direction, and a green sensation in the opponent state.

Hering's arguments were based on the phenomenology of color vision and on psychophysical studies, coupled with assumptions about how the underlying physiology might work. He had no knowledge of the microanatomy and physiology of the retina and visual nervous system. In the 1950s Leo Hurvich and Dorothea Jameson published psychophysical data from color cancellation experiments (mixture experiments, aimed at producing the four Hering chromatic primaries from mixtures of monochromatic lights) to support an opponent-process theory (reviewed in Hurvich 1981, chaps. 5–6). Subsequently, neuroscientists discovered a variety of opponent process neural mechanisms in the retina, lateral geniculate nucleus, and other brain loci. Many researchers now believe that the three types of cone pigment in normal human eyes are linked neurophysiologically into opponent processes (Kaiser and Boynton 1996, chap. 7). Here again, psychophysics and phenomenology led the way to the postulation and subsequent confirmation of neural mechanisms. Peter Kaiser and Robert Boynton, coauthors of the currently standard handbook on human color vision, observe that although anatomy, neurophysiology, and photochemistry have inspired new psychophysical experiments, "the data of the psychophysicist, together with theories developed from such data, provide a framework within which the electrophysiologist conducts his research" (Kaiser and Boynton 1996, pp. 26–27). This remark is similar to Tyler's comment about the relation of psychophysics and neurophysiology in the field of stereoscopic vision, and it supports the general view of the relation between psychology and neurophysiology I presented in section 1. In the two central areas of research on sensory function reviewed here, as indeed in other areas, physiological facts and theories may inspire or confirm research and theory cast in psychological language, but psychological language remains the primary functional vocabulary of research, including neurophysiological research.

3 Are These Teleofunctions?

The cases we have reviewed offer examples in which psychological theory describes a perceptual capacity or inspires speculation about the neural mechanism for a capacity. It is tempting to suppose that the cases of binocular single vision, stereoscopic depth perception, and color matching are instances of proper functions of the visual system. As reviewed above, on the popular Wrightian analysis of (teleo-) function this supposition would entail that these abilities have been procured through natural selection. To back a Wrightian ascription of a teleofunction, it must be supposed that selection has acted or is currently acting to fix or maintain a trait. This requirement poses an epistemological problem for assigning Wrightian teleofunctions. In practice, psychologists rarely have available evolutionary studies of the capacities they investigate, and thus they don't concern themselves with evolutionary considerations (Proffitt 1993). They may indeed assume that evolution stands behind the basic mechanisms they study; but their judgments of what counts as a psychological capacity are made against a background of previous psychological theory and current empirical investigations. So they may not in fact know whether they are dealing with a proper teleofunction or not.

Of the three examples given, the first two are strong candidates for being teleofunctions. The functional significance of binocular single vision and stereopsis, and the pattern of phylogenetic development, strongly indicate that they are adaptations resulting from natural selection. Stereopsis has apparently evolved more than once, including independently in some birds (including owls and falcons) and in mammals (Pettigrew 1986; Tychsen 1992). Nearly all mammals, including the rabbit (with laterally directed eyes), show some stereoscopic vision. Predators show a tendency for frontally directed eyes. Fossil remains indicate that early mammals had laterally directed eyes, but the earliest primates had large, frontally directed eyes and a greatly expanded visual cortex. Anatomical comparisons with the eyes and teeth of living primates suggest that early primates were nocturnal insectivores. Consequently, investigators have proposed (Allman 1977) that frontally directed binocular single vision arose to facilitate frontally pursued predation by allowing the foveal areas to be simultaneously directed at the prey, with stereoscopic depth perception serving to enhance precise spatial localization and to bring background-camouflaged prey into relief (Julesz 1971, pp. 145–146). The question of teleofunctionality is more vexed for cone receptivities, color matching, and opponent color processes. To determine which properties of the color system might be teleofunctionally specified, we need to consider the functions of color vision.

4 Functions of Color Vision

What is color vision for? This question is hopelessly broad, for it takes in all aspects of wavelength discrimination, whether hues are involved or not. Restricting the question to color vision in primates narrows it somewhat. Primates exhibit both dichromatic color vision (some monkeys) and trichromatic color vision (some monkeys, most higher primates, including most human beings). What is (or are) the biological functions(s) of primate sensitivity to differences in light as expressed in the phenomenal colors we experience? We can specify the question more narrowly, and include an evolutionary transition, by focusing on trichromatic color vision by comparison with dichromatic.

We know that color vision varies in systematic ways in relation to wavelengths of light received at the eye, and that various objects exhibit different colors depending on how they reflect the various wavelengths of light. It is therefore tempting to suppose that in color vision the different perceived hues have the function of representing distinct physical properties in objects. At the limit, each hue would specify a unique physical property. For physical objects, a prima facie plausible candidate for the relevant physical property is spectral reflectance distribution of an object's surface: the percentage of incident light it reflects at each wavelength in the visible spectrum. Color scientists have long known that trichromatic vision in primates cannot exhaustively partition the calls of spectral reflectance distributions. There are ranges of photically distinct surfaces (surfaces that absorb and reflect wavelengths quite differently) that nonetheless look the same under ordinary lighting conditions. This class of physically distinct but perceptually identical surfaces are called metamers.

The existence of metamers does not mean that there are no precise laws relating color vision to physical attributes of the stimulus. In fact, perhaps the most precise laws in all of psychology are the psychophysical laws of trichromatic color matching (Kaiser and Boynton 1996, chap. 5). The laws of trichromacy state that for any sample of light, at most three wavelengths are needed to match it. More specifically, an appropriately chosen set of three lights of specified wavelengths (and adjustable intensity) can be used to match any given light. These laws are used to set international standards for dyes, in the engineering of color television sets, and in numerous other industrial applications. The stability of color matching, or at least of hue classification, is so great that the pharmaceutical industry uses color coding for pills, where a mistake in discrimination can cost a life.

The existence of metamers raises trouble for the view that color vision is intended to detect physical properties (see Hatfield 1992). It precludes any one-to-one mapping between hues and unitary physical properties. If the color system is viewed as a physical instrument, then metamerism reveals a dysfunction, a failure of the system to function optimally in partitioning the class of wavelengths, or of surface spectral energy distributions. But this "failure" counts as a dysfunction only if the color system has the teleofunction of detecting unitary physical properties. This returns us to the question of what the function of trichromacy in humans might be.

Instead of assuming that the color system is seeking to solve a physics problem, or is trying to get at the same information that interests a video engineer, let's approach the teleofunction of color vision by asking an evolutionary question: What might trichromacy have allowed animals to do that their dichromatic predecessors couldn't?

There have been two prominent answers to this question. Roger Shepard has argued that trichromacy evolved to serve the needs of color constancy (1992, pp. 501–515). Color constancy is the ability that allows us to see objects as having a stable hue under varying conditions of illumination. Under differing conditions of illumination, an object with unchanging reflectance properties reflects physically different energy distributions to the eye, depending on the characteristics of the ambient light. Shepard considers the changing properties of sunlight near the surface of the earth. Morning light is reddish, high-noon light yellowish. Yet a green object appears (close to) a constant green at both dawn and midday. Shepard argues that trichromacy evolved to compensate for separate red-green and yellow-blue sources of variation in illumination from sunlight, so as to allow constancy. Organisms endowed with color constancy would be able to recognize objects by their precise hues under various lighting conditions.

Jacobs (1981) approaches the function of color vision from a comparative perspective. Looking across the mammals, he suggests that color vision serves three primary functions: (1) to enhance object discriminability; (2) to aid in object recognition; and (3) to signal properties of objects. Visual scientists have long speculated that color vision could help make an object of a certain color more visible against a differently colored background. The example of red, orange, or yellow fruit among green leaves has often been cited (Polyak 1957, pp. 973–974; Wallace 1891, pp. 304–308). Mollon (1989) reviews evidence that color contrast significantly enhances the discriminability of objects that are partially obscured in a dappled manner, as would be ripe fruit seen through leafy foliage. Object recognition is aided if objects of a certain kind are all of the same color; for example, for most varieties, ripe apples are always red. Finally, color may signal a specific property of an object, such as sexual receptivity in some female primates (Hailman 1977, pp. 195–196).

In the examples given by Jacobs, the most is known about the advantages for discrimination of trichromacy over dichromacy. As it turns out, trichromats show greatly enhanced sensitivity for reds and oranges in relation to greens. This enhanced discriminability does not by itself require color constancy or precise color-matching abilities. Red things just need to look markedly different from green things if color is to help discrimination. They needn't always show exactly the same discriminable difference. Moreover, it is difficult to see that the identification or signal properties of color would require precise phenomenal identity across cases. Presumably, it is enough if object kinds and properties are marked by hue values that are strongly similar on various occasions. If we use the red rock for a marker, as opposed to the yellow or green rock, the rock needn't yield precisely the same phenomenal hue in all circumstances; it just has to fall clearly into the same hue category. Industrial uses of color coding make use of contrasting categories of color. It is enough if objects can be stably grouped by color. Thus, although we can discriminate thousands of hues, anyone using color symbolically, to convey typological information to untrained observers, is advised to compress the range of colors to nine (Cushman and Rosenberg 1991, pp. 140–144). Plainly, slight variations in the phenomenal character of the green, yellow, red, blue, white, etc. used in these circumstances will not defeat the intended use.

What then is the teleofunction of the laws of trichromatic color matching? Perhaps none at all. It might well be that these laws simply describe input-output functions that are universal features of normal trichromatic vision, features that result from the fact that three types of cone pigments have become linked to central processors in such a way that they enhance the discriminability of objects and enrich the domain of hue categories available for classifying objects or signaling their properties. Perhaps the laws of trichromacy are purely isofunctional: they reveal a standing property of the color system, but one that is a side effect of the teleofunctional aspects of trichromacy involved in discriminability, object identification, and property signaling, much the way heart sounds are a universal effect of the heart's teleofunction of pumping. We would then say that the color system, in carrying out the teleofunctions identified by Jacobs, draws on cone mechanisms and opponent processing systems that, when tested, in fact yield precise psychophysical laws, including the laws of trichromacy.

The view that the laws of trichromacy don't in themselves reveal a teleofunction receives support from the case of anomalous trichromacy. Anomalous trichromats are individuals in which one or two cones have peak spectral sensitivities that are slightly shifted from those of normals. The result is to shift the relation of perceived hues, such as green and red, in relation to wavelengths. Anomalous trichromats and normal trichromats disagree on which wavelengths produce a pure or "unique" green (and other hues). Depending on the direction of the shift in peak cone sensitivity, anomalous trichromats may have enhanced or slightly reduced powers of discrimination of wavelengths in the red and green portion of the spectrum (Kaiser and Boynton 1996, pp. 443–445). If normal trichromacy is a teleofunction, then anomalous trichromats have defective color vision. But do they? That depends on the standard of functionality. Their color matches differ slightly from those of normals. But ripe red fruit still contrasts markedly with green foliage. Anomalous trichromats perform as well as normals on many other color mediated tasks. The case is quite different for dichromats, for whom a group of objects may look very similar in color, whereas for all trichromats, anomalous or not, the same collection may contain strikingly different colors. My suggestion is that anomalous trichromats don't have deficient color vision; they're just different. And indeed the range of variation within the pigments of "normal" trichromats is greater than had been thought (Neitz et al. 1993).

If we accept this analysis, the strong human concern for precise color matches must still be explained. I would propose that various human cultures have developed practices that place a high value on color-matching ability, and sometimes on inter-subjective reliability of color matches. When the walls of house are painted, one usually wants the wall to appear of a uniform color. A precise match is desired from bucket to bucket of paint. In sharing a decision on the choice of paint, one may want to know that one's partner's color matches are like one's own. In such cases, an iso-function may take on an "instrumental function," that is, it may become an iso-function in the service of a specified end or interest (Godfrey-Smith 1996, p. 17). A candy maker might want to change the dye ingredients used in coloring an outer candy shell in order to reduce biohazard of the dye, but he might want the new candies to have the same color. Here, a metameric match would work. Color science could specify what was needed to obtain a match under usual conditions of illumination for normal trichromats. The science would be precise, but that need not imply that the isofunction in question, the human color-matching function, reveals a teleofunction of the color system. Teleofunction-bearing mechanisms can exhibit precise regularities that are secondary consequences of their proper functions. From this perspective, the mental phenomena described by the laws of trichromacy reveal properties of the machinery of the central nervous system without thereby revealing the teleofunction of the machine.

5 Conclusion

Neuroscientists can, and sometimes do, study the brain independently of the psychological teleofunctions involved. In fact, the early twentieth-century studies of fine-structured neural anatomy were of this type: stain some tissue, and see what looks different from what (Finger 1994, pp. 42–43). In this regard, on could think of taking a naive Baconian approach, so as to study all of the input-output functions open to human imagination: for example, the output of the brain when subject to high voltage shock, the output when an excised brain is frozen and then thawed with a blow torch, or any other relation between "input" (causal influences), internal states (intermediate causal states), and outputs (effects). But naive Baconianism has not been the path toward neuroscientific understanding. That path started from knowledge of mental states and processes gained from reports of phenomenal experience or observations of molar behavior. It proceeded through finer-grained investigation of phenomenal experience and more elaborate measures of molar behavioral performance, which allowed more precise statements of functions performed by neural structures. It continues with new techniques for matching neural activity to psychological process. Still today, after much progress has been made toward characterizing the physical and chemical properties of the brain, the study of the brain's functional organization is guided by descriptions of psychological function (whether iso- or teleofunction). It is difficult to see how things could be otherwise. Descriptions of mind, or of mental activity, condition our functional knowledge of the brain. No matter how you slice it, what we know (of interest) about the brain depends on what we know about mind. Unlocking the mysteries of the brain does not make mentalistic or psychological descriptions go away. Rather, psychology holds the keys to brain science.

Acknowledgments

I thank conference participants for comments on this essay at the meeting of the International Society for the History, Philosophy, and Social Studies of Biology in Seattle, and Tom Meyer for comments on a more recent version.

Notes

1. When psychological functions are regarded as descriptions of neural functioning, it need not be implied that such functions can only be realized by neurons. Pasta rolling is the appropriate functional description of a particular steel pasta roller (and indeed, the properties of the steel surely enter into explanations of how it rolls pasta); but a pasta roller can also be constructed of other materials. That the function of one type of neural structure is rightly said to be a certain psychological function does not imply a type identity between the structure and the function; that is, it leaves open the possibility that other biochemical materials could realize the same psychological function. (See Hatfield 1988, pp. 735–740, for further discussion.)

2. It has been alleged that on a Wrightian view, first instances of a trait don't have a function, since their presence can't be explained by the previous action of natural selection (Boorse 1976, p. 74). But if a propensity interpretation of fitness is accepted (as a friendly amendment to a Wrightian view), then a heritable variation that arises *de novo* by mutation might be assigned a function on the basis of its propensity to increase the survival and reproductive potential of those who have it relative to those who don't (see Bigelow and Pargetter 1987).

3. Even where a neural property can be discovered on its own, such as speed of transmission, it has happened historically that psychological knowledge played an important epistemic role. Helmholtz discovered the speed of transmission incidentally while studying muscle physiology, including the time course of muscular contraction, using frog muscle preparations (Helmholtz 1850/1892–1895, Olesko and Holmes 1993). In the first substantial publication of the results, he appealed to Bessel's famous finding that different sensory modalities (sight and hearing) in an individual show measurable differences in reaction time, citing it as a "certainty" that would lend plausibility to his discovery that the transmission of activity along motor nerves requires a measurable temporal interval (Helmholtz 1850/1892–1895, p. 813). In other words, he referred to a psychological result as the firm basis from which to prepare the way for his physiological finding.

4. A slight complication is that some fish, reptiles, and birds have no decussation in the optic chiasma, but they nevertheless do have binocular vision. In the case of the birds, a decussation occurs further on in the visual pathway, so that decussating nerves project from the thalamus to the cortex (Pettigrew 1986). Newton's functional conjecture again obtains: decussation is needed for single vision (and stereopsis, about which Newton would not have known).

5. This assumption was still not generally accepted in the second half of the nineteenth century, when Helmholtz warned against the tendency to seek anatomical explanations for psychological facts (1866/1925, 3:532).

References

Allman, J. (1977) "Evolution of the Visual System in the Early Primates," *Progress in Psychobiology and Physiological Psychology* 7:1–53.

Barlow, H. B. (1990) "What Does the Brain See? How Does It Understand?" in H. Barlow, C. Blakemore, and M. Weston-Smith (eds.), *Images and Understanding*. Cambridge: Cambridge University Press.

Barlow, H. B., Blakemore, C. and Pettigrew, J. D. (1967) "The Neural Mechanism of Binocular Depth Discrimination," *Journal of Physiology* 193:327–342.

Bigelow, J. and Pargetter, R. (1987) "Functions," *Journal of Philosophy* 84:181–196.

Bishop, P. O. and Pettigrew, J. D. (1986) "Neural Mechanisms of Binocular Vision," *Vision Research* 26:1587–1600.

Block, N. (1980) "Introduction: What Is Functionalism," in N. Block (ed.), *Readings in the Philosophy of Psychology*, 2 vols., vol. 1. Cambridge, MA: Harvard University Press.

Boorse, C. (1976) "Wright On Functions," *Philosophical Review* 85:70–86.

Churchland, P. S. (1986) *Neurophilosophy: Toward a Unified Theory of the Mind/Brain*. Cambridge, MA: The MIT Press.

Cushman, W. H. and Rosenberg, D. J. (1991) *Human Factors Design Analysis*. Amsterdam: Elsevier.

Cummins, R. (1975) "Functional Analysis," *Journal of Philosophy* 72:741–765.

Descartes, R. (1637) *La Dioptrique*. Leiden: I. Maire.

Dretske, F. (1988) *Explaining Behavior: Reasons in a World of Causes*. Cambridge, MA: The MIT Press/Bradford Books.

Finger, S. (1994) *Origins of Neuroscience*. New York: Oxford University Press.

Fodor, J. A. (1974) "Special Sciences, or The Disunity of Science as a Working Hypothesis," *Synthese* 28:97–115.

Fodor, J. A. (1980) "Methodological Solipsism Considered as a Research Strategy in Cognitive Psychology," *Behavior and Brain Sciences* 3:63–109.

Godfrey-Smith, P. (1996) *Complexity and the Function of Mind in Nature*. Cambridge: Cambridge University Press.

Hailman, J. P. (1977) "Communication by Reflected Light," in T. A. Sebeok (ed.), *How Animals Communicate.* Bloomington, IN: Indiana University Press.

Hatfield, G. (1988) "Neurophilosophy Meets Psychology: Reduction, Autonomy, and Physiological Constraints," *Cognitive Neuropsychology* 5:723–746.

Hatfield, G. (1990) *The Natural and the Normative: Theories of Spatial Perception from Kant to Helmholtz.* Cambridge, MA: The MIT Press/Bradford Books.

Hatfield, G. (1992) "Color Perception and Neural Encoding: Does Metameric Matching Entail a Loss of Information?" in D. Hull and M. Forbes (eds.), *PSA 1992,* vol. 1. East Lansing, MI: Philosophy of Science Association.

Hatfield, G. (1993) "Discussion of Dennis Proffitt, A Hierarchical Approach to Perception," in S. C. Masin (ed.), *Foundations of Perceptual Theory.* Amsterdam: Elsevier.

Hatfield, G. (in press) "Attention in Early Scientific Psychology," in R. D. Wright (ed.), *Visual Attention.* New York: Oxford University Press.

Hatfield, G. and Epstein, W. (1979) "The Sensory Core and the Medieval Foundations of Early Modern Perceptual Theory," *Isis* 70:363–384.

Helmholtz, H. (1850/1892–1895) "Messungen über den zeitlichen Verlauf der Zuckung animalischer Muskeln und die Fortpflanzungsgeschwindigkeit der Reizung in den Nerven," in Helmholtz, *Wissenschaftliche Abhandlungen,* 3 vols., 2, 764–843. Leipzig: J. Barth.

Helmholtz, H. (1866/1925) *Treatise on Physiological Optics,* trans. from the 3rd German edition, J. P. C. Southall, 3 vols. Milwaukee: Optical Society of America. The third German edition reprints the first edition of 1866, with added notes.

Hering, E. (1875) "Zur Lehre vom Lichtsinne, VI: Grundzüge einer Theorie des Farbensinnes," in *Sitzungsberichte der Kaiserlichen Akademie der Wissenschaften in Wien. Mathematisch-naturwissenschaftliche Classe,* pt. 3, vol. 70, pp. 169–204.

Hurvich, Leo. (1981) *Color Vision.* Sunderland, MA: Sinauer Associates.

Ibn al-Haytham, A. (1989) *Optics,* vol. 1, trans. A. I. Sabra. London: Warburg Institute.

Jacobs, G. H. (1981) *Comparative Color Vision.* New York: Academic Press.

Julesz, B. (1971) *Foundations of Cyclopean Perception.* Chicago-University of Chicago Press.

Kaiser, P. K., and Boynton, R. M. (1996) *Human Color Vision,* 2nd ed. Washington, D.C.: Optical Society of America.

Kandel, E. R., Schwartz, J. H. and Jessell, T. M. (1991) *Principles of Neural Science,* 3rd ed. Norwalk, Conn.: Appleton and Lange.

Kuhlenbeck, H. (1977) *Central Nervous System of Vertebrates,* Vol. 5, Pt. 1: *Derivatives of the Prosencephalon: Diencephalon and Telencephalon.* Basel: S. Karger.

Matthen, M. (1988) "Biological Functions and Perceptual Content," *Journal of Philosophy* 85:5–27.

Mitchell, D. E., Kaye, M. and Timney, B. (1979) "A Behavioral Technique for Measuring Depth Discrimination in the Cat," *Perception* 8:389–396.

Mollon, J. D. (1989) " 'Tho' She Kneel'd in that Place where They Grew ...' The Uses and Origins of Primate Colour Vision," *Journal of Experimental Biology* 146:21–38.

Neitz, J., Jacobs, G. H. (1990) "Polymorphism in Normal Human Color Vision and Its Mechanism," *Vision Research* 30:621–636.

Neitz, J., Neitz, M. and Jacobs, G. H. (1993) "More than Three Different Cone Pigments among People with Normal Color Vision," *Vision Research* 33:117–122.

Newton, I. (1704) *Optics.* London: Sam. Smith and Benj. Walford.

Olesko, K., and Holmes, F. L. (1993) "Experiment, Quantification, and Discovery: Helmholtz's Early Physiological Researches, 1843–50," in D. Cahan (ed.), *Hermann von Helmholtz and the Foundations of Nineteenth-Century Science.* Berkeley, CA: University of California Press.

Pettigrew, J. D. (1986) "Evolution of Binocular Vision," in J. D. Pettigrew, K. J. Sanderson, and W. R. Levick (eds.), *Visual Neuroscience.* Cambridge: Cambridge University Press.

Poggio, G. F. and Fischer, B. (1997) "Binocular Interaction and Depth Sensitivity in Striate and Prestriate Cortex of Behaving Rhesus Monkey," *Journal of Neurophysiology* 40:1392–1405.

Polyak, S. (1957) *The Vertebrate Visual System.* Chicago: University of Chicago Press.

Posner, M. I., and Raichle, M. E. (1995) "Précis of *Images of Mind,*" *Behavioral and Brain Sciences* 18:327–339.

Priestley, J. (1772) *History and Present State of Discoveries Relating to Vision, Light, and Colours.* London: J. Johnson.

Proffitt, D. (1993) "A Hierarchical Approach to Perception, with Discussion," in S. C. Masin (ed.), *Foundations of Perceptual Theory.* Amsterdam: Elsevier.

Ptolemy, C. (1989) *L'Optique,* ed. and trans. Albert Lejeune. New York: E. J. Brill.

Putnam, H. (1967) "The Mental Life of Some Machines," in H.-N. Castañeda (ed.), *Intentionality, Minds, and Perception.* Detroit, MI: Wayne State University Press.

Shapiro, L. A. (1992) "Darwin and Disjunction: Foraging Theory and Univocal Assignments of Content," in D. Hull, M. Forbes, and K. Okruhlik (eds.), *PSA 1992,* vol. 1. East Lansing, MI: Philosophy of Science Association.

Shapiro, L. A. (1993) "Content, Kinds, and Individualism in Marr's Theory of Vision," *Philosophical Review* 102:489–513.

Shapiro, L. A. (1994) "Behavior, ISO-Functionalism, and Psychology," *Studies in History and Philosophy of Science* 25:191–209.

Shepard, R. (1992) "The Perceptual Organization of Colors: An Adaptation to Regularities of the Terrestrial World?" in J. H. Barkow, L. Cosmides, and J. Tooby (eds.), *The Adapted Mind: Evolutionary Psychology and the Generation of Culture.* New York: Oxford University Press.

Siegel, R. E. (1970) *Galen on Sense Perception.* Basel: S. Karger.

Sober, E. (1990) "Putting the Function Back into Functionalism," in W. Lycan (ed.), *Mind and Cognition: A Reader.* Cambridge, MA: Basil Blackwell.

Stich, S. P. (1978) "Autonomous Psychology and the Belief-Desire Thesis," *Monist* 61:573–591.

Stich, S. P. (1983) *From Folk Psychology to Cognitive Science: The Case Against Belief.* Cambridge, MA: The MIT Press/Bradford Books.

Turner, R. S. (1994) *In the Mind's Eye: Vision and the Helmholtz-Hering Controversy.* Princeton, NJ: Princeton University Press.

Tychsen, L. (1992) "Binocular Vision," in W. M. Hart (ed.), *Adler's Physiology of the Eye.* St. Louis, MO: Mosby.

Tyler, C. W. (1983) "Sensory Processing of Binocular Disparity," in C. M. Schor and K. J. Ciuffreda (eds.), *Vergence Eye Movements: Basic and Clinical Aspects.* Boston, MA: Butterworths.

Tyler, C. W. (1990) "A Stereoscopic View of Visual Processing Streams," *Vision Research* 30:1877–1895.

Wallace, A. R. (1891) *Darwinism: An Exposition of the Theory of Natural Selection.* London: Macmillan.

Wright, L. (1973) "Functions," *Philosophical Review* 82:139–168.

13 Ontogeny, Phylogeny, and Scientific Development

Stephen M. Downes

Biology and psychology intersect in discussions of the development of science. Science is the result of cognitive activity, and accounting for individual cognition is the purview of psychologists. Scientific development has often been likened to an evolutionary process, or more strongly has been characterized as part of the evolutionary process. I propose that the connection between these two discussions can be illuminated by viewing them in terms of the distinction between ontogeny and phylogeny. In the case of scientific development, individual cognitive development is an ontogenetic process and the development of science is a phylogenetic process. I am not the first to have characterized the relation between individual cognitive development and the development of science in these terms, but others who have done so have invariably endorsed a particular view of the relation between ontogeny and phylogeny that is not supported by evolutionary biology. These researchers hold a recapitulationist view of the relation between individual cognitive development and scientific development, which echoes recapitulationism in nineteenth-century biology. This is a view that I will examine in some detail but whose tenability I will reject.

In what follows I will briefly define ontogeny and phylogeny, explain why they are useful concepts to introduce in the discussion of scientific development, and reject the recapitulationist view of their relation. I am interested in conceptual similarities between recapitulationism in biology and in accounts of scientific development, but to illustrate the conceptual overlap a bit of history is in order, as recapitulationism has long been rejected by biologists as an explanatory theory. I will leave a further historical investigation for another occasion: the source of recapitulationist views in developmental psychology.[1]

In the first section of the paper I briefly introduce evolutionary epistemology, as I take any evolutionary account of scientific development to constitute a kind of evolutionary epistemology. In the second section I introduce recapitulationist accounts of the relation between ontogeny and phylogeny in nineteenth-century biology. This is necessary background to making the charge of recapitulationism stick in the case of developmental psychologists accounts of scientific development, which I turn to in the third section. In the fourth section I show that the recapitulationist view of scientific development is undermined by the same problem that undermined recapitulationism in nineteenth-century biology: the lack of a shared mechanism driving ontogeny and phylogeny. I argue that accounts of scientific development should not be written off simply because they are evolutionary accounts, but they can be rejected on the grounds that they are based on a false evolutionary theory. For an evolutionary account of scientific development to be viable, it must be based on a plausible account of evolution.

Before I introduce ontogeny and phylogeny, here is a brief account of some purported relations between evolution and scientific development and my view of the relation. The attempt to understand scientific development in evolutionary terms is referred to as evolutionary epistemology.[2] Thomas Kuhn presents one account:

> The analogy that relates the evolution of organisms to the evolution of scientific ideas can easily be pushed too far.... The process [I describe] as the resolution of revolutions is the selection by conflict within the scientific community of the fittest way to practice future science. The net result of a sequence of such revolutionary selections, separated by periods of normal research, is the wonderfully adapted set of instruments we call modern scientific knowledge. Successive stages in that developmental process are marked by an increase in articulation and specialization. And the entire process may have occurred, as we now suppose biological evolution did, without benefit of a set goal, a permanent fixed scientific truth, of which each stage in the development of scientific knowledge is a better exemplar.   (Kuhn 1970, pp. 172–173)

This puts Kuhn in one camp as regards evolutionary epistemology: the development of science is analogous with evolution. There is a much stronger view, which is that the development of science is part of the evolutionary process in that it is explained by the same mechanisms as evolution. Or as Stein (1996) puts it: "biological evolution [is] the primary cause of the growth of knowledge" (p. 206). This "literal" view of evolutionary epistemology is defended by Michael Ruse (1986), among others. There is an intermediate view articulated by Karl Popper (1982) and by Daniel Dennett at times in his recent *Darwin's Dangerous Idea* (1995). Popper (e.g., 1982), who characterizes evolution very abstractly as a process of conjecture and test, takes scientific development to be continuous with evolution on the grounds that both processes share the same abstract structure. Dennett characterizes evolution algorithmically and tries to show that scientific development and cultural development in general conform to evolution construed as an algorithmic or substrate-neutral process. Whether the development of science is part of the evolutionary process or analogous to it partly hangs on one's conception of the crucial components of evolutionary explanation: the more abstractly or algorithmically one characterizes evolution and hence evolutionary explanation, the more plausible the claim that scientific development is an evolutionary process. The more causally or historically rich one's conception of evolutionary explanation, the less likely one is to view scientific development as continuous with evolution. I hold a more causally and historically rich conception of evolution than Dennett's and Popper's accounts and so I side with Kuhn, and more recently Hull (1982, 1988), in supporting an account of scientific development that draws analogies with evolution. In doing so I am wary of

Hull's general caution about defining the relevant terms: for example, what are the "genes" of scientific development? An even more acute problem than defining the relevant terms arises in the accounts of scientific development I will consider below; this is the problem of drawing one's account from a mistaken evolutionary theory.

1 The Background in Nineteenth-Century Biology

There is a striking similarity between the claims of developmental psychologists studying science and nineteenth-century biologists confronting the relations between ontogeny and phylogeny. Like their counterparts in biology, psychologists hold varying accounts of the relationship between ontogeny, or child cognitive development, and phylogeny, or the history of science. Some developmental psychologists encourage my comparison by directly invoking ontogeny-phylogeny talk (Strauss 1988), whereas others explicitly reject such a connection. The view that children's cognitive development and scientific development call for the same kind of explanation can be traced back to several psychologists in the nineteenth century including Baldwin (see Gould 1977 and Richards 1987). The claim appears in the work of Gestalt psychologists such as Koffka (Gould 1977, p. 144) and notoriously in Piaget (see, e.g., 1971). In what follows I restrict my discussion to the work of contemporary developmental psychologists and will demonstrate the parallels between discussions in nineteenth-century biology and contemporary developmental psychology. My interest is in the plausibility of the psychologists' accounts of scientific development, and I will argue that recapitulationism is a nonstarter as an evolutionary epistemology.

Some nineteenth-century biologists believed that ontogeny recapitulates phylogeny: that an individual organism passes through stages of development that represent the adult stages of its evolutionary ancestors. Here I am relying on fairly standard definitions of ontogeny and phylogeny: ontogeny is the development of an individual organism from a fertilized egg throughout its life and phylogeny is the evolutionary history of a species usually represented as a sequence of adult stages.

Haeckel's biogenetic law, or law of the history of evolution, states that "ontogeny is the short and rapid recapitulation of phylogeny.... During its own rapid development ... an individual repeats the most important changes in form evolved by its ancestors during their long and slow paleontological development" (quoted in Gould 1977, p. 77). If the correlate of phylogeny is the conceptual development of science and ontogeny the child's conceptual development, then, on a strict recapitulationist view, during development, each child would have to acquire all the important concepts in the history of science and reject all those found false. Prior to the development of Haeckel's theory, Von Baer had already pointed out that in the biological context this strict a notion of recapitulation made no sense, because, for example, there were stages missing in ontogeny present as adult forms in the phylogeny of many organisms. But Haeckel's use of the term "important" revealed that he could give a little here. He acknowledged that various stages could be missing from ontogeny and yet his law still apply. Haeckel's view was that ontogeny and phylogeny were to be explained by the same causes or mechanisms (Gould 1997, p. 81). It is this component of his view that is crucial when comparing nineteenth-century biology with developmental psychologists' accounts of science. In his revealing book *Ontogeny and Phylogeny* Gould (1977) explains that Haeckel understands ontogeny and phylogeny as "two steps in a continuum of developmental processes in nature" (p. 80). The developmental psychology literature is filled with references to child development and development in science being the same process or begging for the same explanation. These are the kinds of claims that I am interested in here.

It is worth looking in just a little more detail at the range of recapitulationist views in nineteenth-century biology. On the face of it Haeckel's account is the most straightforward, always being at least caricatured as the claim that ontogeny exactly recapitulates phylogeny. But as we saw in the last paragraph, probably the most important component of Haeckel's view was his idea that ontogeny and phylogeny shared the same mechanism. We can see the frustrating explanatory gap that Haeckel was confronted with: what accounts for development in each generation, and does it also account for the preservation of traits across generations? This problem was confronted by all nineteenth-century biologists, including Darwin. And for all of them embryology provided some or other kind of insight into phylogeny.

In Haeckel's recapitulationist theory each adult stage in phylogeny is represented in ontogeny (see figure 13.1). The adult stages in phylogeny arise by addition at the end of ontogenic development. This account appeals to a principle: the principle of terminal addition. Strict recapitulation by terminal addition presents many problems, but several were confronted by Haeckel and his followers. The first was that with the continual addition of new stages, ontogeny becomes impossibly long. The solution was to propose a second principle, the principle of condensation, which stated that the length of ontogeny must be continually shortened during the subsequent evolution of the lineage. The principle worked either by deletion or acceleration (see figure 13.2). The principle of condensation provided a convenient explanation of missing phylogenetic stages in ontogeny. On this view, recapitulation need not be strict recapitulation, and the discovery of absent stages was not considered empirical evidence against the view.

On Haeckel's view, the relevant stages were whole organism stages, and one of the problems this presented for the view was dubbed "heterochrony" by Haeckel. This is

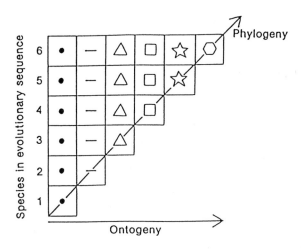

Figure 13.1
Strict recapitulationism (adapted from Gould 1977).

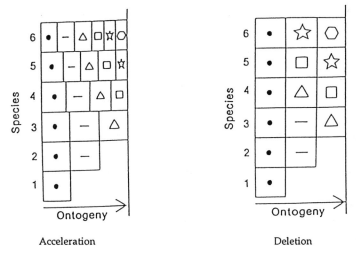

Acceleration   Deletion

Figure 13.2
Condensation in recapitulation (adapted from Gould 1977).

the temporal disruption of recapitulation. Haeckel's oft-repeated example was the early arrival of the heart in mammalian embryology. The whole organism in phylogeny that represents the stage at which the heart appears in ontogeny has no such organ. A different account of the relation between ontogeny and phylogeny arose partly to deal with this problem.

Cope's theory was that the relevant stages are organs, not organisms. Cope invoked a law of parallelism: retrogressive evolution arises by subtraction of terminal stages (for example the adaptation of larvae) and progressive evolution by recapitulation. But this law applied to organs. Cope then introduced a strange account of identity into his notion of recapitulation. A shark is identical to one embryonic stage of a human, on this view, because at the relevant stage the embryo has a heart, a certain configuration of arteries, and cartilage. Different organs accelerate at different rates according to Cope, and so what looked like an exception to Haeckel's view is actually an example of recapitulation on Cope's view (Richardson and Cane 1988).

The main alternative to these various forms of recapitulationism was an account of the relation of ontogeny and phylogeny derived from von Baer. This was the view that Darwin championed but which according to Gould was disregarded for most of the late nineteenth century. The version of von Baer's view that Darwin held was that information about phylogeny was revealed in ontogeny, not because of recapitulated forms, but rather because all species started out with an undifferentiated embryonic form, to which further stages were added higher up the phylogenetic tree. According to Gould, the difference between Darwin's view and the recapitulationists' views was at the level of their differing conceptions of evolution (Gould 1977 p. 73). Darwin's understanding of ontogeny invoked a form of evolutionary conservatism: animals repeat ancestral embryonic stages. In contrast, the recapitulationists believed that the human embryo with gill slits was identical to an adult fish, and the embryo with a heart and cartilage was identical to an adult shark. The problem for the recapitulationists was how these stages arose. As Gould says "If . . . the tiny human fetus with gill slits is (in essence) an adult fish, then we must seek a active mechanism to 'push' the adult shapes of the ancestors into early embryonic stages of descendants" (p. 73). Haeckel's answer, as we have seen, was that ontogeny and phylogeny were driven by the same mechanism, but Haeckel never adequately spelled out quite what that mechanism was. A large part of spelling out a possible mechanism resorted to Lamarckian evolutionary theory. In fact Haeckel's theory of evolution owed at least as much to Lamarck as to Darwin.

What I hope to have illustrated in this section is that there was a variety of recapitulationist views. To talk about ontogeny recapitulating phylogeny need not mean strict recapitulation of every adult stage of phylogeny in ontogeny. What was missing from these accounts was a mechanism to account for the appearance of the relevant stages. What I now go on to discuss further is the variety of views of the relation between child development and scientific development, all of which have a recapitulationist ring to them.

Recapitulationist Accounts of Scientific Development

A significant number of contemporary philosophers and historians of science and developmental psychologists propose that children's cognitive development recapitulates cognitive development in the history of science. Here I focus on the views of developmental psychologists. First, let us look at Piaget, whose work provides a springboard for most contemporary developmental psychology. He presents several implicit views about the relation between children's development and the history of science. He says, for example, that

there is a parallelism between the progress made in the logical and rational organization of knowledge and the corresponding formative psychological processes. With this hypothesis, the most fruitful, most obvious field of study would be the reconstituting of human history—the history of human thinking in pre-historic man. Unfortunately, we are not very well informed in the psychology of primitive man, but there are children all around us, and it is in studying children that we have the best chance of studying the development of logical knowledge, mathematical knowledge, physical knowledge, and so forth. (Quoted in Gould 1977, pp. 144–145)

There are hints of a stronger relation than mere parallel development here. The study of children will apparently adequately substitute for the study of the development of scientific knowledge. Gould's view is that Piaget does not think that phylogeny determines ontogeny in the context we are considering, but Piaget is notoriously ambiguous. He certainly maintains that there is one constraint underlying both child development and scientific development, namely, the structure of the human mind. In some of his work this sounds more like the claim that both types of development involve the same process.

Several contemporary cognitive psychologists hold versions of a recapitulationist view. A recent issue of *Philosophy of Science* featured a presentation and discussion of one such researcher: Alison Gopnik (1996). Gopnik has carried out numerous studies of children's conceptual development and in one article (1988) reflects on the development of object permanence in infants. This is the development of our notion of objects as enduring despite being moved from place to place or being viewed from different perspectives. During this discussion she claims that "there are a number of significant structural similarities between the development of children's knowledge about the world and the development of scientific theories" (p. 197). She then goes on to make the stronger but specific claim that "in the period between 6 months and 18 months we can see all the classical epistemological processes that are involved in theory-formation and change" (p. 209). And, finally, she makes the even stronger and more general claim:

we should think, as Piaget did, of the development of knowledge as a single process beginning in infancy and continuing in its most advanced form in institutional science. Science on this view is best understood as a psychological phenomenon continuous with other phenomena, rather than as an abstract, logical, truth-finding system. (p. 211)

On this view child development and the development of science are one process.

Susan Carey, who has also done numerous studies of child development, proposes a "continuity assumption" (Carey 1991), which is that science is continuous with common sense. The assumption is spelled out in the claims about the particular kinds of conceptual change that occur both in normal conceptual growth and in the development of science (p. 91). The common process is conceptual change. This process occurs in several forms, but all these forms are common to children's conceptual development and the history of science.

One interesting area of research in developmental psychology is the investigation of children's and naive students' reasoning about motion (e.g., McCloskey and Kargon 1988). The naiveté at issue here is naiveté about physics. Here the psychologists' claims are in line with a version of the recapitulationist view, because the work can be understood as the discovery of an early scientific approach to reasoning about motion in children and naive adults. Specifically, children and naive adults have concepts of motion that are closest to medieval impetus theorists. According to investigators, the best way to explain children's responses to various questions about motion is to assume that they have an impetus-type theory (McCloskey and Kargon 1988; Chi 1992). You can think of the discovery of children's use of an impetus theory as analogous to the discovery of gill slits in mammalian embryos. In the nineteenth century, much theorizing about evolution was driven by empirical findings in embryology. In research such as McCloskey's we see empirical findings about conceptual development in children informing theorizing about conceptual development more generally construed.

The developmental psychology literature is full of accounts of the relations between child development and the development of science. All that I want to establish here is that there is a range of views about these relations. This range of views is similar in scope to the range of views about the relation between ontogeny and phylogeny in nineteenth-century biology. What is common to both these discussions of development is that the focus of the researchers is on a common mechanism of development shared by two different developmental processes. Providing an account of this mechanism proved impossible for Haeckel and his contemporaries, and I will now show that the common mechanism underlying child development and scientific development is equally elusive.

2   Problems with the Recapitulationist Account of Scientific Development

Closer examination reveals that few psychologists working on scientific development argue explicitly for a Haeckel-style strict recapitulation. Philip Kitcher provides this diagnosis in a sympathetic review of the literature:

> ... in urging connections between conceptual change in childhood and conceptual change in science I'm not declaring that ontogeny recapitulates phylogeny. There is no suggestion that children have to work through the concepts of Greek science, Medieval science, Renaissance science, and so forth, before triumphantly emerging with our adult conceptual scheme. Rather the claim is that some of the transitions that occur in a child's conceptual development involve similar patterns and processes to those found in the history of science. (1988, p. 217)

Kitcher may be accurately representing his views about the relevant relation, but he is not representing the view of all developmental psychologists. As we have seen there is a wide range of views in the psychology literature most of which are at least as strong as the view Kitcher proposes and all of which I think can be fruitfully compared to recapitulationism in nineteenth-century biology.

One weak claim that developmental psychologists often make is that the study of child development will lead to a better understanding of scientific development. This claim needs some backing up. We usually derive our assumptions that the investigation of one process will shed light on another one from a further theoretical claim about shared processes. Think of the attempts by archaeologists to understand prehistoric cultures through ethno-archaeology (the investigation of existing cultures with practices presumed to be similar to those of our ancestors). What exactly is common to child development and scientific development is not made clear in the literature. Several different claims are made. A crucial question is whether the recapitulationist approach to scientific development rests on postulating shared mechanisms, shared patterns, or simply analogies. Placing the following quotes from Gopnik together illustrates the difficulty:

> There are a number of significant structural similarities between the development of children's knowledge about the world and the development of scientific theories.

[I]n the period between 6 months and 18 months we can see all the classical epistemological processes that are involved in theory-formation and change.

We should think, as Piaget did, of the development of knowledge as a single process beginning in infancy and continuing in its most advanced form in institutional science. Science on this view is best understood as a psychological phenomenon continuous with other phenomena, rather than as an abstract, logical, truth-finding system.

... there are powerful similarities between cognitive development in children and scientific theory change. These similarities are best explained by postulating an underlying abstract set of rules and representations that underwrite both types of cognitive abilities. In fact, science may be successful largely because it exploits powerful and flexible cognitive devices that were designed by evolution to facilitate learning in young children.   (1996, p. 485)

Looking at the similarities between conceptual change in children and in science may yield evidence of a common natural mechanism.   (1996, p. 493)

... investigations of the dynamic aspects of theories in developmental psychology may ultimately have the most to contribute to our understanding of science. Theory change proceeds more uniformily and quickly in children than in scientists, and so is considerably easier to observe, and we can even experimentally determine what kinds of evidence lead to change. In children, we may actually be able to see the "logic of discovery" in action.   (1996, p. 509)

Gopnik presents the relation between children's cognitive development and scientific development as resulting from a set of structural similarities; the same epistemological processes; being two stages in one process; the same set of underlying abstract rules; or a common natural mechanism. These are by no means the same relations. The weakest are the relations of similarity, next are those cashed out in terms of structural similarities or underlying rules, and strongest are those that depend on a shared natural mechanism. Relations of mere similarity are cheap and without a more detailed story do not tell us much. Relations in terms of abstract rules or structural similarities are more interesting, if sustainable. But a very abstract account of evolution, such as Popper's conjecture and test account, renders many prima facie unrelated processes intimates. Relations determined by a shared underlying natural mechanism are easy to come by in evolutionary biology, but there is no evidence so far that the same natural mechanism underlies children's cognitive development and the development of science.

  If there is any consensus among psychologists that a mechanism underlies both processes, a candidate for such a mechanism is "radical conceptual change." Let us review one of the difficulties with this claim. If conceptual change is a process analogous to change in morphology, cognitive psychologists are confronted with problems that appear more intractable than those that confronted Haeckel. Take an example such as the conceptual differentiation between heat and temperature, which constitutes a conceptual change: the move from the possession of the concept of heat alone to the possession of the concepts of both heat and temperature. In what way could this be the same process in the history of science as it is for an individual growing up and learning about heat and temperature? The same psychological process could conceivably occur in a child and an adult, depending on their education and opportunity to learn to apply various concepts (here I mean a different child and adult). It could hardly be the case that the same process would take place in the same individual, unless she suffered terrible amnesia and had to reacquire all her concepts, and the method of reacquisition involved reliving her early conceptual development. So an individual child and scientist could go through the same conceptual process. This may be what McCloskey has in mind when he says that individual scientists and children confront the same stage in theory formation. But if this is the recapitulationists' claim, it is not a claim about conceptual development in science; rather it is a claim about conceptual development in individual scientists.

The general problem is that conceptual change in some scientists may be a necessary condition for conceptual change in science, but the two types of conceptual change are not the same. We can look at this as a confusion between ontogeny and phylogeny: The development of individual scientists may be a continuous process from infancy to maturity, but the development of science seems to involve different processes. When we consider conceptual change in science we usually think of the time from the introduction to the entrenchment of a concept. Such conceptual change involves at least the adoption of the concept by a cross section of the scientific community. For some of those individuals this may involve conceptual acquisition that does not require radical conceptual change. Graduate students or émigrés from an unrelated scientific field may simply acquire the new concept and forge ahead with its application in practice. More can be said here, but so far there is a compelling case that scientific conceptual change and conceptual change in child development are not the same process. The benefit of hindsight and a better theory makes it easy for us to see the falsehood of Haeckel's recapitulationist claims. What I have revealed is that developmental psychologists' recapitulationist claims are on equally shaky ground.

3   Conclusions

We can now see that developmental psychologists' claims about the relation between child development and scientific development are not supported. I used the analogy with accounts of ontogeny and phylogeny in nineteenth-century biology to illustrate some of the problems. So far I can see no evidence for processes shared by the two types of conceptual development or for a mechanism that would make their relation attributable to anything more than a shared pattern. This should come as no surprise, because the problem of the relation between ontogeny and phylogeny in biology, to the extent that it has been addressed, is not fruitfully illuminated by recapitulationism. If we are looking for an account of scientific development that is analogous with evolution, then we ought to rely on the best explanations from current biology.

To reject the developmental psychologists' recapitulationist thesis is not, however, to reject the claim that scientific development is analogous to some kind of evolutionary process. Thomas Kuhn and many others in evolutionary epistemology have made this suggestion, and while I agree with it in spirit, it is important to stress just exactly what kind of evolutionary process is envisaged. Many people agree on the minimal position that scientific development, just as biological evolution, does not converge on any fixed end point and involves variation and a form of selection. But as many philosophers and biologists after Kuhn have illustrated, filling in the details of this story is difficult. The initial insight of the developmental psychologists, that an adequate account of scientific development must include an account of the role of individual scientists' cognitive processes, is on the right track. How this ontogenetic process is related to the phylogenetic process of scientific development is not yet clear. Articulating the relation is one of the challenges facing most varieties of evolutionary epistemology. What I have established here is that pursuing recapitulationism to address these issues is a bad idea, just as it was when applied to analogous issues in nineteenth-century biology.

Acknowledgments

This essay is part of a larger project on ontogeny and phylogeny. The research is sponsored by the Obert C. and Grace A. Tanner Humanities Center at the University of Utah. Don Garrett made helpful comments on an earlier draft of this essay.

Notes

1. Richards (1987) provides some good leads here. A sketch of the trajectory is from Baldwin to Piaget to Kuhn.

2. There are numerous discussions of evolutionary epistemology in the literature and many different views. A good survey of the relevant issues is presented in Plotkin (1982). Stein (1996) also provides a good introduction to the field.

References

Carey, S. (1991) "Knowledge Acquisition: Enrichment or Conceptual Change?" in Carey and Gelman (eds.), op. cit.

Carey, S. (1992) "The Origin and Evolution of Everyday Concepts," in Giere (ed.), op. cit.

Carey, S. and Gelman, R. (eds.) (1991) *The Epigenesis of Mind*. Hillsdale: Lawrence Erlbaum Associates.

Chi, M. (1992) "Conceptual Change within and across Ontological Categories: Examples from Learning and Discovery in Science," in Giere (ed.), op. cit.

Dennett, D. (1995) *Darwin's Dangerous Idea*. New York: Simon and Schuster.

Giere, R. (ed.) (1992) *Cognitive Models of Science*. Minneapolis: University of Minnesota Press.

Gopnik, A. (1988) "Conceptual and Semantic Development as Theory Change: The Case of Object Permanence," *Mind and Language* 3(3):197–216.

Gopnik, A. (1996) "The Scientist as Child," *Philosophy of Science* 63(4):485–514.

Gould, S. J. (1977) *Ontogeny and Phylogeny*. Cambridge, MA: Harvard University Press.

Hull, D. L. (1982) "The Naked Meme," in Plotkin (ed.), op. cit.

Hull, D. L. (1988) *Science as a Process*. Chicago: University of Chicago Press.

Kitcher, P. (1988) "The Child as Parent of the Scientist," *Mind and Language* 3(3):217–228.

Kuhn, T. (1970) *The Structure of Scientific Revolutions*. Chicago: University of Chicago Press.

Little, D. (1991) *Varieties of Social Explanation*. Boulder, CO: Westview Press.

McCloskey, M. and Kargon, R. (1988) "The Meaning and Use of Historical Models in the Study of Intuitive Physics," in Strauss (ed.), op. cit.

Nitecki, M. H. (ed.) (1988) *Evolutionary Progress*. Chicago: University of Chicago Press.

Piaget, J. (1971) *Biology and Knowledge*. Chicago: University of Chicago Press.

Plotkin, H. C. (ed.) (1982) *Learning, Development and Culture*. London: John Wiley and Sons.

Popper, K. (1982) "Of Clouds and Clocks," in Plotkin (ed.), op. cit.

Richards, R. J. (1987) *Darwin and the Emergence of Evolutionary Theories of Mind and Behavior*. Chicago: University of Chicago Press.

Richardson, R. C. and Cane, T. C. (1988) "Orthogenesis and Evolution in the Nineteenth Century," in Nitecki (ed.), op. cit.

Ruse, M. (1986) *Taking Darwin Seriously*. Oxford: Blackwell.

Stein, E. (1996) *Without Good Reason*. Oxford: Clarendon Press.

Strauss, S. (ed.) (1988) *Ontogeny, Phylogeny, and Historical Development*. Stanford, CT: Ablex.

14 Supple Laws in Psychology and Biology

Mark A. Bedau

The nature and status of psychological laws is a long-standing controversy. I will argue that part of the controversy stems from the distinctive nature of an important subset of those laws, what I'll call "supple laws." After reviewing the failures of the two evident strategies for understanding supple laws, I'll turn for inspiration to analogously supple laws in biology. For an emergent-model strategy taken by the new interdisciplinary field of artificial life provides a strikingly successful understanding of supple laws in biology. I'll conclude by inferring what an emergent model of supple laws in psychology should be like.

1 Supple Laws in Psychology

It has long been noticed that the regularities and patterns in our mental lives—what I'll call (without prejudging any questions) "psychological laws"—need ceteris paribus qualifications, that is, qualifications to the effect that the law holds only provided "everything else is equal." Two typical examples, though extremely simplified, clearly illustrate this phenomenon:

Pure Reason: If A believes P and A believes that P implies Q, then ceteris paribus A will infer Q.

Practical Reason: If A wants G and A believes that M will produce G, then ceteris paribus A will do M.

In general, all psychological laws stand in need of similar ceteris paribus qualifications.

A variety of factors bring about the need for ceteris paribus qualifications in psychological laws. People sometimes fail to infer what is implied by their antecedent beliefs because of inattention or illogic, but some exceptions to the law of Pure Reason reflect attentive logical acumen at its best. For example, if agent A who believes proposition P and believes that P implies proposition Q has good antecedent reason to *doubt* Q, A might conclude that it is more reasonable to question P or to question whether P entails Q. Similarly, some exceptions to the law of Practical Reason are due to confusion or weakness of will, but others reflect an apt balance of priorities. To use an example from Horgan and Tienson (1989), if some agent A wants a beer and believes that there is one in the kitchen, then A will go get one—unless, as the ceteris paribus clause signals, A does not want to miss any of the conversation, or A does not want to offend the speaker by leaving in mid-sentence, or A does not want to drink beer in front of his mother-in-law, or *A* thinks he should, instead, flee the house since it is on fire, etc.

The status and source of psychological ceteris paribus laws are quite controversial (see, e.g., Hempel 1965; Dennett 1971, Putnam 1973, 1975; Fodor 1981, 1991; Cartwright 1983, 1995; Horgan and Tienson 1989, 1990, 1996; Schiffer 1991; Dreyfus 1992). (I should note that I'm concerned with the controversies about ceteris paribus laws that *describe* human psychology, not the controversies discussed by Dennett (1984) about how human intelligence *employs* ceteris paribus reasoning.) Misgivings about the status of psychological ceteris paribus laws are quite varied, ranging from worries about whether they are trivially or logically or analytically true, to worries about whether they are falsifiable or usable in scientific explanations, to worries about whether they can be precisely specified, even in principle, in some algorithm. A similarly bewildering variety characterizes the alleged source of the ceteris paribus qualifications; here we find reference to idealizations (Cartwright 1983), implementation-level malfunctions (Dennett 1971; Putnam 1973, 1975; Fodor 1981; Cartwright 1995), the subtle and complex nature of human cognition (Horgan and Tienson 1989, 1990, 1996), as well as "a background of practices which are the condition of the possibility of all rulelike activity" (Dreyfus 1992, p. 57).

A common thread through the bulk of this diversity of opinion is the assumption that ceteris paribus laws have exceptions when they "go wrong." Now, in a trivial sense this is right—an exception, after all, is an exception—but in a deeper sense this is not right. For note that exceptions to psychological ceteris paribus laws fall into two quite different groups, what I'll call "rule-breaking" and "rule-proving" exceptions. Consider the law of Practical Reason: let *A*, *G*, and *M* be particular agents, goals, and actions, and assume that *A* wants *G* and believes that *M* will produce *G*. A *rule-breaking exception* happens when *A* nevertheless performs some other action *M** even though *M* is the most reasonable or sensible thing for *A* to do, given the whole constellation of *A*'s beliefs, desires, capacities, etc. By contrast, a rule-proving exception happens when *A* performs *M** because in this particular situation, given the constellation of the rest of *A*'s beliefs, desires, capacities, etc., *M** is the *more* reasonable or sensible thing for *A* to do. One could say that whereas rule-breaking exceptions involve the "wrong" thing happening in context, rule-proving exceptions involve the "right" thing happening. An exception that proves the rule is appropriate in the context since it achieves the agent's underlying goals better than slavishly following the rule would have and, furthermore, the exception happens *because* it is appropriate in this way.

Those ceteris paribus laws that have exceptions that prove the rule I will call "supple." All ceteris paribus laws are vague in that they describe regularities that hold only for the most part but without delineating what conditions give rise to exceptions. The distinctive feature of supple ceteris paribus laws is that their vagueness has a special source—a certain kind of underlying regularity that explains the supple law. Supple laws have three defining features. The first has to do with how the supple law manifests a deeper, context-sensitive regularity. In *typical* contexts, the underlying regularity is manifested in the pattern of behavior described by the supple law. But in other contexts the same underlying regularity generates exceptions to the supple law. These exceptions "prove the rule" because they reveal the underlying regularity behind the supple law, they indicate the true "meaning" of the supple law.

The second defining feature of supple laws is that their "meaning" is teleological and derives from the telic nature of the underlying regularity. It's not that the underlying regularity has a purpose but that it *describes* the way in which some purpose or function is achieved. The supple law describes how that purpose is achieved in typical contexts and rule-proving exceptions arise in contexts in which some other means better achieves the same purpose. The teleology in supple laws can be mental but it can also be merely biological, as examples below show. (Further details of my preferred understanding of teleology are developed in Bedau 1990, 1991, 1992a, 1992b, 1992c; Bedau and Packard 1992.) Hofstadter (1985) describes mental regularities as "fluid" and Horgan and Tienson (1990) talk of "soft" laws of intentional psychology. But the teleology in supple laws makes them not just "fluid" or "soft" but *aptly* so. The fluidity or softness of supple laws involves the open-ended context-sensitivity with which some purpose of function is achieved.

The third defining feature is exactly this open-ended context-sensitivity of the underlying regularity. For one thing, the purpose in question is achieved in an indefinite number of different ways in an indefinite number of different contexts. But more than this, there may be no rule or algorithm for determining how to achieve the purpose given an arbitrary context; in general, nothing short of trial and error will suffice. Nevertheless, what makes a supple law *supple* is that, in an indefinite number of different contexts, in one way or another, the purpose captured by the underlying regularity *is* achieved. In other words, a law is supple only if the law actually has rule-proving exceptions in (enough of) those contexts in which slavishly following the rule would defeat the purpose in question.

Pure Reason and Practical Reason, our two sample psychological ceteris paribus laws, can plausibly be seen as supple laws, for each can be seen as the manifestation of a deeper regularity that concerns how some specific purpose is achieved in an indefinitely open-ended variety of contexts. Consider Pure Reason first. I assume that it's no accident that people tend to infer the consequences of their antecedent beliefs. Presumably, the purpose served by this process is something like having one's beliefs reflect reality as accurately as possible. (I don't mean that individual agents have this as the conscious intention for fixing their beliefs, of course; moreover I'm not wedded to this particular view of the purpose of forming beliefs.) In addition, underlying the law of Pure Reason is a regularity about reasonable belief formation: ceteris paribus, people infer whatever is most reasonable given what else they believe. In typical contexts, then, when some person A believes both some proposition P and that P implies another proposition Q, then A will infer Q *because* this is the most reasonable inference to make in the context. In this way, the law of Pure Reason is a manifestation of the deeper regularity about reasonable beliefs. Furthermore, the purpose behind the law of Pure Reason derives from, and is the same as, the purpose behind the reasonable belief regularity. Finally, in those contexts in which it is not most reasonable to infer Q, the reasonable belief regularity will be manifested in some other way, such as inferring that P is false or that P does not imply Q. This will constitute an exception to the law of Pure Reason but one that proves the rule.

Similarly, underlying the law of Practical Reason is a regularity about reasonable action: ceteris paribus, people act in whatever way is most reasonable given their beliefs and desires. In addition, the purpose behind the law of Practical Reason derives from, and is identical to, the purpose captured by the reasonable action regularity, which is presumably something like acting so as to best serve one's needs and desires. The reasonable action regularity manifests itself in typical contexts as the law of Practical Reason, but in other contexts the most reasonable action might break the law of Practical Reason.

Note that ceteris paribus clauses appear in the statements of the reasonable belief regularity and the reasonable action regularities, even though these are the regularities that explain the suppleness of Pure and Practical Reason. This should tip us off that the ceteris paribus clauses in psychological laws cover two quite different kinds of exceptions. Rule-breaking exceptions can still remain after all rule-proving exceptions have been removed. As I mentioned above, not all exceptions to supple laws are due to their suppleness. Suppleness is an *aspect* of some ceteris paribus laws but it is not the *full* explanation of ceteris paribus qualifications in any law. Since there are at least two quite different kinds of reasons why ceteris can fail to be paribus there is no such thing as *the* analysis of ceteris paribus laws.

I have been arguing that the suppleness of psychological laws reflects an open-ended dynamic in a mental system's appropriate adaptation to novel contextual changes. In fact, this supple adaptability in psychological processes is a hallmark of their intelligence. Descartes put his finger on precisely this sign of true intelligence in Part V of the *Discourse on Method* when he described the difference between human rationality and the behavior of mere machines:

... although [mere machines] perform many tasks very well or perhaps can do them better than any of us, they inevitably fail in other tasks; by this means one would discover that they do not act through knowledge, but only through the disposition of their organs. For while reason is a universal instrument that can be of help in all sorts of circumstances, these organs require a particular disposition for each particular action; consequently, it is morally impossible for there to be enough different devices in a machine to make it act in all of life's situations in the same way as our reason makes us act.   (1980, p. 30)

Descartes does not just point out that a hallmark of rational creatures is an open-ended flexibility in their ability to act appropriately as contexts change. He also claims that no mere machine could exhibit this suppleness of behavior. I'll argue in a moment that he is both right and wrong about this: right because no *fixed* machine can be supple, and wrong because a suitably *changing* mechanism can be supple. But I wholeheartedly agree with Descartes that suppleness is central sign of genuine intelligence. Ceteris paribus laws are often treated as an embarrassing curiosity, to be ignored or excused. The lesson to be learned from Descartes is that attempts to properly describe and explain supple psychological laws should be celebrated and made a central focus of psychology and the philosophy of mind. A good description and explanation of a supple ceteris paribus law of psychology should have certain virtues. First, it should be precise, specifying when ceteris is not paribus, at least for those exceptions that prove the rule. Second it should be accurate and complete, in the sense that it has no false positives (cases falsely advertised as rule-proving exceptions) and no false negatives (cases falsely advertised as law-conforming). Third, it should be principled, not arbitrary or ad hoc; it should indicate what underlying regularity unifies the law and its rule-proving exceptions. Finally, it should be feasible, consistent with what we know about the natural world, thus allowing us to test empirically the account's accuracy and completeness.

There are two evident strategies for describing and explaining supple psychological laws, and neither is very good. The first strategy—what I'll call the commonsense strategy—engages in hand waving by employing ceteris paribus clauses within the account. One version of the commonsense strategy takes the supple law as a brute fact, thus sacrificing a principled explanation of the supple law. Another variant of this strategy avoids this problem by appealing directly to an underlying regularity, such as the reasonable belief and reasonable action regularities I described above.

The commonsense strategy can achieve a sort of descriptive adequacy, as far as it goes, for its descriptions appeal to the supple law or its underlying regularity and these do obtain. But the strategy lacks all the virtues sought in an account of suppleness. First, no serious attempt is made to indicate under what conditions rule-proving exceptions occur. Appealing to an underlying regularity does provide a minimal explanation of what causes rule-proving exceptions. But if the underlying regularities are like those I sketched above (e.g., ceteris paribus, people infer whatever is most reasonable given what else they believe), they will invoke vague phrases like "whatever is most reasonable" or "as appropriate" and thus will not precisely indicate when to expect a rule-proving exception. This leads to the other flaws with the commonsense strategy. For one thing, a commonsense description and explanation will be too imprecise to be accurate or complete. It might not be able to be convicted of false advertising, but that is because it makes no precise and testable advertisements. Furthermore, its imprecision blocks us from testing whether it fits with our understanding of the natural world, so the commonsense strategy is infeasible.

An alternative strategy for accounting for the suppleness of mental processes is, in effect, to predigest circumstances that give rise to rule-proving exceptions and then specify (either explicitly or through heuristics) how and when they arise in a manner that is precise enough to be expressed as an algorithm. This is exactly the strategy followed by so-called expert systems in artificial intelligence. This expert-systems strategy yields models of supple psychological laws that are explicit and precise enough to be implemented as a computer model, which gives the strategy three important virtues. First, it obviously is precise, since an expert system will indicate precisely when ceteris is not paribus. Second, an expert-systems account is principled, provided the experts' information is. Third, an expert-systems account is feasible (barring some problem with the algorithm), so the account's dynamic behavior can be directly observed and tested for plausibility. There are well-known problems with expert systems, though. (See, e.g., Dreyfus 1992; Hofstadter 1985; Holland 1986; Langton 1989a; Horgan and Tienson 1989, 1990; Chalmers, French, and Hofstadter 1992.) They sometimes work well in precisely circumscribed domains, but they systematically fail to produce the kind of supple behavior that is characteristic of intelligent response to an open-ended variety of circumstances. Their behavior is brittle—they lack the context-sensitivity that is distinctive of intelligence—so they are inaccurate and incomplete. This brittleness shows no evidence of being merely a limitation with present expert systems; attempts to improve matters by amplifying the knowledge base only invite combinatorial explosion. Although precise, feasible, and principled, their brittleness makes expert-systems accounts of supple mental dynamics inaccurate and incomplete. Our experience with expert systems suggest that Descartes was right that the aptness of supple psychological processes is too open-ended to be embodied in any fixed mechanism or algorithm.

I want to promote a third strategy for describing and explaining supple psychological laws. I'm not in a position today to give a concrete illustration of this strategy, so I will do the next best thing and give a concrete illustration of an analogous strategy in an analogous context: emergent artificial life models of supple biological laws. I'll call this approach the emergent-model strategy since its leading idea is that supple macrolevel laws emerge implicitly from an evolving population of explicitly interacting microlevel entities. The result is to view the supple law as produced by a mechanism—a microlevel population of interacting agents—but a mechanism with a capacity to make adaptive changes as the environmental context alters. Emergent models provide a distinctive, indirect but constructive kind of explanation of supple phenomena. My overall conclusion will be that the emergent-model strategy is an adequate way, and the only evident adequate way, to understand supple laws, and my central argument will be an appeal to the analogy with artificial life.

2 Emergent Models of Supple Biological Laws

Evolving populations display various macrolevel patterns on an evolutionary time scale. For example, adaptive innovations that arise through genetic changes tend, ceteris paribus, to persist and spread through the population so as to maximize the population's adaptive fit with its environment. Of course, these patterns are not precise and exceptionless universal generalizations; they hold only for the most part. When their vagueness is due to context-dependent fluctuations of what is appropriate, the macrolevel evolutionary dynamics are supple in the sense intended here. These sorts of supple dynamics of adaptation result not from any explicit macrolevel control (e.g., God does not adjust allele frequencies so as to make creatures well adapted to their environment); rather, they emerge statistically from the microlevel contingencies of natural selection.

The new interdisciplinary field of artificial life is exploring a certain characteristic kind of computer model of evolutionary dynamics. These emergent models (as I'll call them) consist of a microlevel and a macrolevel. (I should stress that I am using "micro" and "macro" in a generalized sense. Microlevel entities need not be literally microscopic; individual organisms are not. "Micro" and "macro" are relative terms; an entity exists at a microlevel relative to a macrolevel population of similar microlevel entities. These levels can be nested. Relative to a population, an individual organism is a microlevel entity; but an individual organism is a macrolevel object relative to, e.g., the microlevel genetic elements that determine the organism's behavioral strategy.) Emergent models generate complex macrolevel dynamics from simple microlevel mechanisms in a characteristic way. This form of emergence arises in contexts in which there is a system, call it S, composed out of microlevel parts.

The number and identity of these parts might change over time. S has various macrolevel states (macrostates) and various microlevel states (microstates). S's microstates are the states of its parts. S's macrostates are structural properties constituted wholly out of microstates; macrostates typically are various kinds of statistical averages over microstates. Further, there is a relatively simple and implementable microlevel mechanism, call it M, which governs the evolution of S's microstates. In general, the microstate of a given part of the system at a given time is a result of the microstates of nearby parts of the system at preceding times. Given these assumptions, I will say that a macrostate P of system S with microlevel mechanism M is emergent if and only if P (of system S) can be explained from X, given complete knowledge of external conditions, but P can be predicted (with complete certainty) from M only by simulating M, even given complete knowledge of external conditions. So, we can say that a model is emergent if and only if its macrostates are emergent in the sense just defined.

Although this is not the occasion to develop and defend this concept of emergence (see Bedau 1997), I should clarify three things. First, "external conditions" are conditions affecting the system's microstates that are extraneous to the system itself and its microlevel mechanism. One external condition is the system's initial condition. If the system is open, then another external condition is the contingency of the flux of parts and states into S. If the microlevel mechanism is nondeterministic, then each nondeterministic effect is another external condition.

Second, given the system's initial condition and other external conditions, the microlevel mechanism completely determines each successive microstate of the system. The macrostate P is a structural property constituted out of the system's microstates. Thus the external conditions and the microlevel mechanism completely determine whether or not P obtains. In this specific sense, the microlevel mechanism plus the external conditions "explain" P. One must not expect too much from these explanations. For one thing, the explanation depends on the massive contingencies under the initial conditions. It is awash with accidental information about S's parts. Furthermore, the explanation might be too detailed for anyone to survey or grasp. It might even obscure a simpler, macrolevel explanation that unifies systems with different external conditions and different microlevel mechanisms. Nevertheless, since the microlevel mechanism and external conditions determine S, they explain P.

Third, in principle we can always predict S's behavior with complete certainty, for given the microlevel mechanism and external conditions we can always simulate S as accurately as we want. Thus the issue is not whether S's behavior is predictable—it is, trivially—but whether we can predict S's behavior only by simulating S. When trying to predict a system's emergent behavior, in general one has no choice but simulation. This notion of predictability only through simulation is not anthropo-centric; nor is it a product of some specifically human cognitive limitation. Even a Laplacian supercalculator would need to observe simulations to discover a system's emergent macrostates.

Norman Packard devised a simple emergent model of evolving sensory-motor agents that demonstrates how supple, macrolevel evolutionary dynamics can emerge implicitly from an explicit microlevel model (Packard 1989; Bedau and Packard 1992; Bedau, Ronneburg, and Zwick 1992; Bedau and Bahm 1994; Bedau 1994; Bedau and Seymour 1994; Bedau 1995). What motivates this model is the view that evolving life is typified by a population of agents whose continued existence depends on their sensory-motor functionality, that is, their success at using local sensory in-formation to direct their actions in such a way that they can find and process the resources they need to survive and flourish. Thus information processing and re-source processing are the two internal processes that dominate the agents' lives, and their primary goal—whether or not they know this—is to enhance their sensory—motor functionality by coordinating these internal processes. Since the requirements of sensory-motor functionality may well alter as the context of evolution changes, continued viability and vitality require that sensory-motor functionality can adapt in an open-ended, autonomous fashion. Packard's model attempts to capture an espe-cially simple form of this open-ended, autonomous evolutionary adaptation.

The model consists of a finite two-dimensional world with a periodically replen-ished resource-distribution and a population of agents. An agent's survival and reproduction are determined by the extent to which it finds enough resources to stay alive and reproduce, and an agent's ability to find resources depends on its sensory-motor functionality—that is, the way in which the agent's perception of its contingent local environment affects its behavior in that environment. An agent's sensory-motor functionality is encoded in a set of genes, and these genes can mutate when an agent reproduces. Thus, on an evolutionary time scale, the process of natu-ral selection implicitly adapts the population's sensory-motor strategies to the envi-ronment. Furthermore, the agents' actions change the environment because agents consume resources and compete with each other for space. This entails that the mixture of sensory-motor strategies in the population at a given moment is a significant compo-nent of the environment that affects the subsequent evolution of those strategies. Thus the fitness function in Packard's model—what it takes to survive and reproduce—is constantly buffeted by the contingencies of natural selection and unpredictable changes (Packard 1989).

All macrolevel evolutionary dynamics produced by this model ultimately are the result of explicit microlevel mechanisms acting on external conditions. The model starts with a population of agents with randomly chosen sensory-motor strategies, and the subsequent model dynamics explicitly control only local microlevel states: resources are locally replenished, an agent's genetically encoded sensory-motor strategy determines its local behavior, an agent's behavior in its local environment determines its internal resource level, an agent's internal resource level determines whether it survives and reproduces, and genes randomly mutate during reproduction. Each agent is autonomous in the sense that its behavior is determined solely by the environmentally sensitive dictates of its own sensory-motor strategy. On an evolutionary time scale these sensory-motor strategies are continually refashioned by the historical contingencies of natural selection. The model generates macrolevel evolutionary dynamics only as the indirect product of an unpredictably shifting agglomeration of directly controlled microlevel events (individual actions, births, deaths, mutations). The model has no provisions for explicit control of macrolevel dynamics. Moreover, macrolevel evolutionary dynamics are typically emergent in the sense that, although constituted and generated solely by the microlevel dynamics, they can be derived only through simulations.

Packard's model is not intended as a realistic simulation of some actual biological population. Rather, it is an "ideal" model, aiming to capture the key abstract principles at work in evolving systems generally. Packard's model is in effect a thought experiment—but an *emergent* thought experiment (Bedau 1998). As with the armchair thought experiments familiar to philosophers, Packard's model attempts to answer "What if *X*?" questions, but what is distinctive about emergent thought experiments is that what they reveal can be discerned only by simulation.

I will illustrate the emergent supple dynamics in Packard's model as seen in some recent work concerning the evolution of evolvability. The ability to successfully adapt depends on the availability of viable evolutionary alternatives. An appropriate quantity of alternatives can make evolution easy; too many or too few can make evolution difficult or even impossible. For example, in Packard's model, the population can evolve better sensory-motor strategies only if it can test sufficiently many sufficiently novel strategies; in short, the system needs a capacity for evolutionary "creativity." At the same time, the population's sensory-motor strategies can adapt to a given environment only if strategies that prove beneficial can persist in the gene pool; in short, the system needs a capacity for evolutionary "memory."

Perhaps the simplest mechanism that simultaneously affects both memory and creativity is the mutation rate. The lower the mutation rate, the greater the number of genetic strategies remembered from parents. At the same time, the higher the mutation rate, the greater the number of creative genetic strategies introduced with children. Successful adaptability requires that these competing demands for memory and creativity be suitably balanced. Too much mutation (not enough memory) will continually flood the population with new random strategies; too little mutation (not enough creativity) will tend to freeze the population at arbitrary strategies. Successful evolutionary adaptation requires a mutation rate suitably intermediate between these extremes. Furthermore, a suitably balanced mutation rate might not remain fixed, for the balance point could shift as the context of evolution changes. One would think, then, that any evolutionary process that could continually support evolving life must have the capacity to adapt automatically to this shifting balance of memory and creativity. So, in the context of Packard's model, it is natural to ask whether the mutation rate that governs first-order evolution could adapt appropriately by means of a second-order process of evolution. If the mutation rate can adapt in this way, then this model would yield a simple form of· the evolution of evolvability and, thus, might illuminate one of life's fundamental prerequisites.

Previous work (Bedau and Bahm 1994) with fixed mutation rates in Packard's model revealed two robust effects. The first effect was that the mutation rate governs a phase transition between genetically ordered and genetically disordered systems. When the mutation rate is too far below the phase transition, the whole gene pool tends to remain frozen at a given strategy; when the mutation rate is significantly above the phase transition, the gene pool tends to be a continually changing plethora of randomly related strategies. The phase transition itself occurs at a characteristic mutation rate. The second effect was that evolution produces maximal population fitness when mutation rates are around values just below this transition. The upshot of these two effects is that evolutionary adaptation tends to be maximized when the gene pool is "at the edge of disorder."

In light of our earlier suppositions about balancing the demands for memory and creativity, this work suggests that evolutionary memory and creativity are balanced at the edge of genetic disorder. To test this balance hypothesis, Packard's model was modified so that each agent has an additional gene encoding its personal mutation rate (Bedau and Seymour 1994). In this case, two kinds of mutation play a role when an agent reproduces: the child inherits its parents' sensory-motor genes, which mutate at a rate controlled by the parent's personal (genetically encoded) mutation rate; and the child inherits its parents' mutation rate gene, which mutates at a rate controlled by a population-wide metamutation rate. Thus first-order (sensory-motor) and second-order (mutation rate) evolution happen simultaneously. So, if the balance hypothesis is right and mutation rates at the critical transition produce optimal conditions for sensory-motor evolution because they optimally balance memory and creativity, then we would expect second-order evolution to drive mutation rates into the critical transition. It turns out that this is exactly what happens.

Examination of many simulations confirms the pattern predicted by the balance hypothesis: Second-order evolution tends to drive mutation rates to the edge of disorder, increasing population fitness in the process. If natural selection is prevented from shaping the distribution of mutation rates in the population, the mutation rates wander aimlessly due to random genetic drift. But the mutation dynamics are quite different when natural selection operates. Although the population is initialized with quite high mutation rates well into the disordered side of the spectrum, as the population becomes more fit (i.e., as it more efficiently gathers resources) the mutation rates in the population drop into the ordered side of the mutation spectrum.

If the balance hypothesis is the correct explanation of this second-order evolution of mutation rates into the critical transition, then we should be able to change the mean mutation rate by dramatically changing where memory and creativity are balanced. In fact, the mutation rate does rise and fall along with the demands for evolutionary creativity. For example, when we randomize the values of all the sensory-motor genes in the entire population so that every agent immediately forgets all the genetically stored information learned by its genetic lineage over its entire evolutionary history, the population must restart its evolutionary learning from scratch. It has no immediate need for memory (the gene pool contains no information of proven value); instead, the need for creativity is paramount. Under these conditions, we regularly observe a striking sequence of events: (a) the residual resource in the environment sharply rises, showing that the population has become much less fit; (b) immediately after the fitness drop, the mean mutation rate rises dramatically as the mutation rate distribution shifts upward; (c) by the time that the mean mutation rate has risen to its highest point the population's fitness has improved substantially; (d) the fitness levels and mutation rates eventually return to their previous equilibrium levels.

These results show that the mutation rate distribution shifts up and down as the balance hypothesis predicts. A change in the context for evolution can increase the need for rapid exploration of a wide variety of sensory-motor strategies and thus dramatically shift the balance toward the need for creativity. Then, subsequent sensory-motor evolution can reshape the context for evolution in such a way that the balance shifts back toward the need for memory.

This all provides evidence for the following supple law of second-order evolution (at least in the modified Packard model):

Edge of Disorder: Mutation rates evolve ceteris paribus to the edge of disorder.

The Edge of Disorder law is supple because it is the manifestation of a deeper regularity that concerns how some specific purpose is achieved in an indefinitely open-ended variety of contexts. The underlying regularity here is that, ceteris paribus, mutation rates evolve in such a way that evolutionary memory and creativity are optimally balanced for successful adaptability. This point at which the mutation rate balances evolutionary memory and creativity is typically at the edge of genetic disorder, but an indefinite variety of environmental contingencies can shift the point of balance. In other words, the Edge of Disorder law is vulnerable to exceptions that prove the rule. Not only are there rule-breaking exceptions in which evolutionary memory and creativity are not balanced (e.g., microlevel stochasticity, a break environment [Bedau 1994, 1996b]). Finally, the concern in the bulk of existing connectionist modeling is with equilibrium behavior that settles onto stable attractors. By contrast, partly because the microlevel entities are typically always reconstructing the environment to which they are adapting, the behavior of the emergent models I have in mind would be characterized by a continual, open-ended evolutionary dynamic that never settles onto an attractor in any interesting sense. Neuroscientists sometimes claim that macrolevel mental phenomena cannot be understood without seeing them as emerging from microlevel activity. Churchland and Sejnowski (1992), for example, argue that the brain's complexity forces us to study macrolevel mental phenomena by means of manipulating microlevel brain activity. Their position is superficially similar to the emergent models perspective, but there is an important difference between the two. For Churchland and Sejnowski, manipulating the mind's underlying microlevel activity is merely a temporary practical expedient, a means for coming to grasp the mind's macrolevel dynamics. Once the microlevel tool has illuminated the macrolevel patterns, it has outlived its usefulness and can be abandoned. No permanent, intrinsic connection binds our understanding of micro- and macrolevels. By contrast, my thesis is that the mind's macrolevel dynamics can be adequately described or explained only by making essential reference to the microlevel activity from which it emerges. The microlevel mechanism in the emergent model is a complete and compact description and explanation of the macrolevel dynamics. Since these global patterns are supple, they inevitably have rule-proving exceptions. Thus, to get a precise and detailed description of the macrolevel laws, there is no alternative to simulating the model. In this way, the microlevel model is ineliminably bound to our understanding of the emergent supple laws.

At this stage of development, the emergent-model strategy for understanding supple psychological processes raises at least as many questions as it answers. Our final judgment of it must await the time when we have concrete models to explore. But we can conclude today that this strategy shows some striking promise. This is encouraging, for the suppleness of psychological processes is at once both enigmatic and essential to the intelligence of life and mind.

Acknowledgments

For valuable discussion, thanks to Terry Horgan, Karen Neander, Norman Packard, Kim Sterelny, and audiences at my 1997 ISHPSSB presentation in Seattle, at my cognitive science seminar at the University of California at Los Angeles, and at my philosophy colloquia at the University of California at San Diego and the University of Newcastle.

References

Ackley, D., and Littman, M. (1992) "Interactions between Evolution and Learning," in C. Langton, C. Taylor, D. Farmer, and S. Rasmussen (eds.), *Artificial Life II*. Reading, MA: Addison-Wesley.

Anderson, J. A., and Rosenfeld, E. (eds.). (1988) *Neurocomputing: Foundations of Research*. Cambridge, MA: The MIT Press/Bradford Books.

Bedau, M. A. (1990) "Against Mentalism in Teleology," *American Philosophical Quarterly* 27:61–70.

Bedau, M. A. (1991) "Can Biological Teleology Be Naturalized?" *The Journal of Philosophy* 88:647–655.

Bedau, M. A. (1992a) "Where's the Good in Teleology?" *Philosophy and Phenomenological Research* 52:781–805.

Bedau, M. A. (1992b) "Goal-directed Systems and the Good," *The Monist* 75:34–49.

Bedau, M. A. (1992c) "Naturalism and Teleology," in S. J. Wagner and R. Warner (eds.), *Naturalism: A Critical Appraisal*. Notre Dame: University of Notre Dame Press.

Bedau, M. A. (1994) "The Evolution of Sensory-motor Functionality," in P. Gaussier and J.-D. Nicoud (eds.), *From Perception to Action*. Los Alamitos, CA: IEEE Computer Society Press.

Bedau, M. A. (1995) "Three Illustrations of Artificial Life's Working Hypothesis," in W. Banzhaf and F. Feckman (eds.), *Evolution and Biocomputation: Computational Models of Evolution*. Berlin: Springer.

Bedau, M. A. (1996a) "The Nature of Life," in M. Boden (ed.), *The Philosophy of Artificial Life*. New York: Oxford University Press.

Bedau, M. A. (1996b) "The Extent to which Organisms Construct Their Environments," *Adaptive Behavior* 4:476–482.

Bedau, M. A. (1997) "Weak Emergence," in J. Tomberlin (ed.), *Philosophical Perspectives: Mind, Causation and World*, Vol. 11. New York: Blackwell.

Bedau, M. A. (1998) "Philosophical Content and Method in Artificial Life," in T. W. Bynam and J. H. Moor (eds.), *The Digital Phoenix: How Computers Are Changing Philosophy*. New York: Blackwell.

Bedau, M. A. and Bahm, A. (1994) "Bifurcation Structure in Diversity Dynamics," in R. Brooks and P. Maes (eds.), *Artificial Life IV*. Cambridge, MA: The MIT Press/Bradford Books.

Bedau, M. A., and Packard, N. (1992) "Measurement of Evolutionary Activity, Teleology, and Life," in C. Langton, C. Taylor, D. Farmer, and S. Rasmussen (eds.), *Artificial Life II*. Reading, MA: Addison-Wesley.

Bedau, M. A., Ronneburg, F., and Zwick, M. (1992) "Dynamics of Diversity in a Simple Model of Evolution," in R. Manner and B. Manderik (eds.), *Parallel Problem Solving from Nature* 2. Amsterdam: Elsevier.

Bedau, M. A. and Seymour, R. (1994) "Adaptation of Mutation Rates in a Simple Model of Evolution," in R. Stonier and X. H. Yu (eds.), *Complex Systems: Mechanisms of Adaptation*. Amsterdam: IOS Press.

Belew, R. K., McInerney, J., and Schraudolph, N. N. (1992) "Evolving Networks: Using the Genetic Algorithm with Connectionist Learning," in C. Langton, C. Taylor, D. Farmer, and S. Rasmussen (eds.), *Artificial Life II*. Reading, MA: Addison-Wesley.

Brooks, R. and Maes, P. (eds.) (1994) *Artificial Life IV*. Cambridge, MA: The MIT Press/Bradford Books.

Cartwright, N. (1983) *How the Laws of Physics Lie*. New York: Oxford University Press.

Cartwright, N. (1995) "Ceteris Paribus Laws and Socioeconomic Machines." *The Monist* 78:276–294.

Chalmers, D. J., French, R. M., and Hofstadter, D. R. (1992) "High-level Perception, Representation, and Analogy," *Journal of Experimental and Theoretical Artificial Intelligence* 4:185–211.

Churchland, P. S. and Sejnowski, T. J. (1992) *The Computational Brain*. Cambridge, MA: The MIT Press/Bradford Books.

Cliff, D., Harvey, I., and Husbands, P. (1993) "Explorations in Evolutionary Robotics," *Adaptive Behavior* 2:73–110.

Dennett, D. C. (1971) "Intentional Systems," *The Journal of Philosophy* 68:87–106

Dennett, D. C. (1984) "Cognitive Wheels: The Frame Problem of AI," in C. Hookway (ed.), *Minds, Machines, and Evolution: Philosophical Studies*. Cambridge: Cambridge University Press.

Descartes, R. (1980) *Discourse on Method and Meditations on First Philosophy*. Trans. by D. A. Cress. Indianapolis, IN: Hackett.

Dreyfus, H. (1992) *What Computers Still Cannot Do* (Rev. ed.). Cambridge, MA: The MIT Press.

Farmer, J. D., Lapedes, A., Packard, N., and Wendroff, B. (eds.) (1986) *Evolution, Games, and Learning: Models for Adaptation for Machines and Nature*. Amsterdam: North Holland.

Fodor, J. A. (1981) "Special Sciences," in J. A. Fodor (ed.), *Representations*. Cambridge, MA: The MIT Press/Bradford Books.

Fodor, J. A. (1991) "You Can Fool Some of the People All of the Time, Everything Else Being Equal: Hedged Laws and Psychological Explanations," *Mind* 100:19–34.

French, R. M. (1995) *The Subtlety of Sameness: A Theory and Computer Model of Analogy-making*. Cambridge, MA: The MIT Press/Bradford Books.

Hempel, C. (1965) "Aspects of Scientific Explanation," in C. Hempel (ed.), *Aspects of Scientific Explanation and Other Essays in the Philosophy of Science*. New York: Free Press.

Hofstadter, D. R. (1985) "Waking Up from the Boolean Dream, Or, Subcognition as Computation," in D. R. Hofstadter (ed.), *Metamagical Themas: Questing for the Essence of Mind and Pattern*. New York: Basic Books.

Holland, J. H. (1986) "Escaping Brittleness: The Possibilities of General-purpose Learning Algorithms Applied to Parallel Rule-based Systems," in R. S. Michalski, J. G. Carbonell, and T. M. Mitchell (eds.), *Machine Learning II*. Los Altos, CA: Morgan Kaufmann.

Horgan, T. and Tienson, J. (1989) "Representation without Rules," *Philosophical Topics* 17:147–174.

Horgan, T. and Tienson, J. (1990) "Soft Laws," *Midwest Studies in Philosophy* 15:256–279.

Horgan, T. and Tienson, J. (1996) *Connectionism and the Philosophy of Psychology*. Cambridge, MA: The MIT Press/Bradford Books.

Langton, C. (1989a) "Artificial Life," in C. Langton (ed.), *Artificial Life*. Reading, MA: Addison-Wesley.

Langton, C. (ed.) (1989b) *Artificial Life*. Reading, MA: Addison-Wesley.

Langton, C., Taylor, C. E., Farmer, J. D., Rasmussen, S. (eds.) (1992) *Artificial Life II*. Reading, MA: Addison-Wesley.

Mitchell, M. (1993) *Analogy-making as Perception*. Cambridge, MA: The MIT Press/Bradford Books.

Packard, N. (1989) "Intrinsic Adaptation in a Simple Model of Evolution," in C. Langton (ed.), *Artificial Life*. Reading, MA: Addison-Wesley.

Parisi, D., Nolfi, N., and Cecconi, F. (1992) "Learning, Behavior, and Evolution," in F. Varela and P. Bourgine (eds.), *Toward a Practice of Autonomous Systems*. Cambridge, MA: The MIT Press/Bradford Books.

Putnam, H. (1973) "Reductionism and the Nature of Psychology," *Cognition* 2:131–146.

Putnam, H. (1975) "The Nature of Mental States," in H. Putnam (ed.), *Mind, Language, and Reality*. Cambridge: Cambridge University Press.

Putnam, H. (1991) *Representation and Reality*. Cambridge, MA: The MIT Press/Bradford Books.

Rumelhart, D. E., and McClelland, J. L. (1986) *Parallel Distributed Processing: Explorations in the Microstructure of Cognition*, 2 Vols. Cambridge, MA: The MIT Press/Bradford Books.

Schiffer, S. (1991) "Ceteris Paribus Laws," *Mind* 100:1–17.

Todd, P. M., and Miller, G. F. (1991) "Exploring Adaptive Agency II: Simulating the Evolution of Associative Learning," in J.-A. Meyer and S. W. Wilson (eds.), *From Animals to Animats: Proceedings of the First International Conference on the Simulation of Adaptive Behavior*. Cambridge, MA: The MIT Press/Bradford Books.

Varela, F., and Bourgine, P. (eds.). (1992) *Toward a Practice of Autonomous Systems*. Cambridge, MA: The MIT Press/Bradford Books.

Werner, G. M., and Dyer, M. G. (1992) "Evolution of Communication in Artificial Organism," in C. Langton, C. Taylor, D. Farmer, and S. Rasmussen (eds.), *Artificial Life II*. Reading, MA: Addison-Wesley.

VI PARALLELS BETWEEN PHILOSOPHY OF BIOLOGY AND PHILOSOPHY OF PSYCHOLOGY

15 Genes and Codes: Lessons from the Philosophy of Mind?

Peter Godfrey-Smith

Do genes really *code* for biological traits? Of course genes have an important *causal* role in development and the production of traits of organisms, but is this causal role a matter of genes coding for their effects?

Some would say that there is not much left to argue about. The view that the relation between DNA and some traits is a coding relation is part of basic textbook biology. A philosopher might disagree with the textbook view, but then that is a rejection of some very well-established science—not something for a philosopher to do lightly.

On another view, the talk of codes and programs in molecular biology has no genuine theoretical role. Although this talk appears constantly in textbooks and popularizations, and even in research articles for illustrative purposes, it is not a real part of the theory. Rather, talk of codes and programs is just a picturesque way of talking about certain causal relations (or perhaps correlations) between genes and traits. This talk could be dropped or denied without loss of explanatory power. So according to this second view, genes do not really code for traits, but to say this is not to break with biological orthodoxy. Philip Kitcher holds a view of this kind.

My own view is opposed to both of these. In contrast to Kitcher, I believe that the idea that some genes code for some traits is a real part of current biological theory. According to the standard picture, as I understand it, both genes and environmental conditions have causal effects on phenotypic traits, but only the genes code. And in contrast to the first view outlined above, I do not think that the idea that genes code is off-limits for philosophical discussion. It would certainly be folly for a philosopher to deny the standard biological account of the chemistry of DNA and the mechanisms through which it affects the production of proteins. But the concept of genetic coding seeks to add something to that basic picture; it seeks to add a claim about the special nature of some kinds of genetic causation, and a theoretically important analogy between these genetic processes and processes involving symbols and messages in everyday life. Further, the idea of coding itself—both in general and in genetics—is not a straightforward concept that everyone understands in the same way. Dissenting voices within both biology and philosophy have claimed that it is a mistake to see genes as coding.[1] A philosophical discussion could be useful at least as a contribution to clarity on this issue, and perhaps to help settle the question.

The question of coding has recently become linked to debates around "developmental systems theory" (Oyama 1985; Griffiths and Gray 1994). Advocates of developmental systems theory claim, among other things, that the idea that genes code for traits is part of a picture that assigns to genes a false causal priority in development and evolution. Developmental systems theory opposes the idea that some of the factors that contribute to development are sources of information or form, while other factors are mere background support, or raw material. For the developmental systems view, it is a mistake to think that genes have a causal role that is different in kind from that of nongenetic factors. At least in principle, both kinds of factor can carry information, both can be inherited, and so on. So developmental systems theory is associated with various claims of *symmetry* for genetic and nongenetic factors.

As far as possible, in this chapter I will discuss coding in isolation from questions about the causal priority of genes.[2] Someone could claim that genes really do code for traits while denying that genes have preeminent causal importance. The other combination of views (genes as preeminent, but not coding) is possible also. But the more general issue of symmetry between the causal roles of genetic and nongenetic factors will be discussed in some detail.

If genetic coding is recognized in some cases, it need not be recognized in all. One also has to decide exactly what messages genes can contain. Do genes only code for proteins, or can a gene also code for penicillin resistance, if the protein produced has a key role in producing this resistance? That question is a question about "how far out" coding can reach. A different but related question arises when a single gene makes a difference to a complex trait that can only be built with many other genes and developmental factors (a trait like delayed sexual maturity, perhaps). The first of these two questions (the "how far out" question) will be discussed in this chapter; the second (involving complex traits) I leave for another day.

The question of whether genes code would be easier if there were a widely accepted theory of coding and representation that we could apply to the problem. There is, indeed, a widely accepted framework in which some questions about messages can be addressed—the mathematical theory of information. And some writers have applied information-theoretic concepts directly to the problems involving genes (Maclaurin forthcoming). But I will argue that this approach is unlikely to furnish a full solution to the problem.

Within philosophy, perhaps the most promising place to look is the literature on physicalist theories of representation in the philosophy of mind.[3] These theories aim to state, in physicalistically acceptable terms, necessary and sufficient conditions for an internal state of an organism to be a representation of some specific object or state of affairs in the organism's environment. If this debate had produced a clear winner we might be able to apply this winning theory (or adaptations of it) to a range of other problems involving representation and meaning—including the problem of the relation between genes and traits. The discussion has not, however, produced a consensus.

Even though there is no consensus, perhaps this work can at least provide some clues as we wrestle with the genes. I will attempt to make some connections in this chapter.

This chapter does not, however, give a full solution to the problem of genetic coding. It does not even reach a firm decision on whether genetic coding is real or whether the idea is useful. This is an exploratory discussion. Much of the aim is just to outline some of the available options.

1   The Genetic Code and the "Gene for" Concept

In this section I will discuss some common ways of talking about the genetic code within biology and then look at how some recent philosophical literature has handled both the question of coding and the idea that a gene can be "for" a phenotypic trait. The first part of this section will also give a refresher on basic facts about the role of DNA in the manufacture of proteins.

We start with two statements from biology textbooks about the role of DNA.

The information dictating the structures of the enormous variety of protein molecules found in living organisms is encoded in and translated by molecules known as *nucleic acids*.   (Raven, Evert, and Eichhorn 1992, p. 59)

[DNA] contains a coded representation of all of the cell's proteins; other molecules like sugars and fats are made by proteins, so their structures are indirectly coded in DNA. [DNA] also contains a coded set of instructions about when the proteins are to be made and in what quantities.   (Lodish et al. 1995, p. 10)

These are the sorts of statements I take as evidence for the claim that contemporary biology attributes to DNA a special set of properties that are described in semantic terms. Although standard views in contemporary biology certainly see these coding relations as fundamentally causal, these views also hold that among the various causal relations involved in development and metabolism, some causal relations are special in that they involve the interpretation of a message or the "expression" of coded instructions.

Let us look briefly at the relevant biological processes, and the terminology used for them within biology. Two main steps are distinguished in the causal chain between DNA and a protein. "Transcription" is the process in which DNA gives rise to mRNA ("messenger RNA"). Then "translation," which occurs at the ribosomes, generates the protein itself. The molecule of mRNA produced during transcription is formed using a stretch of DNA directly as a template, and the mRNA contains a sequence of "bases" that corresponds, by a standard rule, to the sequence of bases in the DNA from which it was derived. In organisms other than bacteria, the mRNA is usually processed (in ways I will discuss later) before it is used in translation. Then, at the ribosomes, the processed mRNA is used to direct the formation of a chain of amino acids—a protein.

In this process of translation, a crucial role is played by another kind of RNA molecule, tRNA (or "transfer RNA"). Molecules of tRNA bind to particular amino acids (of which there are twenty kinds), and at the ribosomes these tRNA molecules bind to specific three-base sequences in the mRNA. So each triplet of bases in the mRNA is associated, via the chemical properties of tRNA, with a particular amino acid. The "genetic code" is, strictly speaking, the rule linking RNA-base triplets with amino acids, and this interpretation of the RNA determines the interpretation of the DNA from which the mRNA is derived.

The ribosome moves along the mRNA chain, and as it does a chain of amino acids is assembled, with the sequence of amino acids corresponding exactly to the sequence of bases in the mRNA by the rule comprising the genetic code.[4] As there are four bases in the mRNA (almost the same four as in DNA) there are sixty-four possible triplets. Of these, sixty-one specify particular amino acids; some amino acids are specified by as many as six different triplets. The three remaining triplets are "stop" signals. The chain of amino acids folds (and may be processed in other ways) to produce a finished protein. Protein structure is described at four different levels, of which the *primary* and *tertiary* are the most important for our purposes. The primary structure of a protein is its sequence of amino acids; the tertiary structure is the three-dimensional folded shape of a single amino acid chain. The causal role of proteins depends greatly on their tertiary structure.

There is much more to all these processes, of course, and on some views the extra detail is essential to an understanding of what is going on. But on more standard views, the information given above outlines the core of the process by which the genetic message is expressed.

Philosophically, the term "translation" seems a strange one, even within the standard picture of the genetic code. Translation, in ordinary usage and in philosophical theory, takes a message from one symbol system or language to another. But although the standard view sees the DNA sequence as a sort of language, amino acids and proteins are not usually understood as coding for anything (unless that is their job elsewhere in the body). On standard views, DNA and RNA are messenger molecules, but the series of messages ends when the protein is formed. So the process of "translation," as it is usually conceived, would be more accurately described as "interpretation." And sometimes biological discussions do use that term, although it is not nearly as standard as "translation."[5]

Turning to recent philosophical discussion of the relation between genes and traits, a natural place to start is with a work that does not make any claims about coding at all. Sterelny and Kitcher (1988) give an analysis of what is involved in some gene's being a "gene for" a particular trait, but they do not express any part of the analysis in terms of coding. Roughly, to talk of a "gene for X" in the sense of Sterelny and Kitcher (1988) is to talk of a reliable correlation, in normal genetic and nongenetic environments, between the gene and the trait.[6]

Griffiths and Gray (1994), arguing for the explanatory symmetries characteristic of developmental systems theory, claim that an analysis of "gene for X" in the style of Sterelny and Kitcher must allow that environmental conditions or cytoplasmic factors can be "for" particular traits in the same sense that genes can. This is because when we call a gene a gene for X, we hold certain environmental factors constant, as a background condition. But we can also hold genetic factors fixed, as a background condition, and speak of environmental or cytoplasmic factors "for" traits. Griffiths and Gray took Sterelny and Kitcher to be seeking an analysis that would retain the idea that genes *code* for traits as well as causing them, so Griffiths and Gray took this symmetry between genes and environment to be a problem for Sterelny and Kitcher.

In their responses to Griffiths and Gray, Sterelny and Kitcher diverge in interesting ways. Kitcher (forthcoming) accepts Griffiths and Gray's point about the explanatory symmetry of genetic and nongenetic factors and accepts that there are "environments for" as well as "genes for" traits. He claims that his and Sterelny's original reconstruction of the concept of a "gene for X" is entirely compatible with this move. But when Griffiths and Gray suggest that Kitcher would not want to say that an environmental feature *codes* for a trait, Kitcher dismisses their talk of coding as "a rhetorical flourish irrelevant to the discussion" (forthcoming p. 19).

When I objected to Kitcher that standard views in genetics do see genes as coding whereas environmental conditions cannot code, Kitcher replied (in personal correspondence) that there is no need to make literal sense of claims about genes coding for traits. It is just a colorful mode of talk that has no role in the explanatory structure.

Where Kitcher steers clear of coding and all properties akin to it, Sterelny takes the opposite route in his response to the symmetry arguments of the developmental systems literature. Sterelny, Smith, and Dickison (1996) accept that genes can be ascribed semantic properties as well as causal properties. Their preferred term is a philosophically strong one: "the genome does *represent* developmental outcomes" (1996, p. 387, emphasis in original). In response to the symmetry arguments, Sterelny, Smith, and Dickison claim that some nongenetic factors in development have the same kinds of properties that genes are usually taken to have; there are both genetic and nongenetic replicators. Replicators have been shaped by selection for their developmental role, and replicators "represent phenotypes in virtue of their function" (p. 387). I will discuss Sterelny, Smith, and Dickison's view in more detail in section 4 below.

Sterelny, Smith, and Dickison also talk a good deal about *information*, as many others do. Griffiths and Gray, who oppose many standard ways of talking about genes, do not object to the idea that genes contain information, so long as the use of informational concepts is not restricted in its application to genes and used to "privilege" genes over other developmental resources (p. 283). I will discuss the possible role of the concept of information in the next section.

2   Indicative and Imperative

As we have seen, philosophers and biologists use a range of semantic expressions when talking about what genes do. Sometimes it is said that genes carry information about traits, or represent the outcomes of developmental processes. Sometimes genes contain coded instructions. We find picturesque terms such as "blueprint," and far more neutral terms such as "specify."[7] The differences between these formulations are important, as some claims are empirically more plausible than others, and some attributions of meaning or content raise more philosophical problems than others.

A detailed analysis of how exactly genes code can be expected to choose some one of these formulations as best. As I will not defend any particular analysis in this chapter, I leave some of these issues fairly open. But some preliminary points can be made. In this section I will discuss one important distinction between different kinds of semantic properties that genes might have.

Ruth Millikan, adapting older terminologies, distinguishes in her general theory of signs between "indicative" and "imperative" representations (1984). Roughly, an indicative representation is supposed to describe how things are, and an imperative one is supposed to bring something about. So a declarative sentence is an example of an indicative representation, and a command is an example of an imperative representation.

If genes are representations at all, which kind are they? I claim they must be seen as *imperative* representations. Their role is to prescribe rather than describe. So within the family of semantic terms that philosophers and biologists have used about genes, the most appropriate ones are those that suggest imperative rather than indicative contents. Viewing genes as containing "coded instructions," as "prescribing" or "dictating," has more chance of being right than viewing them as "describing" anything.[8]

The difference between assigning indicative and imperative contents is readily seen in cases of error, noncompliance, and misrepresentation. Suppose you order a pizza, but what arrives is pasta. In such a situation, your message is not faulty or erroneous. The mistake was made by the people who received the message and filled your order. If, on the other hand, you did not order but *predicted* (or guessed, or claimed) that they would bring pizza, and they brought pasta, then your claim is where the error lies.

In the case of genes, which party is "at fault," according to the standard picture, if the protein produced on some occasion does not match the specifications of the DNA? I claim that on standard views about genes and coding, it is not the gene that is at fault for misdescribing the protein (for guessing pizza when they brought pasta). Rather, the interpreting mechanisms in the cell are at fault, for failing to comply with what the DNA instructed.

So even before the details of an analysis of coding properties of genes have been worked out, it is clear that the aim should be an analysis of DNA's capacity to carry messages with imperative semantic content. DNA, if it contains a message at all, contains instructions rather than descriptions.

Partly because of this, one of the most popular ways to ascribe semantic properties to genes—a way that uses the concept of *information*—is not a good approach to the problem.

There is a variety of ways in which the concept of information is used in describing what genes do. It is common to say that genes carry within them information, in coded form, about the proteins made by the organism, and perhaps information about whole complex traits the organism exhibits. Here the genes are playing a role like that of a message. At other times the genes are said to be a *source* of information used by the cell or by the organism in development. This may (although it need not) mean something different than the idea that the genes are a message.

I will argue that the only proper role the concept of information has here is a weaker, less interesting role than it is often taken to have. There is a weak sense of "information" in which anything is a source of information if it can occupy a variety of possible states. And in this sense of information, if the states of some X are reliably correlated with the states of some Y, then X carries information about Y.[9] This is the sense of information dealt with in the mathematical theory of information (Shannon 1948) and refined into a semantic theory by Dretske (1981). Information in this sense has also been discussed more informally by philosophers as "natural meaning."

Information, or natural meaning, is everywhere. It does indeed connect DNA with proteins and phenotypic traits, but it connects them in both directions and connects both of these to environmental conditions as well. DNA sequences have many possible states, as do proteins. Given background conditions (which define a "channel") the genes carry some information about the proteins produced by a cell. The proteins produced also carry some information about the genes responsible for them. In both directions the transmission of information is imperfect, for a variety of reasons.

So we can regard the environment as a background condition against which genes carry information about phenotypic traits. But as defenders of developmental systems theory insist, we can also view genetic conditions as background conditions or part of the "channel." Against such a genetic background, we can see environmental conditions as carrying information about phenotypes. And we can also see phenotypes as carrying information about environmental conditions.

If any of these attributions of informational properties is acceptable, then all of them are. In some cases and in some directions there will be more information carried than there is in others, but that does not affect the basic point about the ubiquity of information.

Attributions of informational properties of this kind cannot be used to analyze the special role played by concepts of coding in molecular biology. This is clear from the fact that although information is ubiquitous and runs in these cases in both directions, the coding relationships discussed in molecular biology are not. Coding is (i) specific to the relationship between genes and phenotypes, and (ii) asymmetric, as genes code for phenotypes but not vice versa.

This is not an argument that talking about information can have no useful role in molecular biology. Talk of information is often a useful way of picking out correlations and causal relationships of various kinds. The argument is just that this concept of information gives no grounding for the asymmetry expressed by the idea that whereas genes and environments both causally affect phenotypes, only the genes have their effects by coding for phenotypic features.

Compare another usage of "information" in molecular biology. Biologists sometimes talk about the information that genetic variation within and between populations carries about phylogenetic relationships. We can learn from analysis of DNA sequences the order in which various species split off from one another, and other historical facts of this kind. Different parts of organisms' genomes carry different amounts of information about these historical relationships ("junk" DNA is especially helpful). Sometimes one kind of analysis will be useful for more recent history whereas another analysis will be used to reconstruct more distant events. All of this is often described in terms of the "information" carried by patterns of genetic variation. But it is obviously only information that *we* use, not information that is part of any explanation of the causal role that genes play in development or evolution. There is no more to this kind of information than correlation or "natural meaning"; the genes are not *trying to tell us* about their past, Genes in this sense are like tree rings (a standard example of natural meaning). Similarly, the fact that *we* might sometimes be able to read the composition of a protein off a DNA sequence does not imply that the *cell* literally reads the composition of the protein off the sequence. If the role played by the concept of genetic "information" in explanations of development is something more important than the trivial role it plays in phylogenetic reconstruction, then a novel and richer concept of information-transmission must be developed.

I do not say that this is impossible; there may be other, richer concepts of information that biologists and philosophers could develop and apply here, and some analyses might use the concept of information as one component in a theory along with others (as Dretske does in his philosophy of mind). I am not claiming that no analysis of the semantic content of DNA that uses the idea of "information" can possibly succeed.[10] But I do think this is an unpromising way to proceed. It is unpromising first because the idea that genes "carry information about" proteins or phenotypic traits is naturally understood in terms of indicative rather than imperative content.[11] And it is unpromising also because it will always be hard to keep a richer concept of information distinct from the original sense—a sense with genuine usefulness in biology—in which information exists whenever there is reliable correlation.

3   Analyses in Terms of Developmental Role

This section and the next will look at two ways to develop an analysis in which genes are assigned coding properties. On the first approach, discussed in this section, genes code in virtue of their role in developmental and metabolic processes.[12]

In this chapter I understand "development" as a process that takes place strictly within a single generation. Development is set of local causal processes through which early stages give rise to later stages, something that could in principle be described without any reference to evolutionary history. Genes, as everyone agrees, play a causal role in such processes. One possible view on the question of coding is that the causal role DNA plays in developmental processes is one that can fairly be described in terms of its bearing coded instructions; the content of specific DNA sequences is determined by a rule of interpretation that derives from this causal role. I will call these "developmental role" theories. They are loosely analogous to "conceptual role" theories of meaning in the philosophy of mind (Block 1986).

As I envisage views of this kind, the peculiar characteristic of DNA that justifies its being treated as a code lies in the fact that its sequence is physically *read* by the cell during the construction of proteins. The cell first creates an mRNA molecule whose sequence corresponds to the sequence of bases in the DNA, and then part of the cell's machinery physically moves along the mRNA molecule, at each step interacting with the base sequence, producing with each step a chain of amino acids whose linear structure corresponds, by a standard rule, to the linear structure of the mRNA. This process, according to an analysis of coding in terms of developmental role, is one in which the mRNA is directly read by the ribosomal machinery, and the whole process is one in which the DNA is read as well.

On the developmental role view, what makes genes into coded messages is not just the *specificity* of their causal role—the fact that particular DNA sequences reliably give rise to particular products. The causal specificity of DNA is one important feature but not the only one. This is because "specificity" is a property that applies to a wide range of causal relations. A cutting enzyme might be highly specific in what it cuts. A raw material might only be usable in one specific building project. So specificity alone is not the issue; it also matters how this specificity arises. What makes the "genetic code" into more than just a set of causal associations is the nature of the processes that underlie those associations. DNA is causally specific through being read; other factors (like enzymes) have causal specificity of different kinds and for different reasons. A family of enzymes might have its causal role described by a general principle linking particular enzymes to particular reactions, but this will not be an *interpretation* rule in the case of the enzymes, because enzymes are not read by these processes.

As I stressed at the beginning of this chapter, the idea that genes code because of their developmental role may or may not be coupled with a claim about the preeminent causal importance of genes. One might hold that genes code without holding that "genes are destiny." The point of the concept of coding, on a developmental role analysis, is to pick out one particular causal role among many. Within developmental and metabolic processes there are raw materials (like amino acids), cutters and joiners (enzymes), stores of energy (like ATP), readers and assemblers (ribosomes)—and there are coded instructions as well (the genes). Raw materials and stores of energy might be just as important as messages, but they are different *kinds* of causal players.

An analysis along these lines will face a variety of challenges. One philosophical line of objection proceeds by claiming that cells cannot possibly "read" anything because the concept of "reading" is one that only has a place in a system of human conventions of public symbol use. So it is nonsense to say that genes contain a message read by the cell.

The ordinary sense of "reading" may well be one that requires that the reader be an agent with mental states. But machines that "read" in extended senses, like bar-code readers in supermarkets, are all around us. A biologist can reply that though cells read in an extended sense, this is a sense that has a useful role in our understanding of many kinds of machines. Possibly the "reading" status of supermarket bar-code readers depends on the machines' being the products of human design, so the genes and ribosomes are in this respect even further from the ordinary use. But it can be argued that "reading" in this sense *is* still a distinctive kind of causal process, one with real similarities to ordinary human reading and interpretation. Reading as done by supermarket machines is physically different from weighing and imprinting; different also from "guessing" and other actions that machines might perform in extended senses.

So a biologist will very likely accept that cellular mechanisms only read things in an extended sense of "read." That would imply that DNA only contains coded instructions in an extended sense of "coded instruction." But this concession could reasonably be regarded as a minor one by a biologist; it is consistent with the claim that the processes in which DNA is involved have a remarkable and theoretically important similarity to ordinary processes of reading and interpretation. And that is what many biologists might regard as the important point—it does not matter exactly how the genetic senses of "reading" and "interpretation" are connected to everyday senses.

A more important challenge derives from biological considerations. DNA is not "read" by the cell in exactly the ordinary sense, admittedly, but is it appropriate to see these processes as akin to reading at all? The suggestion I made about reading embodies quite a controversial claim about the causal processes in which DNA is involved. It might be objected that it is more accurate to see the DNA sequence as having the role of a *template*, rather than something that is read by the cell. Talk of "reading" might be harmless in certain informal contexts, but the processes of protein synthesis are not of a kind that will support the linguistic analogy, if one looks closer.

When we look closer, we see that the sequence of bases in DNA is transferred onto an mRNA sequence by a process in which the DNA acts directly as a template for the synthesis of mRNA. Then the mRNA itself acts as a template along which a chain of amino acids forms. The ribosomes where the protein is formed are not much like readers; rather, they are an elaborate kind of scaffolding where certain reactions take place spontaneously. So according to this biological objection, the idea that DNA functions as a template is an *alternative* to the view that DNA contains instructions that the cell reads. Using a template and using instructions are two different ways in which structure can be transferred or conveyed. On such a view, the standard rule linking DNA-base triplets to amino acids is seen, again, as describing a set of causal specificities and no more.

To this objection the advocate of genetic coding has two replies. It might be replied that a description in terms of reading is more accurate than a description in terms of mere templates, perhaps because of the *combinatorial* properties of the RNA–amino acid relationship. Alternatively, perhaps the two descriptions are compatible. Watson et al. (1987), the king of molecular biology textbooks, puts a lot of theoretical stress on the concept of a template but evidently does not see this as antithetical to the idea of coding. Both issues are hard to settle and I will not attempt to solve the problem here.

Another feature of the developmental role approach to coding is that there will be interesting differences between different organisms, with respect to how appropriate it is to see genes as containing a code. All views on this matter must accept and make room for a range of phenomena that cause trouble for the most simplistic views of the coding properties of genes. These phenomena are sometimes specific to certain kinds of organisms. Viruses sometimes have the sequence of one gene completely embedded inside that of another, which is read differently. Both viruses and bacteria have genes whose sequences partially overlap. At the other end of the scale of economy, genomes of eukaryotes (roughly, all organisms except bacteria) contain *introns*. This is DNA sequence that is transcribed into RNA but then removed before translation. So the protein produced typically does not correspond to the sequence of any contiguous stretch of DNA. There can even be alternative ways of splicing together pieces of mRNA, to form different products from the same initial "primary" transcript.[13] But prokaryotic genomes do not contain introns.

Should a view that analyzes coding in terms of developmental role concede that prokaryote genomes more clearly or unproblematically contain coded messages than eukaryote genomes? For some this would be a problematic conclusion. If so, this suggests that developmental-role analyses should not require too strict a correspondence between protein structure and DNA sequence.

Interestingly, Sarkar (1996) thinks that a different conclusion should be drawn; he thinks that the most important objections to the idea that DNA contains a code only apply in the case of eukaryotes. If what was true for bacteria was also true for elephants, Sarkar says, this would "make the linguistic view of genetics palatable" (1996, p. 860). So as I read Sarkar, he thinks that bacterial genes might reasonably be seen as coding for traits, as a consequence of their developmental role, whereas genes in eukaryotes do not contain a code.[14] He bases this argument not just on the presence of introns and other nonfunctional DNA, but on a variety of other complexities peculiar to eukaryotes. In particular, RNA is sometimes "edited" in ways that go beyond the removal of introns; sometimes the editing involves substitutions or additions of bases. In eukaryotes the sequence of amino acids in a protein need not correspond exactly by the standard interpretation rule to any stretch (contiguous or not) of the DNA sequence.

My own view is cautiously opposed to Sarkar's on this point. It seems to me unlikely that the differences between eukaryotes and prokaryotes are such that coding properties might be found in the DNA of one but not the other. And in his treatment of these issues, Sarkar seems sometimes to require far more transparency of interpretation for genetic codes than one would normally require for ordinary public messages. For example, Sarkar says "natural languages do not contain large segments of meaningless signs interspersed with occasional bits of meaningful symbols" (p. 863), whereas eukaryote DNA does contain such junk. But surely a great deal of symbolic interaction in everyday life is interspersed with meaningless extra elements. Admittedly, to reach a ratio of ninety-five percent junk to five percent real information (as is often claimed for human DNA) we might have to look at some unimpressive regions of social life (I leave it to readers to insert their preferred examples), but ordinary concepts of meaning and representation certainly do not preclude messages' containing in their physical structure much that has to be edited out.

Within philosophy of mind, many theories of representation have difficulty accounting for the possibility of error and false representation.[15] This is thought important because it is taken to be essential to the concept of representation that wherever there is representation there is the possibility (in principle) of misrepresentation. So theories of genetic coding that rely purely on developmental role must face the problem of specifying how misrepresentation, error, and noncompliance can exist. The problem is not one of showing how the DNA can err in description; the problem is showing how the cell might misread the DNA and thereby fail to do what is instructed. An analysis drawing only on the actual causal role of DNA can distinguish between the common and the uncommon—in some particular case a process involving the genes might give rise to a product that differs from the normal product. But philosophers of mind have labored long and hard over the fact that the ordinary concept of misrepresentation seems to be one in which the distinction between proper and erroneous in semantic contexts is *not* the same as the distinction between the common and the rare. It is also not the same as the distinction between the beneficial and the harmful. So for an analysis of the genetic code purely in terms of developmental role, either whatever is common is what is "supposed" to happen, or

else there is no sense in which some things are supposed to happen whereas others
are not.

Once the issue of error and malfunction is raised, some philosophers of mind will
think that what is needed at this point is an appeal to a concept of "proper functioning" based in evolutionary history. There are several ways in which this might be
done. A minimal way would be to add an evolutionary concept of normal or proper
functioning to an analysis that relies in all other respects on the developmental role
of DNA. This might be quite a promising way to proceed. But another possible
response is to think that more extensive use of evolutionary concepts is needed. Such
a move would build a more substantial bridge between genetics and the philosophy
of mind.

4   Analyses in Terms of Evolutionary History

If genetic coding is analyzed in terms of evolutionary history, nothing about the
pattern of interaction between DNA and proteins, considered just as a physical process, makes a DNA sequence into a message. Rather, the key fact is that the evolutionary history of these mechanisms is one that has given to DNA certain *biological
functions*. DNA has the function of coding for amino acids, or the function of
instructing the cell which proteins to produce (or some other functional property of
that kind). This approach is taken by Sterelny, Smith, and Dickison in "The
Extended Replicator" (1996).

A view of this kind is able to draw on a range of influential ideas in the philosophy
of mind. Millikan's view (1984), which I have drawn on earlier, is the most elaborate
evolution-based theory of representation, and the most potentially adaptable to the
case of DNA. Others include the theories of Papineau (1993) and Sterelny (1990). In
philosophy of mind these are sometimes called "teleofunctional" theories.

Two principles are basic to teleofunctional theories, and both principles might be
applied to genes. One is the idea that a rich concept of function is essential to an
understanding of representation and other semantic properties. The other is the idea
that functions are to be analyzed in terms of evolutionary history: functions are
effects or capacities that have been selected for.[16]

DNA and the mechanisms that interact with it are certainly products of evolution,
and some of their effects have been selectively important. So it should be possible to
assign functions to these structures. And such an assignment of functions might provide a semantic analysis of what the genes are telling the cell.

There are many ways to develop the specifics of such a view. One dimension on
which historical theories can be compared concerns how much of a role the theory

gives to factors other than evolutionary history. Some theories might hold that genes code simply because of the special properties of their selective histories; others might see the appeal to history as one factor used along with others to analyze coding. I will discuss examples of both approaches below. Sterelny, Smith, and Dickison's view puts almost total weight on historical properties in explaining why genes code. But in response to problems with that view, some might add nonhistorical factors to the story.

Another distinction has to do with how important the *systematic* properties of DNA are taken to be. Suppose we are seeking to assign biological functions to a DNA strand. There are two ways in which this could be done.

Option A: The *particular* DNA strand might have its own biological functions, as a consequence of the success under natural selection of strands with the same sequence (or almost the same sequence) that gave rise to the present strand. That is, we can look to the selective history of particular DNA sequences. Their interpretation is determined by the effects they have had that have led to their selective success.

Option B: It might be that DNA *in general* has the function of specifying proteins, as a consequence of the selective history of the entire machinery of protein synthesis. Then the interpretation of any particular sequence of DNA would be determined not by its own past effects, but by a standard interpretation rule that applies to all DNA, a rule determined by the evolutionary history of DNA in general.

Option A has the consequence that a new mutation has no natural interpretation; whatever the cell does with it, there is no reading/misreading distinction. On Option B, a new mutation does have a natural interpretation, as a consequence of the general rules of DNA interpretation, and a new mutation could in principle be *mis*interpreted even on its debut. I imagine that many readers will regard this as an argument against Option A. The DNA-reading mechanisms are supposed to react in particular ways to any bit of DNA, new or old. That suggests that the new sequence does have a natural interpretation.

In general I think Option B is the right approach for a view that analyses coding in terms of evolutionary history. However, there might be some place in the picture for the reasoning behind Option A. Suppose a new mutation appears which for some reason always interferes with the reading mechanisms. The mutation does have effects on protein synthesis, but effects that are not part of the usual pattern of DNA interpretation. However, the effects are useful and the mutation is successful under natural selection. Call the protein it produces by this nonstandard route "protein Z."

Does this sequence of DNA *code* for protein Z? Inducing the cell to make protein Z is the sole effect the gene has that explains its success under natural selection. So on many theories of biological function, the gene will have the function of producing protein Z. But the causal route of this production is not one that involves the usual pattern of DNA interpretation. In a sense, this gene is being selected for despite being always misinterpreted. In a causal and functional sense, this is a *gene for Z*, but the gene does not *code* for Z. Some readers might find this an odd conclusion to draw, and I will discuss it further below.[17]

Incidentally, this case is not as biologically far-fetched as it might sound. "Frame-shift" events sometimes occur during translation; the ribosomal machinery reads a quadruplet as a triplet, or backs up and reads a base twice, and then carries on reading triplets after that anomalous event. This happens through a variety of causes, but in some organisms there are "slippery" mRNA sequences that act, in the context of the cell, to *induce* frame-shifting, and this frame-shifting can be required for normal development.[18] I don't know if there are cases of frame-shifting where a sequence is *always* read out of frame, however.

My response to the protein Z case is guided not just by a preference for Option B, but by a general assumption about the relationship between biological functions and semantic properties. The assumption is that not everything with a biological function in DNA replication and expression has semantic properties. This assumption about functions and representation is surely true in many contexts—legs are for walking, but they do not represent walking. They do not (usually) represent anything, despite having biological functions. Something can have the function of producing a certain object or outcome in biological processes, without representing or coding for that object or outcome. I apply the same assumption to DNA.

I highlight this assumption about functions because the solution to the coding question defended by Sterelny, Smith, and Dickison (1996) works by denying this assumption for a special class of cases. Sterelny, Smith, and Dickison claim (as I interpret them) that an analysis of the biological functions of DNA in developmental processes suffices to determine a semantic interpretation of genes; all factors that have biological functions within development *represent* the outcomes of those developmental processes. Although Sterelny, Smith, and Dickison do not claim this tight connection between function and representation for all biological functions (so they would accept my claim above about legs), they view developmental functions as a special case. Here the connection between function and representation is especially tight.[19]

I do not think this is a satisfactory solution. There are entities that have the function of playing a certain role in development, where it seems quite implausible to assign them representational properties. For example, an enzyme such as an *aminoacyl-tRNA synthetase* has a certain function in protein synthesis, and this function can be understood in terms of a selective history. These enzymes have the function of attaching particular amino acids to particular tRNA molecules. That is the effect they have that explains why they are there. Though that is the function of these enzymes, and the enzymes are very causally specific, the enzymes do not code for or represent their products. There might be no representation without function, but there is function without representation.

Consequently, in understanding how it might be that genes code for traits, it is not enough to have an analysis that assigns them the biological function of having a certain causal role in the production of proteins. Something more is needed.

Here is one suggestion, a friendly amendment to Sterelny, Smith, and Dickison's proposal. Perhaps DNA codes because the DNA sequence is supposed to help produce a product *with a certain abstract relation to its own structure*. DNA sequences have historically entered into causal processes in which a certain mapping relation between DNA sequences and amino acid sequences has been important in the evolutionary history of these mechanisms. That "mapping relation" is, of course, the familiar set of relations usually described as comprising the interpretation of the genetic code. A certain pattern, in which DNA sequences have generated (with the help of other cellular machinery) proteins with a certain abstract relationship to themselves, has been central to the evolutionary success of these mechanisms.

Would this move provide Sterelny, Smith, and Dickison with a way to forge the link they want between function and representation? This amendment would certainly give them a way to respond to the problem with enzymes that I raised above. Although an aminoacyl-tRNA synthetase is supposed to have a certain causal role, this role is not one that involves production of products with abstract mapping relations to itself (except in trivial senses).

This suggestion might not take us far enough, however. I do not know of a real case that might provide a counterexample, but here is a hypothetical case. Suppose there is a family of enzymes whose function is to join identical protein subunits into larger structures. One enzyme joins pairs of the subunits together, and it has two binding sites at which the subunits attach when being joined together. Another enzyme makes units out of three of the subunits, and it has three binding sites. Another joins four subunit molecules ... and so on. Here we have enzymes whose function is to manufacture certain products, where the products are supposed to have a certain abstract relation to the enzymes (an enzyme with two sites produces a double molecule, and so on). But would this be a case where the enzymes represent or code for their products? If intuitions are to be trusted in this arena, it seems to me that these enzymes would not represent their products, in any ordinary sense of "represent." Again we have function without representation.

What might be the next step down the same road? The previous suggestion was inspired by parts of Ruth Millikan's semantic theory (1984). The concept described above is related to Millikan's concept of an "imperative intentional icon." However, it is not quite the same as Millikan's concept. Should we move still closer to her view?

If so, the most important further step would be to introduce what Millikan calls "producers and consumers" of the representation. For Millikan (and on some other views) all representations *mediate* between two specific kinds of functionally characterized things.[20] "Producers" are supposed to produce the representation, in the performance of their functions. "Consumers" are supposed to have their activities modified by the state of the representations they receive from the producers. In particular, in the case of imperative signs, the sign affects its consumers in such a way that the consumers acquire the function to produce a certain state of affairs in the world. This state of affairs is the sign's compliance condition, its content.

If we introduced this idea, we could rule out the hypothetical case involving enzyme-binding sites that I described above. The only problem is: who are the producers and consumers, in the case of DNA?

There are various options for answering this question, but none of them seems attractive. The consumers are probably easier to handle than producers. The most likely initial candidates for consumers are the ribosomes, the machinery where protein synthesis actually occurs.

A first problem with this idea is that it is mRNA, not DNA, that interacts with the ribosomes and functions in the assembly of proteins. So applying Millikan's view in this way, we could see the mRNA as instructing the ribosomes, and then see the DNA in the nucleus as the producer of the representation. In some ways this makes sense, but it is not a solution that vindicates standard views about coding. Those views, recall, assign content to the DNA in the nucleus, not just to the mRNA. So perhaps we might see the ribosomes as consumers of both the DNA and the mRNA—the mRNA is an intermediary.

Another problem is that regarding the ribosomes as "consumers" in Millikan's sense might be giving them too much credit. If there are to be consumers, it probably makes more sense to see them as comprising a set of interacting factors, including the ribosomes themselves, the tRNA molecules, and the various proteins and other molecules that initiate, control, and fuel the process.

So to see the DNA as interpreted by a Millikan-style "consumer," we must identify this consumer with a range of factors, including those that produce mRNA and those that use mRNA in building proteins. I do not see this "distributed" nature of the consumer as a problem for such an analysis.

But who is the *producer* of the DNA message? It might be the entire previous cell that gave rise to this one, in which case messages would get passed indefinitely back through cells across the generations. But then it is harder to see why these "producers" are giving particular instructions to the machinery of the new cell. We should not say (as people sometimes do) that the DNA is a message passed from one generation to the next, as the DNA is part of what *gives rise* to the next generation. The next generation is a *product* of the DNA (plus other inherited resources) rather than its consumer.

I will not follow up these specific ideas here. Some might think that I have unfairly dismissed a viable alternative, or that there are options I have not discussed that will do the trick. (Sterelny, Smith, and Dickison say at one point, for example, "a gene can have the function of telling *the developmental program* how to build hemoglobin molecules" [1996 p. 338, emphasis added]. Maybe something like this could work.)

In general terms, though, if one accepts that Sterelny, Smith, and Dickison's original (1996) view is unsatisfactory but on the right track, the obvious thing to do is to constrain the *kinds* of causal role that are associated (when suitably historically embedded) with representational properties. The problem will be to do this without forcing a misleading or ill-fitting set of distinction onto our empirical picture of the causal role of genes. Insisting that we locate a set of "producers" and "consumers" might, for instance, be forcing onto the cell a framework derived from elsewhere and with no empirical motivation in this context, just to retain the idea that genes code for something.

Recall, in contrast, the first amendment I made to Sterelny, Smith, and Dickison's view above. This was the suggestion that DNA codes for proteins because the DNA sequence is supposed to help make a product with a certain abstract relation to its own structure. Even though this view might not make all the right intuitive distinctions, nonetheless this proposal does make use of *empirically* well-motivated distinctions. We might decide to retain only a modified or weakened concept of coding, one that can be understood in terms of this view.

5 Distality: Amino Acid Sequences, Folded Proteins, and Traits

I will discuss one more topic before drawing some conclusions. This topic is: *what* exactly can the genes code for? As far as *causal* relations are concerned, we can trace a chain as far as we please, from the proximal effects of genes (amino acid sequences), to folded proteins, and then further on to traits such as camouflage, blue eyes, penicillin resistance (in bacteria), and musical ability. All will agree that genes have *some* causal role in even the most complex traits. But as I have argued in earlier sections, the question of what is coded for by genes is not the same as the question as what is caused by genes.

One common view in this area is a permissive one, seen in Dawkins (1982) and Sterelny and Kitcher (1988). According to this permissive view, the standard concept of "gene for *X*" recognizes no natural boundary beyond which it is false to associate effects with particular genes. If a gene has a systematic association with a trait that is complex and far removed causally from particular proteins—even if it is manifested in *another organism* from the one bearing the gene, as in some of Dawkins's favorite cases—still there is no problem in principle with saying that the gene is a gene "for" that trait or effect.

I suggest that if coding is taken seriously then this permissive view cannot be the whole story. If coding is a real relation linking genes and their effects, then not every causal consequence of a gene will be coded for. There is no indefinitely extended phenotype in the sense involving coding, even if there is in the sense involving causing.[21]

To see the point, consider some everyday cases involving messages with imperative content. Here we certainly do not see all the effects of a message, even the reliable and systematic ones, as necessarily coded for by the message. Suppose I know that if I order the extra-large pizza, this will have the consequence that the delivery arrives late. This fact does not imply that when I order the pizza I am also ordering them to make the delivery late. The likely effects of a message, even an imperative message, are not all part of the content of the message. Not everything caused by a message is coded for.

So the question of how "distal" the content of genetic instructions might be has to be settled by specific analyses of coding. Suppose a gene produces an amino acid that folds to produce an enzyme that catalyzes a reaction that produces a pigment that makes the organism camouflaged from its predators. The amino acid is the most proximal of these effects of the gene, camouflage the most distal. Here there are no fewer than four possible degrees of distality that the content of the DNA might have. Even on the assumption that there is a simple causal chain from amino acid sequence to enzyme to pigment to camouflage, there is the further question of which of these products is coded for. (The problem obviously gets harder in the case of traits that involve large numbers of genes and many nongenetic causes.) Different theories of coding will answer this question in different ways; in each case there are constraints on the possible contents of genetic messages deriving from factors used to analyze those messages.

In many cases the theories I have considered will apparently preclude assigning very distal contents to genetic messages. Theories relying on the developmental role of DNA, in particular, will have this consequence. If what makes DNA a message is the fact that it is read, along with the facts concerning the specificity of base triplets to amino acids, then apparently the only thing DNA can code for is the sequence of amino acids (the primary structure of a protein). Not even the folded state of the protein is coded for, even in cases (if there are such) where the pattern of folding is fully determined by the amino acid sequence.

Within the family of theories of coding that make use of evolutionary history, there is not so definite a verdict. If an analysis of coding gives an important role to the "consumer" of the message, as in a Millikan-style theory, then the content of the code is constrained by the possible scope of the biological functions of the consumer. If the consumer of the genetic message is the ribosomal/tRNA machinery, then the genetic message can only have as a content something that this machinery can have the function of bringing about. Probably then, the content of the message is no more distal than instructing the production of a protein. (This claim might be contested, I realize, by people with ambitious views about functions.) At the end of the section on evolutionary theories of the genetic code I discussed the possibility of a theory that stays closer to the outlines of Sterelny, Smith, and Dickison's (1996) view and does not try to accommodate all intuition-driven counterexamples. I suggest that these views, too, will tend to have the consequence that the content of genetic instructions can be no more distal than the production of a protein. But here again, there might be other possible views.

Biologists, incidentally, do not exhibit consensus on this issue of distality. Some specifically restrict coding to the specification of the amino acid sequence (Sarkar quotes Crick saying this, for example, see [1996], p. 858). In contrast, recall this sentence from the Lodish et al. textbook: "[DNA] contains a coded representation of all of the cell's proteins; other molecules like sugars and fats are made by proteins, so their structures are *indirectly coded in DNA*" (1995, p. 10, emphasis added).

The question of distality is made more complex by the role of the common phrase "gene for...." Does this phrase imply a relation involving coding, or just some relation of causal or statistical relevance? In earlier sections I discussed cases that suggest that there can be a useful concept of "gene for X" that does not imply a coding relation, and I think that many will agree that *if* coding is to be viewed as a real part of the theoretical structure of molecular biology, then it will be useful to also have a sense of "gene for X" that does not imply that the gene codes for X.

It might even be useful to recognize several different senses of "gene for X" that do not imply coding. Along with the coding sense, there may be distinct *statistical*, *causal*, and *teleofunctional* senses. In the statistical sense, a gene is a gene for X if it displays a certain pattern of correlation with X. A causal sense of "gene for X" will require that the connection between the gene and the trait have the right causal properties (there is likely to be much dispute about which properties these are).[22] And in the teleofunctional sense, a gene will be a gene for X if it has been maintained under natural selection because of its association with trait X.

I suggest that those who take coding seriously should want to recognize at least one of these other "gene for" concepts, and perhaps will find a use for all three.

If biologists recognize genetic coding along with one or more concepts of "gene for X" that do not imply coding, there will be a range of cases where we have a gene for X that does not code for X. For example, on a "gene for X" concept like Sterelny and Kitcher's, it is straightforward to have a gene for camouflage, penicillin resistance, or even reading. A sense of "gene for X that requires tighter causal connections might recognize genes for camouflage and penicillin resistance but not for reading—there are various ways the details of such views could be developed. But most reasonable theories of genetic coding will probably not hold that there can be genes coding for camouflage or penicillin resistance, and almost certainly not for reading.

So many views will recognize genes for X that do not code for X. Is the converse possible? Could there be a gene that codes for X that is not gene for X? On some views this is a possibility. I discussed a semihypothetical case in section 4 where a gene is systematically misinterpreted by translation mechanisms but nevertheless succeeds under selection. So we have a causal, statistical, and even naturally selected association between a gene and a protein, where the standard interpretation of the gene does not associate that gene's sequence with the protein's primary sequence. The gene does not code for the protein, on any theory of coding that gives a central place to the standard rule linking base triplets with amino acids, but in every other sense the gene is associated with that protein.[23]

A central topic of this paper has been an asymmetry within standard views in biology: both genes and environmental conditions have causal effects on development, but only genes code for (some of) their effects. My various proposals for analyzing genetic coding have all been developed with this constraint in mind. What is the status of the three alternative concepts of "gene for X" with respect to this asymmetry? The statistical and causal concepts of a "gene for X" will clearly be as applicable in principle to environmental conditions and nongenetic inherited factors as to genes—here I agree with Griffiths and Gray (1994). The status of the teleonomic sense is less clear; whether there can be "environmental conditions for X" in this sense will depend on the details of the analysis of evolutionary processes. On standard views, environmental conditions are not shaped by selection in the same sense that genes are, and there will be no "environmental conditions for X" in the teleonomic sense. But the "extended replicator" view developed by Sterelny, Smith, and Dickison (1996) would certainly treat genes and environments symmetrically in this respect. Other unorthodox views might do the same.

To close this section I will introduce yet another concept of "gene for X," as it has a combination of features that some might find useful. Suppose it is accepted that genes can only code for the primary structure of proteins. Is it possible to describe a concept of "gene for X" that (i) includes coding and (ii) is hence restricted to genes and not environments, but (iii) is more distal in the values it allows for "X"? One way would be to say that a gene is a gene for X if it *codes* for a protein that *causes* the distal trait. To pick a simple case, if a gene in a bacterium codes for an enzyme that causes penicillin resistance, that is a gene for penicillin resistance. This also provides a way to make some sense of the quote from Lodish et al. that I gave above: "... other molecules like sugars and fats are made by proteins, so their structures are indirectly coded in DNA" (1995, p. 10).

This last concept might be helpful to orthodox views, but I stress that if it is true that genes cannot code for anything other than amino acid sequences, that is an important fact that should be highlighted more than it often is by biologists. The importance of combating the mistaken idea that "genes are destiny" requires that the concept of coding, with which so many errors can be made, be kept strictly in its place.

6 Summary

I have no firm conclusions to draw, but here is a summary of some of my main points.

(i)   A central feature of the coding problem is the status of the claim that whereas genes, environmental conditions, and other factors can all *causally* affect traits, genes are distinguished from other factors in that only they *code* for some of their effects.

(ii)   The existence of a standard rule mapping amino acids to RNA and DNA base triplets does not solve the coding question. This rule could be seen as describing a set of causal specificities, without giving a rule of interpretation.

(iii)   If DNA sequences have semantic content, this content is imperative rather than indicative.

(iv)   An appeal to the concept of information, as understood in the mathematical theory of information, is unlikely to solve the problem, especially as such an approach is ill equipped to solve the asymmetry problem described in (i) above.

(v)   Two approaches to the problem of coding are analyses based on the *developmental role* of DNA and analyses based on *evolutionary history*.

(vi)   Views based on developmental role fight their main battles over the concept of "reading." They may also fight philosophical battles over the distinction between reading and misreading. If the idea that the cell reads DNA sequences can be defended, a satisfactory analysis might be possible by combining developmental role with a minimal appeal to a historical concept of function, to deal with the problem of misreading.

(vii)   Within the evolutionary approach, the analysis of Sterelny, Smith, and Dickison (1996) has problems with counterexamples, because it casts its net too widely. One way to avoid counterexamples is to move closer to Millikan's view in philosophy of mind, but this risks forcing an empirically unmotivated framework onto the biology.

(viii)   All the genes can code for, if they code for anything, is the primary structure (amino acid sequence) of a protein.

(ix)   If coding is taken seriously, there are good reasons to recognize one or more concepts of "gene for X" that do not imply that the gene *codes* for X. These concepts of "gene for X" might be analyzed in statistical, causal, or teleofunctional ways.

There are clearly some things genes can do that environmental conditions cannot do—act as a template in the construction of amino acid chains, most notably. But despite the enthusiasm of biologists, whether or not this role is best understood in terms of a coding relationship is a harder, and to my mind unresolved, issue.

Acknowledgments

Much of this essay emerged from discussions with Richard Francis. Thanks also to Lori Gruen, Philip Kitcher, Susan Oyama, Kim Sterelny, and Kritika Yegna-shankaran for helpful correspondence and discussion.

Notes

1. For various kinds of dissent and unease, see Oyama (1985), Lewontin (1991), Moss (1992), Sarkar (1996), Griesemer (forthcoming), and Francis (forthcoming). In this chapter I will use the language of coding in standard ways when I discuss examples.

2. This is a move whose importance has been stressed to me by Kritika Yegnashankaran.

3. The literature is large. Landmarks include Dretske (1981), Millikan (1984), and Fodor (1987). See Stich and Warfield (1994) for a collection of key papers and Sterelny (1990) for a review of the options.

4. I say "exactly" but there are exceptions in particular cases—see below in section 4.

5. So I see this sentence, from one of the textbooks cited earlier, as making a very strange claim, even within a strongly symbolic view of molecular biology: "The synthesis of protein is known as translation because it involves the transfer of information from one language (nucleotides) to another (amino acids)" (Raven, Evert and Eichhorn 1992, p. 144). I see claims like this as some evidence for a view like that of Kitcher, who holds that talk of coding makes no real contribution to molecular biology.

6. They add a causal constraint as well, but in a somewhat indirect way. An allele A is a gene for X if it is at a *locus* that causally affects X, and individuals with A (in normal genetic and nongenetic environments) have X. (Perhaps it would be better to build the causal constraint into the relation between A and X, or not have it at all.) Sterelny and Kitcher also stress that there are several different ways of handling the fact that a gene's role is dependent on its environment—they do not think there is just one way to understand "normal environment."

7. See Oyama (1985), chapter 5, for a menagerie of such phrases and formulas.

8. For "prescribing" see Lodish et al. (1995) p. 101; for "dictating" see the quote from Raven et al. at the start of section 1.

9. Some analyses require that the correlation be based in natural law (Dretske 1981).

10. For one attempt, see Maclaurin (forthcoming). In my view, this proposal has problems deriving from his not taking the content of DNA to be imperative.

11. Griffiths and Gray (1994, p. 281) give an interesting quote from Konrad Lorenz in which Lorenz says both that genes (i) contain a blueprint, and (ii) give the organism descriptive information about its environment ("a rival is red underneath"). I will not tackle this second kind of content-attribution here. Certainly not all claims about genetic coding can be understood as providing environmental information. I'm not sure how many could be handled this way.

12. For most of this discussion I will just say "developmental," not "developmental and metabolic," because the philosophical literature on this point is mostly concerned with development.

13. For a discussion of some of these phenomena that draws interesting philosophical conclusions, see Neumann-Held (unpublished).

14. I say "might reasonably" because Sarkar's overall view is that the concept of coding plays little positive role in molecular biology, and although it is much less problematic in the case of prokaryotes, on balance we might be better off without the concept.

15. See Fodor (1984) for an influential discussion.

16. For a discussion of the exact relations between functions and evolutionary history, see Godfrey-Smith (1994). Dretske's view (1988) is the best representative of theories that use a biological concept of function for part of the analysis but use other concepts as well.

17. Sarkar also discusses an interesting case. The RNA coding for part of an enzyme (NADH dehydrogenase subunit 7) in a parasite (*Trypanosoma brucei*) has hundreds of "U" bases inserted and some deleted before it is translated into protein. So there is no piece of DNA (contiguous or not) whose sequence corresponds, by the standard rule, to the amino acid sequence of the finished protein. Consequently, Sarkar says, "the DNA segment encoding the primary transcript can hardly be considered a *gene for* NADH dehydrogenase subunit 7" (1996, p. 861, emphasis added). I prefer to say this is a gene *for* NADH dehydrogenase subunit 7, but not a gene that *codes* for it.

18. On frame-shifting see Watson et al. (1987), p. 458; Lewin (1997), pp. 237–249

19. Sterelny, Smith, and Dickison's main discussion of these issues is around pp. 387–389 of their paper. For example: "One element of the developmental matrix exists only because of its role in the production of the plant lineage phenotype. That is why it has the function of producing that phenotype, and hence why it represents that phenotype" (1996, p. 388).

20. I use the term "representation" here although Millikan (1984) calls these signs "icons" and reserves "representation" for a richer concept.

21. Related to this is the fact that causal relations are transitive in a way that coding relations are not. If A codes for outcome B and B codes for outcome C, that does not imply that A codes for C.

22. For a good discussion of the differences between some of the relevant causal and statistical concepts in this area, see Block (1995) and a classic discussion in Lewontin (1974).

23. Some might say that even the presence of introns generates this consequence, but I do not agree with that. Then there is the case discussed by Sarkar (note 17 above), and cases of frame-shifting. Whether Sarkar's case fits might depend on the details of how the editing process is caused.

References

Block, N. (1986) "Advertisement for a Semantics for Psychology." Reprinted in Stich and Warfield 1994, op. cit.

Block, N. (1995) "How Heritability Misleads about Race," *Cognition* 56:99–128

Dawkins, R. (1982) *The Extended Phenotype*. Oxford: Oxford University Press.

Dretske, F. (1981) *Knowledge and the Flow of Information*. Cambridge, MA: The MIT Press.

Dretske, F. (1988) *Explaining Behavior*. Cambridge, MA: The MIT Press.

Fodor, J. A. (1984) "Semantics, Wisconsin Style." Reprinted in Stich and Warfield 1994, op. cit.

Fodor, J. A. (1987) *Psychosemantics*. Cambridge, MA: The MIT Press.

Francis, R. (forthcoming) *Genes, Brains, and Sex in the Information Age.*

Godfrey-Smith, p. (1994) "A Modern History Theory of Functions," *Noûs* 28:344–362.

Griesemer, J. (forthcoming) "The Informational Gene and the Substantial Body: On the Generalization of Evolutionary Theory by Abstraction," in N. Cartwright and M. Jones (eds.), *Varieties of Idealization*. Amsterdam: Rodopi.

Griffiths, P. and Gray. R. (1994) "Developmental Systems and Evolutionary Explanation," *Journal of Philosophy* 91:277–304.

Kitcher, P. S. (forthcoming) "Battling the Undead: How and How Not to Resist Genetic Determinism," to appear in R. Singh, C. Krimbas, D. Paul, and J. Beatty (eds.), *Thinking about Evolution: Historical, Philosophical, and Political Perspectives*. Cambridge: Cambridge University Press.

Maclaurin, J. (forthcoming) "Reinventing Molecular Weismannism: Information in Evolution," to appear in *Biology and Philosophy*.

Lewin, B. (1997) *Genes VI*. Oxford: Oxford University Press.

Lewontin, R. C. (1974) "The Analysis of Variance and the Analysis of Cause." Reprinted in R. Levins and R. C. Lewontin, *The Dialectical Biologist*, Cambridge MA: Harvard University Press, 1985.

Lewontin, R. C. (1991) *Biology as Ideology: The Doctrine of DNA*. New York: Harper.

Lodish, H., Baltimore, D., Berk, A., Zipursky, S. L., Matsudaira, P., and Darnell, J. (1995) *Molecular Cell Biology*, 3rd edition. New York: Freeman.

Millikan, R. G. (1984) *Language, Thought, and Other Biological Categories*. Cambridge, MA: The MIT Press.

Moss, L. (1992) "A Kernel of Truth? On the Reality of the Genetic Program," *PSA 1992* 1:335–348.

Neumann-Held, E. (unpublished) "Let's De-BlackBox the Gene!" Presented at ISHSSPB Conference, Seattle 1997.

Oyama, S. (1985) *The Ontogeny of Information.* Cambridge: Cambridge University Press.

Papineau, D. (1993) *Philosophical Naturalism.* London: Blackwell.

Raven, P. H., Event, R. F., and Eichhorn, S. E. (1992) *Biology of Plants*, 5th edition. New York: Worth.

Sarkar, S. (1996) "Decoding Coding—Information and DNA," *BioScience* 46:857–864.

Shannon, C. E. (1948) "A Mathematical Theory of Communication," *Bell System Technical Journal* 27:379–423, 623–656.

Sterelny, K. (1990) *The Representational Theory of Mind: An Introduction.* London: Blackwell.

Sterelny, K. and Kitcher, P. S. (1988) "The Return of the Gene," *Journal of Philosophy* 85:339–361.

Sterelny, K., Smith, K, and Dickison, M. (1996) "The Extended Replicator," *Biology and Philosophy* 11:377–403.

Stich, S. P. and Warfield, T. A. (eds.) (1994) *Mental Representation: A Reader.* Oxford: Blackwell.

Watson, J., Hopkins, N., Roberts, J., Steitz, J. A., and Weiner, A. (1987) *The Molecular Biology of the Gene*, 4th edition. Menlo Park, CA: Benjamin/Cummins.

16 Understanding Biological Causation

Charbel Niño El-Hani and Antonio Marcos Pereira

Despite the pervasive influence of the reductionist program, several philosophers and scientists still hope to avoid both radical dualism, as it challenges one of the main tenets of the scientific worldview, namely, ontological physicalism, and reductionism, since it would be most desirable that any description of the natural world should preserve the relative independence of the diverse levels of organization and, thus, the autonomy of sciences other than physics. This middle road between radical dualism and reductionism often combines the notions of supervenience and emergence, claiming that higher-level phenomena are both grounded in and emergent from a lower-level material structure.[1] In this chapter, our purpose is to examine how the understanding of biological causation is affected by the concepts of supervenience and property emergence.

1 Kim's Dilemma

1.1

We take as a starting point a polemic against Kim concerning the intelligibility of combining the notion of higher-level causation with that of supervenience. The supervenience of biological properties on physical properties[2] can be stated as follows:

For every biological property B there is a physical property P such that necessarily whenever something instantiates P at t it instantiates B at t (P is called a "base" or "subvenient" property of B). Moreover, nothing can instantiate B at t unless it instantiates some physical base property of B at t.[3]

Kim argues, in an essay on the problem of mental causation, that the concept of supervenience leads to a dilemma.[4] The argument begins with a tautology:

(i)   Either the supervenience of biological properties on physical properties fails or it holds.

As to the first horn of this tautology, it can be said:

(ii)   If the principle of supervenience fails, there is no way of understanding the possibility of biological causation (if the premise of the causal closure of the physical domain is held).

The acceptance of the unintelligibility of biological causation given the failure of the supervenience of biological properties on physical properties is supported by the premise that if the causal relations of any physical event are traced, they will never take us outside the physical domain. But if the principle of the causal closure of the physical domain is rejected, nonphysical explanations of biological processes would be acceptable, that is, biological causation would be intelligible even if the principle of supervenience was refuted. Nevertheless, we consider biological causation within the physicalist perspective and so agree with Kim's conclusion: "If you reject [the] principle [of the causal closure of the physical domain], you are ipso facto rejecting the possibility of a complete and comprehensive physical theory of all physical phenomena."[5]

The second horn of the tautology can be pursued by the following line of argument (see figure 16.1):

(iii)   Suppose that an instance of biological property B causes another biological property B^* to be instantiated.

Applying the concept of supervenience, we obtain:

(iv) B^* is supervenient on a physical base property P^*.

This proposition poses the question of how does the biological property B^* get instantiated. Two answers are possible:

(v) B^* is instantiated (a) because B caused B^* to be instantiated, or (b) because P^*, the physical base property of B^*, is instantiated.

Now the concept of supervenience poses the following problem: B^* always occurs because its base property P^* occurs; hence, if P^* occurs, B^* must occur, no matter the preceding events. This casts doubt on the claim that B^* is caused by B. But ... what caused P^* to be instantiated? The answer to this question allows us to reconcile the claim that B caused B^* with the premise that B^* is supervenient on P^*:

(vi) B causes B^* by causing P^*.

A general postulate can be then stated: "To cause a supervenient property to be instantiated, one must cause its base property to be instantiated."[6]

The concept of supervenience, however, implies that:

(vii) B itself has a physical base property P.

Now, if we compare B and P in regard to their causal status concerning P^*, we conclude that P preempts the claim of B as a cause of P^*:

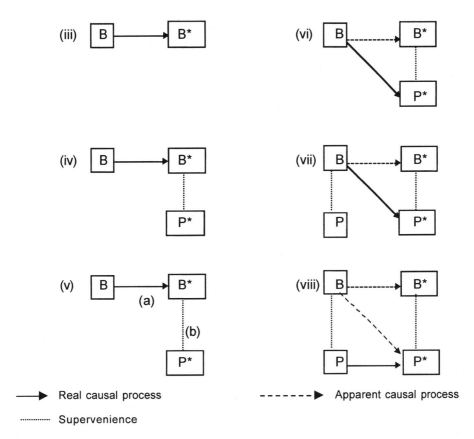

Figure 16.1
Kim's line of argument.

(viii)   *P* causes *P**, and *B* supervenes on *P*, and *B** supervenes on *P**.

Thus, in the case of the presumed *B*-*B** causation, the situation is rather like a series of shadows cast by a moving car. The moving car represents the real causal process, and the series of shadows it casts, no matter how regular and lawlike it seems, does not constitute a causal process:[7]

(ix)   The *B*-*B** and *B*-*P** causal relations are only apparent, arising out of a genuine causal process from *P* to *P**. Biological causation collapses into physical causation.

We can now enunciate Kim's dilemma, restating it as regards biological causation:

(x)   If the supervenience of biological properties on physical properties fails, biological causation is unintelligible; if it holds, biological causation is unintelligible. Hence biological causation is unintelligible.

1.2

The acceptance of the unintelligibility of biological or mental causation, even when the notion of supervenience holds, is like an invitation to nonphysical explanations. Kim pursues, then, a way out of his dilemma: if mental causation collapses into physical causation, the reduction of mental to physical processes would provide an explanation compatible with physicalism. When Kim opens the section on reduction as a way out, he poses the question of whether his dilemma applies to all supervenient properties: "One good question to raise about the foregoing argument is this: Wouldn't the same argument show that all properties that supervene on basic physical properties are epiphenomenal, that their causal efficacy is unintelligible?"[8]

He maintains that although biological properties, for instance, are also supervenient on physical properties, there is no reason to worry about biological causation, since "with properties like biological and chemical properties, we are much more willing to accept a reductionist solution to the problem. That is, if the 'higher-order' properties can be reduced to basic physical properties ... there is no *independent* problem of the causal efficacy of the reduced properties."[9]

As we intend, here, to challenge the idea that biological properties are as easily reducible as Kim assumes, we have now to present (1) arguments supporting the thesis that biological descriptions cannot be simply replaced by physical descriptions, and (2) a way out of Kim's dilemma that sustains the intelligibility of biological causation when the notion of supervenience is accepted.

2   Property Emergence, Biological Meaningfulness, and the Irreducibility of Biological Descriptions

2.1

The notion of emergence has a clear intuitive appeal, since it seems to account for the easily observable differences between the several levels of organizational complexity that one can perceive in the material world. When we try to capture the sort of macro-micro relationship that appears, say, between biological and chemical systems, or between mind and brain, or between social systems and the biological basis of behavior, we are commonly compelled to articulate a middle road between the extremes of substance dualism and reductionism. And this middle road frequently consists in the claim that the phenomenon in question is at once grounded in and yet emergent from the underlying material structure with which it is associated.

In spite of its intuitive appeal and its key role in the reductionism/antireductionism debate, many philosophers and scientists regard the notion of property emergence with suspicion. O'Connor indicates two reasons for such a distrust: (1) various formulations of the notion of emergence have been imprecise and are not obviously reconcilable with one another; (2) this notion seems to violate the maxim that you can't get something from nothing.[10] A third problem lies in the claim that the idea of downward causation, held by emergentists, amounts to a violation of the causal closure of the physical domain.[11]

As to the first problem, it is quite clear that it does not require a rejection of the notion of property emergence, but rather a critical appraisal of its previous formulations. And if we consider in some detail the idea that downward causation entails a breach of the causal closure of the physical domain, we will be able to answer the last two objections.[12]

2.2

The notion of emergence can be allied to a description of the evolution of matter as a process that passes through successive and hierarchical orders of complexity.[13] Indeed, emergentism was the first systematic formulation of the multilayered model of the world.[14] In this model, evolution is both continuous and discontinuous: if, on the one hand, a commitment to an evolutionist worldview entails the belief that functional systems of higher complexity are the product of the evolution of simpler systems, on the other hand, this evolutionary process passes through discontinuous barriers between levels of organization. If the continuity of the evolutionary process results in the supervenience of the complex system's properties on properties characterized in simpler systems and hence, in a refusal of substance dualism and an acceptance of ontological physicalism, its discontinuity, by its turn, implies that this supervenience cannot be translated into reducibility and hence gives rise to a rejection of reductionist ontologies of properties and a defense of nonreductive physicalism. Nonreductive physicalism combines ontological physicalism with property dualism, claiming that higher-level properties constitute an autonomous domain that resists reduction to the physical domain:[15] the properties of the diverse levels of organization would be *identical in nature* (given that the evolution of matter is, in a sense, continuous) but *different in complexity* (given the discontinuity between the hierarchical levels of organization).[16] That is, the properties of the relatively more complex levels of organization would be properly seen as emergent properties, *higher-level equivalents* of the properties of the relatively simpler levels.

Once a material system attains a sufficient level of structural complexity, or, in other words, once a higher level of organization arises in the evolutionary course, it acquires, as a matter of nomological but not metaphysical necessity, at least an emergent quality.[17] These emergent qualities are obviously supervenient on the properties of the composing elements, but must be taken as the outcome of the coordinated assembling of these elements in a new system. They can be seen as the product of both an upward and a downward causation; that is, they arise from a line of causality running both from parts to whole and from whole to parts. For this reason, they are irreducible to the properties of and relations between the parts. These emergent qualities introduce a change in the behavior of a system that possesses them, and, consequently, the higher-level processes cannot be fully explained in terms of the laws that govern its components in the absence of such organizational complexity.[18]

2.3

Kim properly maintains that emergentism can be thought of as consisting of the following three doctrines:

(1)   All that exists in the space-time world are the basic particles recognized by physics and their aggregates. (Ontological physicalism)

(2)   When aggregates of material particles attain an appropriate level of organizational complexity, genuinely novel properties emerge in these complex systems. (Property emergence)

(3)   Emergent properties are irreducible to, and unpredictable from, the lower-level phenomena from which they emerge.[19] (Irreducibility of emergents)

Downward causation is, as Kim recognizes, another fundamental tenet of emergentism.[20] He claims, however, that the notion of downward causation breaches the causal closure of the physical domain and, therefore, cannot be accepted by any physicalist:

The nonreductive physicalist, like the emergentist, is committed to irreducible downward causation, causation of physical processes by nonphysical properties, and this of course means that the causal closure of the physical is breached. The emergentist perhaps will not be troubled by it, but the nonreductive physicalist, insofar as he is a physicalist, should be.... [T]o abandon the physical causal closure is to retrogress to the Cartesian picture that does not allow, even in principle, a complete and comprehensive physical theory of the physical world.... This is something that no serious physicalist will find palatable.[21]

It is our contention here that downward causation does not entail a violation of the physical causal closure, and that, besides, Kim's reasoning misconstrues the emergentist view. Is it reasonable to claim that emergentism gives way to substance dualism and, furthermore, that an emergentist might not be troubled by a breach of the causal closure of the physical and, hence, admit an appeal to nonphysical causal agents to account for physical phenomena? First of all, it should be remembered that one of the main tenets of emergentism is ontological physicalism. As is any other physicalist, an emergentist is not inclined to accept nonphysical explanations of natural phenomena.

But then we come to the problem posed by Kim: how are we to make sense of downward causation without being committed to a violation of the causal closure of the physical domain? We understand that the notions of property emergence and downward causation are not incompatible with ontological physicalism, since they do not deny the claim that material particles constitute everything that exists. Emergentist hypotheses do not propose that living beings or minds bring with them nonphysical causal processes, but only stress the difference between events that take place in purely physical systems and phenomena connected with highly organized aggregates of physical particles, identical to physical systems in nature but different from them in complexity.

Kim uses in his reasoning a too narrow notion of "physical." It does not follow from the fact of physical causal closure that any explanation that is not supposed to breach it must include only the basic physical particles as captured in lower-level theories. Instead, this principle can be read broadly as a claim against nonmaterial causation, in a sense that obviously is not breached by downward causation. If we maintain that downward causation lies in the coordinated assembling of the parts in highly complex aggregates of basic physical particles, it is clear that it cannot amount to a breach of physical causal closure, for it has nothing to do with the causation of physical processes by nonphysical causal agents, as we see, for instance, in substance dualism. An emergentist conceives that all systems, no matter their complexity, are material, constituted by the very same basic physical particles, and the eye-catching difference between them lies not in any sort of substance diversity, but rather in their different levels of organizational complexity.

Furthermore, Kim is not right when he claims that according to the emergentist, why a higher-level property emerges from lower-level properties will forever remain a mystery and must be accepted as an unexplainable brute fact.[22] Instead, the notion of downward causation, defined as the coordination of the relations between the parts in largely replicable and stable patterns, can provide us with a rational basis for the understanding of property emergence.

2.4

Downward causation turns out to be a problem only for an ontology that allows only strict efficient causation. We claim, however, that the nature of downward causation is properly interpreted not as an "effective" top-down causation, but in a more Aristotelian sense, as are other causal modes, such as a kind of "formative" and "functional" causation.[23] We shall use here the four Aristotelian types of causality as reinterpreted by Emmeche, Køppe, and Stjernfelt:[24]

(1)   Efficient causality can be defined as a cause-effect relation involving an interactional exchange of energy and resulting in a temporal sequence of states being causally interrelated;

(2)   Material causality refers to immanent properties in the entities of a given level, which are composed, in a physicalist picture, by the entities of the immediately lower level;

(3)   Formal causality relates to the pattern or form into which the component parts of a higher-level entity are arranged;

(4)   Functional causality refers to the role played by a part within an integrated processual whole, or the purpose of a behavior as seen from the perspective of a system's chance of remaining stable over time.

Emmeche, Køppe, and Stjernfelt distinguish three versions of downward causation:[25]

(1)   *Strong downward causation*: this version of downward causation is related to the claim that there is a substantial difference between higher- and lower-level entities, and there can be a strict efficient causality from entities or processes at a higher level to a lower one. Even though the organizational aspect can be regarded as a necessary condition of the higher level, it is not organization but an ontological change in substance that is the distinctive feature of the several layers discernible in the world. This version of downward causation way be invoked, for instance, by substance dualism in the philosophy of mind and by vitalism in biology.

(2)   *Medium downward causation*: here, the higher-level entity, as a real substantial phenomenon in its own right, acts as a constraining condition (a kind of formal cause) for the activity of lower-level entities. Medium downward causation does not amount to a direct efficient causation from an independent higher-level entity to a lower-level one, as in the strong version. Instead, the control of the part by the whole is seen as a sort of functional causation, based on efficient and formal causation in a

multinested system of constraints. An example of this version of downward causation is found in Roger Sperry's "interactionism."[26]

(3) *Weak downward causation*: in this version, the higher level is seen not as a substance but rather as an organizational level, characterized by the pattern or form into which the component parts are arranged. The higher-level entity consists of entities belonging to the lower level, but the forms of the higher level are believed to be irreducible. Or, in our own terms, there is both an identity of nature and a difference of complexity between higher- and lower-level entities.

If we take the crucial difference between medium and weak versions to be a greater stress, in the former, on the functional aspects of downward causation,[27] we may conclude that our remarks on this issue are more related to the medium version. We claim that higher-level entities have, as crucial features, both the entanglement of matter and form, and the ascription of specific roles for the components as their relational properties are constrained and controlled by the irreducible form of a higher-order structure. A complex system shows a decisive dependence on the spatial and temporal control of the possible actions of its components for its own stability and survival. In higher-level systems, such as living beings and minds, formal, functional, material, and efficient causation are fundamentally interdependent. Yet, although the first two causal modes are better understood as top-down causation, effective causal processes still play a role in the underlying material structure that constitute the system. Thus, although the components and low-level effective causal connections are material causes of any complex system, form and function in such a system constrain the possible effective causal connections the components can enter into.

2.5

The notion of downward causation evokes the following question: If the whole is nothing but an aggregate of interrelated parts, how can it exert any influence transcending the relations between its components?[28] If we want to ascertain the nature of such a downward causation, we must simply engage in a comparison between, say, a mouse and a mass of mouse cells or molecules inside a test tube. Although a random collection of cells or molecules is in some sense undeniably plenty of relational properties, no one in his right mind could deny the remarkable difference between that collection and the organism. We can reasonably claim that this difference lies in the coordination of the relations between the molecules, cells, or any other of its components by the biological system, and that such a coordination can be seen as an irreducible line of causality running from the whole to the parts.

A higher level of complexity, for example, living matter, differs from the preceding level, for example, inanimate matter, due not only to the interactions between its elements, but rather to a new mode of coordinating these interactions. Parts randomly gathered also display relational properties, but there is no higher-order system coordinating these relations.

Following Baas, we can make use of type theory to understand why formal and functional top-down causation cannot be sufficiently described in terms of relational properties of a system's parts. He claims that going from one level to the next, a basic problem in studying multilevel systems, is analogous to type theory in logic.[29] Although Russell's theory of types is currently seen as an excessively drastic and unnecessary solution to the problem of paradoxes,[30] it is worth considering if it can be of any help when we strive to understand the relation between systems and their components. In type theory, the universe of individuals (objects that are not sets) and sets is arranged in a hierarchy of levels or types: the individuals are at level 0; the sets whose members are the individuals, at level 1; the sets whose members are sets of level 1, at level 2; and so on. All classes must be homogeneous as to the type: we cannot ascribe any meaning, in the context of the theory, to a set composed by objects pertaining to different types. Suppose, then, that a is an individual. The expression "$\{a, \{a\}\}$" is grammatically incorrect, since it denotes a set with two elements pertaining to distinct types.[31] As Ayer sums up, in Russell's theory of types, what can be said truly or falsely about objects of one type cannot be meaningfully said about objects of a different type.[32]

According to Baas, type theory can be suitably applied to the passage from physical or biological types to a higher-order type. The systemic organizing principle coupled with such a passage can be thought of as pertaining to a higher logical type than the relations between the components of the system, since it acts over these relations, transforming them by means of their coordination. Nothing can be an element of itself: the set of all umbrellas, for instance, clearly has the property of *not being an umbrella;*[33] similarly, the set of all relational properties that constitutes the form and function of a higher-level system has the property of *not being a relational property*. The coordination of the relations between the components that sets biological systems apart from physical systems should not be regarded as mere relational properties of the parts, but rather as a property that emerges only in the set of relational properties, that is, in the biological system as a whole.

2.6

We can make sense of this higher-order coordination as a *pattern* of processes that characterizes, in a proper level of organizational complexity, emergent systems in the evolution of matter, such as living beings or minds. But why should it be enough to refrain from the denial of biology or psychology as legitimate and autonomous domains of the scientific investigation, if we are just alluding to patterns of interactions between the parts when we make use of the notion of coordination? Why should not be enough to refer only to the relational properties of the basic physical particles and hence consider all higher-level properties as ultimately reducible to the predicates of the fundamental physical theories?[34] To answer these questions, we must consider if emergent qualities can be said to relate to nonreducible patterns that come into play in systems characterized by the appropriate level of organizational complexity. This also amounts to making our commitments explicit with regard to the nature of the explanation we intend to offer of phenomena. As long as we remain committed to an essentialist point of view, to the notion that our discursive practices may somehow be vehicles by which we grasp the "real nature of nature," we cannot avoid being stuck in a dilemma such as Kim's. And we do not seem to be able to overcome it unless we are willing to give up one of our previously stated commitments, namely, the causal closure of the physical domain.

Nonetheless, once we give up the commitment to essentialism and its reductionist consequence, we shall feel confident to claim that the putative relation between our descriptions of phenomena in terms of physics is not a matter of comparing and checking the way our descriptions relate to little chunks of the world, but that it is a matter of how we relate one description of the world in reductionist terms with another—that we are prone to endorse—in antireductionist terms.[35] This meta-philosophical stance we propose has to be made explicit in order to allow us to drop the notion that the vocabulary of physics is the most accurate vocabulary to describe any phenomena, and to drift toward the perspective that different ways of talking about the phenomena at stake are more philosophically interesting than just a unidimensional reductionist description.

If we accept, with Davidson and Rorty, that there is no fundamental distinction between "the way the world really is" and "convenient, but metaphorical, ways of talking about the world,"[36] we can frame our problem in a different manner: when we relate property emergence in biological systems to a higher-order coordination that can be understood as a *pattern* of processes, we are relying upon a sustainable, or, in other words, convenient level of description—or can such a pattern be sufficiently described in terms of the relational properties of chemical or physical particles?

Although it seems to be an obvious claim, we should call attention to the "fact" that without the complex network that makes an organism an organism, most causal links or relational properties in general, connecting cells, molecules, and so on that are ordinarily seen in living systems, can hardly make any sense. Even if one conceives that the relatively stable patterns of interactions between the parts within an integrated processual whole can be described in the vocabulary of physics, this will not suffice for a commitment to reductive physicalism. This is so because if we deny the autonomy of living systems—and hence of biological theories—and consider them merely as an outcome of the relational properties of their parts, we will fail to retain the higher-order, cohesive system that, by coordinating the physical interactions peculiar to living matter, makes them possible. This argument is grounded in a notion that may be called "biological meaningfulness"[37] and can be captured in the following conditional:

If P-P^* is a relational property of chemical or physical particles that occurs on account of the organic network that coordinates events inside an organism B, P-P^* does not make any sense and cannot be described sufficiently in the absence of the organizational complexity that characterizes B.

The notion of biological meaningfulness entails the irreducibility of biological theories to chemical and physical theories, no matter if one believes or not that the biological macroproperties can be given a reasonable account in terms of the relational properties of chemical or physical particles. We are dealing here with two different levels of description, and the higher-level description cannot be simply discarded if we are to be able to capture the meaning of the relational properties of the chemical or physical particles themselves within the coordinated networks that characterize organisms.[38] On the other hand, although a higher-level theory can be regarded as irreducible to lower-level theories, this claim does not entail, as Kincaid observes, the conclusion that lower-level theories are not relevant to explain higher level phenomena.[39] Rather, microlevel accounts are required for an adequate understanding of the phenomena pertaining to higher-order structures.

Let us return once again to the striking difference between a random mass of molecules and an organism. As to the former, it is clearly enough to describe the *simple relational properties* exhibited by lower-level chemical systems have only relational properties with no lasting and reliably replicable patterns, at least not in the same degree that can be seen in living beings. On the other hand, it seems to be necessary to describe—as we may call them—the *patterned relational properties* that characterize biological systems while considering the organism as a whole, or, in other words, to sustain a higher-level of description as convenient for capturing the entanglement of matter and form that constrains the possible interactions between the constituting particles so as to guarantee the regulation and stability of the complex system's behavior.

Enzymatic catalysis, for instance, can certainly be given a proper description at a molecular level, but this does not suffice to deny the necessity of a higher-level, biological description. This is so because enzymatic activity is always regulated, both temporally and spatially, and this regulation cannot be seen as a mere relational property of the enzymes, but must be rather regarded as a result of an intricate interplay of organic systems that coordinates the enzymes' actions at the level of the organism itself. DNA transcription and replication are also molecular processes and thus can be described in terms of chemical or physical theories. Nonetheless, their temporal and spatial coordination, which guarantees that cells neither replicate their DNA at the same time nor express the same genes, is a property emerging from the interdependence of organic systems, and despite its supervenience on molecular interactions, must be also described at a biological level. And so on.

2.7

We can draw a useful parallel between the notion of biological meaningfulness and Davidson's claim that "reduction" is a relation between linguistic items, and not between ontological categories.[40] Following his line of argument, we can maintain that even if one accepts that a biological event can be described in terms of chemical or physical theories, this will not entail the conclusion that such a biological event is nothing but a chemical or physical event. Relational properties of chemical or physical particles cannot be substituted for the organism in all true sentences about organisms, while preserving the truth of these sentences.[41] Even worse, a sentence concerning, for instance, the causal relationship between two biochemical processes that takes place in the context of the network coordinating events inside an organism is completely meaningless if it tries to capture the sense of the relationship merely in chemical or physical terms. A meaningful sentence about a biological event, such as the relation between the synthesis of a hormone and a controlled change in the behavior of an organism, cannot be simply restricted to the terms of chemical or physical descriptions; a biological level of description is certainly essential for its meaningfulness, since the sentence must refer to the very organizational complexity that makes the mentioned relation possible.[42] If we translate a biological meaningful sentence of biology into a meaningful sentence of physics or chemistry, this will not capture all that is meaningful in the sentence of biology.

In short, the notion of biological meaningfulness suggests that the organism must be preserved if we want to formulate true sentences about most relations that take place between its components, and this means, of course, that the level of description that characterizes biological theories must be also retained. Although we can accept —as we do— the value of microstructural descriptions of living beings (or minds), no elimination of macrolevel descriptions follows from such an acceptance, since they are a sine qua non for the lower-level descriptions to have meaning.

The complex nature of biological systems lends them an independence that cannot be captured by reductionist explanations.[43] The problem of causal explanation in biology demands the recognition that even if the outcomes of biological systems can be properly explained in terms of their components, these reductionist accounts are not sufficient for warranting the dismissal of biological descriptions. We must alter the way we conceive of the relation between living matter and its components in view of the awareness that a better understanding of the biological processes can be obtained if we combine descriptions at the level of the biological systems themselves with any account concerning the properties of and relations between their parts.

3   Emergence as a Way Out

3.1

Kim's dilemma would be easily solved if we could accept a reductionist way out. If so, the collapse of mental causation into physical causation would be an obvious consequence of the reducibility of mental properties. But Kim denies the very possibility of a escape route through reductionism. He claims that qualia, the phenomenal, qualitative characters of our experiences, cannot be construed as extrinsic or relational properties and hence cannot be reduced to their physical supervenience base.[44] He ends his essay circumscribed by the dilemma he was led to by the concept of supervenience: no matter if mind-body supervenience fails or holds, mental causation is in any case unintelligible. Someone with a propensity toward nonphysical explanations can but rejoice with this outcome, for it suggests that every theory aiming at a complete understanding of mind must necessarily invoke nonphysical factors.

Similarly, once we acknowledge that biological properties are not as easily reducible to physical properties as Kim assume, we are back with the problem of the collapse of biological into physical causation. Kim's dilemma is maintained: if biology cannot be reduced to physics, there seems to be no solution that makes biological causation intelligible in light of supervenience physicalism. We face a paradox: on the one hand, we sustain that $B \neq P$, namely, that biological processes are irreducible to molecular phenomena; on the other hand, if we follow Kim's line of argument, we will be compelled to claim that biological causation collapses into physical causation.

3.2

We hope to put forward here another way out of this dilemma, one which was not examined by Kim in the essay we have been discussing: an escape route through emergence that eschews the collapse of biological and also of mental causation into physical causation, preserving the intelligibility of both when the concept of supervenience is accepted.[45] Granted, Kim has addressed the issue of emergence in several of his essays.[46] Nonetheless, for our present purposes, it is enough to notice that he has not used this concept as the basis for an alternative way out of his dilemma.[47] He has addressed only the escape route through reduction and, since it has failed, closed both the article we have been examining and his book *Philosophy of Mind* entrapped in the very same dilemma:

> It is not happy to end to book with a dilemma, but we should all·take it as a challenge, a challenge to find an account of mentality that respects consciousness as a genuine phenomenon that gives us and other sentient beings a special place in the world and that also makes consciousness a causally efficacious factor in the workings of the natural world. The challenge, then, is to find out what kind of beings we are and what our place is in the world of nature.[48]

It seems to be necessary to use both the concepts of supervenience and emergence in a persuasive defense of nonreductive physicalism, or else we will be caught in the dilemma that, as Kim shows, apparently follows from the concept of supervenience, and we will have either to accept, as a way out, reductive physicalism or to refuse the idea that all higher-level properties supervene on physical properties, thereby embracing substance dualism.[49] Not surprisingly, Kim himself seems to be annoyed by the dilemma posed by the concept of supervenience. For instance, after advocating the irreducibility of qualia as a crucial argument against mind-body reductionism, he writes: "before we rejoice at the thought that reductionism has been defeated, remember this: if our considerations on emergentism and downward causation are generally correct, then if reductionism goes, so does the intelligibility of mental causation."[50]

The challenge posed by Kim at the end of his book is yet another motivation for us to engage in an attempt to bring forth a more vigorous formulation of the intuitively appealing notion of property emergence, in order to support our belief that both living matter and mind are irreducible material entities endowed with distinctive causal powers.

3.3

The crucial difference between Kim's line of argument and the escape route through emergence, as we conceive it, lies in the assumptions concerning the nature of downward causation. At a certain step in his argument, Kim claims that the realization of the higher-level property B^* by its physical base property P^* can be reconciled with the claim that B causes B^* to be instantiated if we suppose that B causes B^* by causing P^* (figure 16.1, vi). Notice, however, that downward causation is being depicted in Kim's argument as an effective top-down causation. Consequently, Kim's conclusion is inevitable: as B itself has a physical base property P, biological causation seems to collapse into physical causation.

But suppose we understand downward causation in a rather different sense, not as strict efficient causation but as a kind of formal and functional causation. It can then be claimed that Kim's dilemma will be avoided if the notions of supervenience and property emergence are combined in an apt formulation of nonreductive physicalism. A formal reading of an escape route through emergence runs as follows:

(i)   The concept of supervenience holds. Every higher-level property Q is supervenient on some physical property P, but this relation only entails that since Q and P are *identical in their material nature*, nothing can instantiate Q at t unless it instantiates P at t.

(ii)   The notion of property emergence also holds. Due to a *difference of complexity* between higher- and lower-level entities, or, in other words, an irreducibility of higher-order form and behavior, Q is an emergent property or *higher-level equivalent* of P.

We can develop the argument through a concrete example (see figure 16.2).

(iii)   Suppose that the visual perception of a predator activates an escaping behavioral response in a prey. Formally, an instance of biological property B causes another biological property B^* to be instantiated.

It follows from our premises (i) and (ii) that:

(iv)   The escaping behavior (B^*) is supervenient on adrenalin synthesis (P^*), but it can be seen, due to a difference in complexity, as a higher-level equivalent of P^*.

We can pursue, then, the possible origins of the biological property B^*:

(v)   The escaping behavior (B^*) was instantiated (a) because the visual perception of the predator (B) caused B^* to be instantiated, or (b) because adrenalin synthesis (P^*), the base property of B^*, was instantiated.

Since adrenalin synthesis (P^*) is the physical supervenience base of the escaping behavior (B^*), the following claim is the only way to reconcile the proposition that

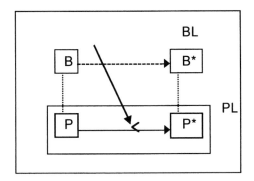

`----▶` apparent causal relation arising out of subvenient molecular events

`——▶` causal relation between retinal photochemical events and adrenalin synthesis in the context of a network of organic systems that ascribes meaning to this very relation (see Figure 16.3)

`——▶` coordination of sensory, cerebral and hormonal events that ascribes meaning to the relation between retinal photochemical events and adrenalin synthesis (biological top-down formal and functional causation)

`............` supervenience relation

Figure 16.2
The escape route through emergence. B = visual perception of a predator by a prey; B* = escaping behavior elicited in the prey; P = photochemical events in the retina of the prey; P* = adrenalin synthesis; PL = chemical/physical level of organization; BL = biological level of organization.

B^* was caused by the visual perception of the predator (B) with the conclusion, entailed by the concept of supervenience, that B^* can be only instantiated when its base property P^* is instantiated:

(vi)   The visual perception of the predator (B) causes the escaping behavior (B^*) by causing adrenalin synthesis (P^*).

Indeed, we see no problem in accepting, as a general principle, that any same-level biological causation is apparent, since it always depends on the instantiation of subvenient molecular processes. Nonetheless, such an instantiation is necessarily linked to the biological level of organization, or, in other words, it cannot be meaningfully captured by a mere reduction to a molecular description, since it depends on the peculiar coherence and coordination seen in biological systems. In our example, the visual perception of the predator can only result in the escaping behavior seen in the prey if it causes adrenalin synthesis to be instantiated, but the very connection between visual perception and adrenalin synthesis can hardly make any sense outside an organic network that coordinates sensory, cerebral, and hormonal events (see figure 16.3).

Applying once more the notions of supervenience and property emergence, we obtain:

(vii)   The visual perception of the predator (B) is supervenient on a physical base property, namely, the photochemical events that take place in the retina of the prey (P), but it can be seen, due to a difference in complexity, as a higher-level equivalent of P.

On Kim's line of argument, the supervenience of visual perception (B) on retinal photochemical events (P) leads to the conclusion that the latter cause adrenalin synthesis (P^*) and hence its supervenient property, the escaping behavior observed in the prey (B^*): the real causal process would take place between P and P^*, and the relation between B and P^* would be only apparent. Thence it follows that the causal relationship between the visual perception of the predator (B) and the escaping behavior of the prey (B^*) could be sufficiently characterized in the terms of chemical and, ultimately, physical theories, as a causal link between retinal photochemical events (P) and adrenalin synthesis (P^*).

This is far from being an indisputable conclusion. On the contrary, it is a rather doubtful one, since it ignores that the relation between P and P^* only makes sense if understood in the context of the coordination that acts over the relations between organic systems.[51] Both the molecular and the organismic (biological) levels of description must be retained if we want to formulate meaningful sentences about most phenomena pertaining to biological systems. We simply cannot make sense of the relationship between the retinal photochemical events and adrenalin synthesis if we hope to capture its meaning only in chemical or physical terms, avoiding reference to higher-level, biological descriptions. In spite of the explanatory gain associated with a linguistic reduction to molecular mechanisms, no ontological reduction that could deny the autonomy of biological theories and their objects follows from the relevance of such microdescriptions.

One should recognize that an irreducible biological influence lies in the organic network that coordinates events inside an organism, ascribing meaning to relational properties of chemical and physical particles that could not be sufficiently described without reference to the organizational complexity that characterizes the organism. This high-level organizing principle constrains the possible low-level interactions of

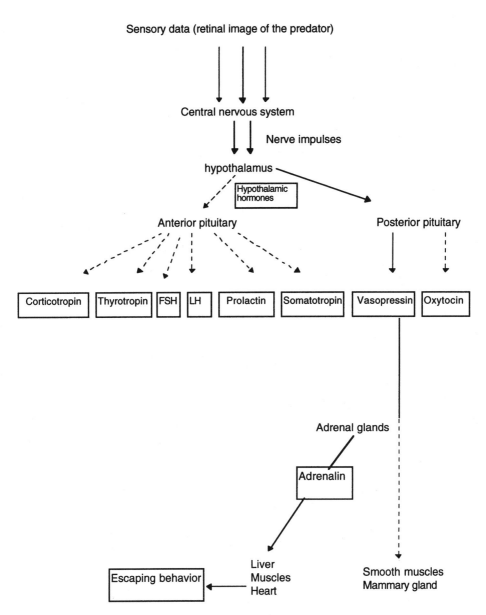

Figure 16.3
Organic network coordinating sensory, cerebral and hormonal events involved in the relationship between the sensory perception of a predator and the escaping behavior of the prey (adapted from Lehninger 1990, p. 514).

constituting particles, modifying their behavior so as to preserve the regulation of the system as a whole.

Furthermore, although biological properties supervene on physical properties, biological causation does not collapse into physical causation and is intelligible, not as a kind of strict efficient causation, but rather as a formal and functional top-down causation. If, on the one hand, it is true that no new efficient causal powers can magically accrue to B over and above the efficient causal powers of P, on the other, a nonreducible formal and functional causal influence does emerge at the higher-level entity:[52]

(viii)   The retinal photochemical events (P), the physical supervenience base of the visual perception of the predator (B), causes, *in the context of an organic network* coordinating sensory, cerebral, and hormonal events, adrenalin synthesis (P^*) to be instantiated, and P^*, in its turn, instantiates its supervenient property, the escaping behavior observed in the prey (B^*).

Hence:

(ix)   The escape route through emergence is not blocked and Kim's dilemma is dissolved.

Acknowledgments

We are indebted to Claus Emmeche and Kelly Smith for their incentive and their helpful suggestions. We are grateful to Jaegwon Kim for kindly sending us several of his papers. We want also to thank the Area of Mathematics and Science Teaching, Faculty of Education, University of São Paulo, for giving to one of us (C. N. E.) a travel award that made it possible to present a previous version of this essay at the last meeting of the International Society for the History, Philosophy and Social Studies of Biology, at the University of Washington, Seattle, July 1997. Finally, we are grateful to Valerie Hardcastle, organizer of the session on Connections between Philosophy of Biology and Philosophy of Psychology in that meeting. Research partially supported by grants from PICDT-CAPES (C. N. E.) and PIBIC-UFBA/CNPq (A. M. P.)

Notes

1. But compare O'Connor (1994).

2. "Physical," here, is construed broadly to include chemistry as well as physics (Kim 1996). Similarly, Carnap (1991, p. 395) takes "physics" as a common name for the nonbiological fields of science, including chemistry, mineralogy, astronomy, geology, and meteorology.

3. Adapted from Kim (1995a), p. 3.

4. Kim (1995). The dilemma presented by Kim for mind-body supervenience will be adapted here to the supervenience of biological properties on physical properties. Kim denies that this dilemma applies to the relationship between biological and physical properties, as we will see later. We disagree with him in this respect. The same argument is presented by Kim in his book *Philosophy of Mind* (1996, pp. 230–232).

5. Kim (1995a), p. 4. Notice that Kim's conclusion admits the possibility of a rejection of physicalism. In this case, it would be argued that every theory aiming at an understanding of physical phenomena based only on physical events is incomplete, and nonphysical causal agents should be always invoked. Since the physical causal closure itself is rejected by a "nonphysicalist," it is both obvious and desirable for her that the possibility of a complete physical theory of physical phenomena must be also refused.

6. Kim (1995a), p. 6.

7. Kim (1995a), pp. 7–8.

8. Kim (1995a), p. 8.

9. Kim (1995a), p. 9.

10. O'Connor, *op. cit.*, p. 91.

11. Kim (1996), pp. 232–233.

12. For a more detailed account, see El-Hani and Pereira (in prep.).

13. Novikoff (1945); El-Hani and Pereira (in press). This description of the evolution of matter, supported by the notion of levels of organization, amounts to an ontological claim. Nevertheless, we should highlight that if we believe, with Davidson and Rorty, that there is no relation between nonsentences and sentences called "making true," it follows that such an ontological claim cannot be seen as a proposition about the way the world really is, but rather as a convenient but metaphorical way of talking about the world (Rorty 1991, p. 116).

14. Kim (1996), p. 226.

15. Kim (1996), p. 212.

16. If we admit that all explanations of higher-level predicates, given the identity of nature sustained throughout the evolutionary course, must be physicalist, or, in other words, can never disobey the principle of physical causal closure, we have here a starting point for an argument showing that the notion of property emergence does not violate the maxim that one cannot get something from nothing.

17. O'Connor, *op. cit.*, p. 92; Feibleman (1954); Odum (1988).

18. O'Connor, *op. cit.*, p. 92.

19. Kim (1996), pp. 227–228.

20. Compare Kim (1996), p. 228. In another essay (1993, p. 350), he writes: "downward causation is much of the point of the emergentist program."

21. Kim (1996), pp. 232–233.

22. Kim (1996), p. 229.

23. Emmeche (in prep.), p. 23.

24. Emmeche, Køppe, and Stjernfelt (in prep.).

25. Emmeche, Køppe, and Stjernfelt, *op. cit.*; Emmeche, *op. cit.*, pp. 23–24.

26. Sperry (1983).

27. Emmeche, Køppe, and Stjernfelt, *op. cit.*

28. This question is roughly equivalent to the claim that property emergence cannot be accepted since it violates the maxim that you cannot get something from nothing. El-Hani and Pereira (in prep.) claim that if we take due account of form and function in higher-level systems, we will see that genuinely novel qualities in those systems appear not out of nothing, but rather from a new arrangement, in a multinested system of constraints, of the very same basic physical particles.

29. Baas (1996), p. 635.

30. See Miller (1996), pp. 69, 73–74.

31. Miller, *op. cit.*, pp. 72–73.

32. Ayer (1982), p. 30.

33. Miller, *op. cit.*, p. 75.

34. A similar question is raised by O'Connor (*op. cit.*) when he examines Alexander's identification, in his book *Space, Time, and Deity*, of emergent qualities with the idea of "configurational patterns."

35. In fact, a case has already been made in favor of this antireductionist perspective from inside philosophy of biology itself—two examples being Oyama (1985) and Varela (1979). Some contemporary philosophers have also been proposing very interesting and seemingly fruitful venues for exploration along these lines; see Putnam (1987) and Rorty (*op. cit.*).

36. Rorty, *op. cit.*, p. 116.

37. We owe this expression to Claus Emmeche.

38. Oyama (*op. cit.*) explicitly addressed this question in several different ways in her book (see especially pp. 142–149).

39. Kincaid (1988).

40. See Davidson (1986). See also Rorty, *op. cit.*

41. We must highlight that the notion of biological meaningfulness stands in opposition to the claim (found in Rorty, *op. cit.*, p. 115) that "to reduce the language of Xs to the language of Ys one must show either (a) that if you can talk about Ys you do not need to talk about Xs ...". In our argument, we claim that we can—and even must, if we accept the notion of supervenience—talk about the micromechanisms underlying a macrolevel property, but even so this microlevel account does not suffice for an elimination of the macrolevel description. On the other hand, the second criterion mentioned by Rorty regarding the reduction of a higher-level to a lower-level language follows from the notion of supervenience: "... or (b) that any given description in terms of Xs applies to all and only the things to which a given description in terms of Ys applies."

42. See section 3.

43. Smith (1994), p. 122.

44. Kim (1995a), p. 12. See also Kim (1996), pp. 236–237.

45. Edelman's theory of mind and consciousness can provide us with a formulation of both the emergence of mental properties and formal or functional downward causal influence of mental events on the lower-level cerebral processes and entities from which they emerge. See El-Hani and Pereira (in prep.). Brief accounts of Edelman's theory are found in Edelman and Tononi (1995) and Sacks (1995). Detailed accounts are found in Edelman (1987, 1988, 1989).

46. See Kim (1993, 1995b, 1996).

47. It is not difficult to understand why Kim has not considered a potential escape route through emergence. He is not inclined to accept emergentist hypotheses, as we can see in his remarks concerning downward causation (see section 2).

48. Kim (1996), p. 237.

49. Many philosophers prefer supervenience to emergence when they try to characterize the dependence relation between entities or properties at different levels (Emmeche, *op. cit.*, p. 17). We do not agree with this viewpoint, and, rather, both consider emergence as a species of supervenience and claim that, if we want to assume a nonreductive physicalist stance, supervenience is not enough to support it. We need to combine both concepts to make sense of the dependence relation between a higher-level system and its components (see El-Hani and Pereira, in prep.). Emergence seems to demand a specific kind of supervenience relation, since nonemergent properties are also grounded in an underlying material structure (O'Connor, *op. cit.*; Emmeche, *op. cit.*).

50. Kim (1996), p. 237.

51. There is no room for a reductionist counterargument emphasizing that figure 16.3 portrays a set of molecular interactions. It would be enough to highlight, in response, another property present in the figure, the biological coordination of those molecular interactions.

52. This stands in opposition to the following claim: "There are no new causal powers that magically accrue to M [in our example, B] over and beyond the causal powers of P. The approach to mental causation last pictured, therefore, is essentially reductionist: No new causal powers emerge at higher levels, and this goes against the claim of the emergentist and the nonreductive physicalist that higher-level properties are novel causal powers irreducible to lower-level properties" (Kim 1996, p. 232). Notice that the arrow representing biological causation in figure 16.2 has its origin not in the higher-level property B (as in figure 16.1), but in the biological level of organization itself.

References

Ayer, A. J. (1982) *Philosophy in the Twentieth Century*. London: Phoenix.

Baas, N. A. (1996) "A Framework for Higher-order Cognition and Consciousness." in S. R. Hameroff, A. W. Kaszniak, and A. C. Scott (eds.), *Toward a Science of Consciousness: The First Tucson Discussions and Debates*. Cambridge, MA: The MIT Press.

Carnap, R. (1991) "Logical Foundations of the Unity of Science," in R. Boyd, P. Gasper, and J. D. Trout (eds.), *The Philosophy of Science*, pp. 393–404. Cambridge, MA: The MIT Press.

Davidson, D. (1986) *Inquiries Into Truth and Interpretation*. Oxford: Clarendon Press.

Edelman, G. M. (1987) *Neural Darwinism: The Theory of Neuronal Group Selection*. New York: Basic Books.

Edelman, G. M. (1988) *Topobiology*. New York: Basic Books.

Edelman, G. M. (1989) *The Remembered Present: A Biological Theory of Consciousness*. New York: Basic Books.

Edelman, G. M. and Tononi, G. (1995) "Neural Darwinism: The Brain as a Selectional System," in J. Cornwell (ed.), *Nature's Imagination: the Frontiers of Scientific Vision*, pp. 78–100. Oxford: Oxford University Press.

El-Hani, C. N. and Pereira, A. M. (in press) "A Survey of Explanatory Methodologies for Science Teaching, I: Reductionism, Antireductionism, and Emergence," *Annals of the History, Philosophy and Science Teaching Conference, "Toward Scientific Literacy."*

El-Hani, C. N. and Pereira, A. M. (in prep.) "Higher-level Descriptions: Why Should We Preserve Them?" in P. B. Andersen, N. O. Finnemann, P. V. Christiansen, and C. Emmeche (eds.), *Downward Causation*. Dordrecht: Kluwer.

Emmeche, C. (in prep.) "Defining Life, Explaining Emergence."

Emmeche, C., Køppe, S., and Stjernfelt, F. (in prep.) "Levels, Emergence and Three Versions of Downward Causation." in P. B. Andersen, N. O. Finnemann, P. V. Christiansen, and C. Emmeche (eds.), *Downward Causation*. Dordrecht: Kluwer.

Kim, J. (1993) *Supervenience and Mind*. Cambridge: Cambridge University Press.

Kim, J. (1995a) "What Is the Problem of Mental Causation?" 10th *International Congress of Logic, Methodology and Philosophy of Science*. Florence, Italy.

Kim, J. (1995b). "Emergent Properties," in T. Honderich (ed.), *The Oxford Companion to Philosophy*, p. 224. Oxford: Oxford University Press.

Kim, J. (1996) *Philosophy of Mind*. Boulder, CO: Westview Press.

Kincaid, H. (1988) "Supervenience and Explanation," *Synthese* 77:251–281.

Lehninger, A. L. (1990) *Princípios de Bioquímica*. São Paulo: Sarvier.

Miller, D. (1996) "Russell, Tarski, Gödel: Um Guia de Estudos," *Ciência & Filosofia* 5:67–105.

Novikoff, A. B. (1945) "The Concept of Integrative Levels and Biology," *Science* 101(2618):209–215.

O'Connor, T. (1994) "Emergent Properties," *American Philosophical Quarterly* 31(2):91–104.

Odum, E. P. (1988) *Ecologia*. Rio de Janeiro: Guanabara.

Oyama, S. (1985) *The Ontogeny of Information*. New York: Cambridge University Press.

Putnam, H. (1987) *The Many Faces of Realism*. LaSalle, IL: Open Court.

Rorty, R. (1991) *Objectivity, Relativism, and Truth: Philosophical Papers, Vol. 1*. Cambridge: Cambridge University Press.

Sacks, O. (1995) "A New Vision of the Mind," in J. Cornwell (ed.), *Nature's Imagination: the Frontiers of Scientific Vision*, pp. 101–121. Oxford: Oxford University Press.

Smith, K. C. (1994) *The Emperor's New Genes: The Role of the Genome in Development and Evolution*. Durham, NC: Duke University (Ph.D. Thesis).

Sperry, R. (1983) *Science and Moral Priority: Merging Mind, Brain, and Human Values*. New York: Columbia University Press.

Varela, F. (1979) *Principles of Biological Autonomy*. New York: North Holland.

17 The Individual in Biology and Psychology

Robert A. Wilson

Individual organisms are obvious enough kinds of things to have been taken for granted as the entities that have many commonly attributed biological and psychological properties, both in common sense and in science. The sorts of morphological properties used by the folk to categorize individual animals and plants into common sense kinds (that's a *dog*; that's a *rose*), as well as the properties that feature as parts of phenotypes, are properties of individual organisms. And psychological properties, such as believing that taxes are too low, and remembering the last seven digits you read in the phone book, are likewise properties of individual organisms.

Yet the individual has played a more controversial and I think more interesting role in a number of debates in both biology and psychology with a philosophical edge. In philosophical thinking about psychology and cognition over the last twenty years or so the individual has been viewed not merely as the subject of psychological predication (to re-express the point in the previous paragraph), but also as a sort of boundary beyond which psychology either should not or need not venture. This is the central, general idea of Jerry Fodor's (1980) thesis of *methodological solipsism* or what Tyler Burger (1979) called *individualism*. It is the idea that psychological or mental properties ought not to presuppose the existence of anything beyond the head of the individual who has those properties, an idea shared in various guises by philosophers of earlier times as different from one another as are Descartes, Brentano, and Carnap. Individualism is expressed in contemporary materialist philosophy of mind in terms of the technical notion of *supervenience*: mental properties must supervene on the intrinsic, physical properties of the individuals who have them. This is to say that individuals identical in their intrinsic, physical properties must also be identical in their psychological properties; this is so no matter how different their environments.

In the philosophy of biology, one of the more substantial debates involving the status of individuals has been that over the *units of selection*. Darwin and the bulk of nineteenth-century evolutionary thought took natural selection to operate on individual organisms via the sorts of phenotypic traits with respect to which individual organisms both within and between species can vary. Here the individual has served as a sort of unit of selection by default. But units both larger than the individual—the group or even the species—and smaller than the individual—most famously, the gene or small genetic fragment—have been viewed as alternative units of selection over the last fifty years. Wynne-Edwards's *Animal Dispersion in Relation to Social Behavior* (1962) became a symbol of the dangers of group selectionist thinking following George Williams's influential critique of group selection in his *Adaptation and*

Natural Selection (1966), a book that also contained the seed of the idea that Richard Dawkins championed in both the title and substance of his *The Selfish Gene* (1976).

In what follows I want to concentrate on just three themes that interact with this pair of debates, with particular focus on illustrating the mutual relevance of the biological and psychological discussions. I will not have much to say here about methodological solipsism in psychology or the units of selection debate in evolutionary biology per se, but I hope that the connections of my three themes to these more general topics are clear. The themes I shall briefly explore are: Dawkins's notion of the extended phenotype and its relation to nonsolipsistic or *wide* views of psychology; the closely related metaphors of causal powers and encoding in both biology and psychology; and the idea of individuality itself and its relation to topics such as complexity and the locus of control.

1 The Extended Phenotype

It is not Dawkins's *The Selfish Gene* that I want to focus on but a much less-discussed book of Dawkins's, written, as Dawkins says in his preface, for his "professional colleagues, evolutionary biologists, ethologists and sociobiologists, ecologists, and philosophers and humanists interested in evolutionary science" (p. v). This is Dawkins's *The Extended Phenotype* (1982), not only written but also received very much in the shadow of its more widely-read predecessor. At a general level, the conclusion of this section of the paper is simple to state: the received perspective on the central idea of the book—the idea of the extended phenotype itself—limits the plausibility of that idea, in much the way that individualistic construals of the computational theory of mind impose a constraint on that theory that it need not and should not bear.

The chief idea of *The Extended Phenotype* is that the phenotypes that express particular genes or genetic fragments do not stop at the boundary of the organism, but extend into the world at large. Shells that are found are no less part of the phenotype of hermit crabs than are shells that are grown by other crabs, and the web morphology of a given species of spider (or even individual spiders) is as much a phenotype of that species (or individual) as are the length of its legs or the distribution of pigment on its body. Phenotypes might belong to a given individual organism even though they reach beyond the boundary of the body of that organism.

We can put this in terms of the distinction between *replicators* and *vehicles* that Dawkins introduced in *The Selfish Gene*, and which David Hull (1984) generalized in his discussion of *replicators* and *interactors*. Replicators are entities that are capable of making copies of themselves, and that do so with enough reliability to be represented from generation to generation. Vehicles, by contrast, are what replicators lodge themselves in, what house one or more replicators. Replicators and vehicles (or interactors) may but need not be the very same entity, a point that Hull emphasized. Dawkins's thesis in *The Selfish Gene* is that genes are replicators, and individuals are vehicles. Thus, since replicators are what natural selection operates on, genes are the units of selection. Using this same distinction, we can say that Dawkins's thesis in *The Extended Phenotype* is that a replicator's phenotype need not be restricted to the vehicle that replicator happens to occupy.

A corollary of this view, one prompted by Dawkins's own probing questioning of the past focus on organisms in evolutionary biology, is that organisms are simply convenient ways of packaging many phenotypic characters: *their* existence is also a result of the extended reach of the gene on the world at large, since packaging biological matter in this way has proven to be mighty effective in preserving replicators across evolutionary time. Although the "convenient way of packaging" expression comports with the general view of individual organisms in *The Selfish Gene*, the final chapter of *The Extended Phenotype*, entitled "Rediscovering the Organism," treats the emergence of individuality itself more seriously, taking up suggestions in the work of John Bonner on development and phylogeny that I will return to briefly later in this chapter.

In championing the extended phenotype, Dawkins saw himself as liberating the phenotype from the bounds of the individual organism, and with it the crucial notion of phenotypic differences between organisms within a population. The idea that phenotypes can be and sometimes are *extended* in the sense that Dawkins intends seems to me both true and important, though we should not overemphasize this importance. Dawkins's own "wildest daydream ... that whole areas of biology, the study of animal communication, animal artefacts, parasitism and symbiosis, community ecology, indeed all interactions between and within organisms, will eventually be illuminated in new ways by the doctrine of the extended phenotype" (1982, p. 7) has not be realized, and the practice of what Dawkins called an "extended genetics" (1982, p. 203), which would supplement conventional genetics by following the effects of genes out into the world beyond the individual organism, has hardly developed over the last fifteen years.

I shall focus below on three ways in which the idea of the extended phenotype as Dawkins presents and defends it is significantly more controversial than what we might think of as the bare-bones extended phenotype. I will say why this is so, and in so doing propose a divorce between the bare-bones extended phenotype—the idea of the extended phenotype in itself—and that idea as Dawkins develops it.

First, Dawkins presents the extended phenotype as a natural consequence of his defense of the selfish gene. (This is one reason that what in the preface he calls the "heart of the book," articulating the idea of the extended phenotype, is to be found in three chapters that follows ten others devoted to cleaning up misunderstandings about and objections to the idea of the selfish gene.) Since genes are, to a good approximation, the only or best replicators in the evolutionary process, they are the units of selection and their differential survival is what matters in evolution. They replicate via the phenotypes they express, of course, but only traditional bias leads us to think of these as strictly *bodily* or *organismic* manifestations of the gene—thus, the extended phenotype. Those with qualms about the selfish gene view will see this defense of the extended phenotype as not much more than an interesting exercise in reasoning.

By contrast, Dawkins's most forceful arguments, in my view, for embracing the extended phenotype are *parity* arguments that rely only incidentally on the selfish gene view. In these arguments Dawkins uses widely accepted views of what sorts of things count as phenotypes and the relation between genes and phenotypes, arguing that since there is no relevant difference between these paradigms and phenotypes that extend beyond the boundary of the organism, such parity offers a defense of the extended phenotype. If you are prepared to accept something that grows as part of an organism as a phenotype—a shell, perhaps—why not accept something that it acquires through its interaction with the world—another shell—as a phenotypic expression of its genes? If behaviors—such as stalking in lions—can be phenotypes, as the ethologists convinced us long ago (prior to the sociobiology of the 1970s), then why not behavior that reaches into the body, and the behavioral repertories of other organisms—such as that of parasitized or otherwise manipulated hosts? Similarly, in Dawkins's own words, since "we are already accustomed to phenotypic effects being attached to their genes by long and devious chains of causal connection, . . . further extensions of the concept of phenotype should not overstretch our credulity" (1982, p. 197).

Dawkins's basic point is that there is nothing in the concept of a phenotype restricting it to the boundary of the organism, and this point stands independent of the selfish gene view. As he says in several places (e.g., 1982, pp. 198, 214), he is making a "logical point" about the concept of a phenotype, and as such the point has little to do with significantly more controversial views of the unit of selection. This implies, of course, that one could augment the traditional, individual-centered view of natural selection and adaptation, the idea that the individual is "the" unit of selection (or at least, in these heady pluralistic days, *a* unit of selection) with an extended conception of the phenotype. In fact, there would seem little to bar one from incorporating the extended phenotype into a pluralistic view of the units of selection that embraced forms of group selection, such as David Sloan Wilson and Elliott Sober (1994, Sober and Wilson 1998) have recently defended.

Second, and relatedly, Dawkins often talks of the "extended phenotypic effects" (1982, p. 4) that replicators have, the "phenotypic effects of a gene" (1989, p. 238), and of phenotypes as the "bodily manifestation of a gene" (1989, p. 235). This creates the impression that Dawkins thinks of phenotypes as properties *of genes* (as in "the long reach of the gene"), and so obscures the point that phenotypes are, in the first instance, properties *of individual organisms*. Genes certainly have phenotypic effects (extended or otherwise), in the sense of playing a significant causal role in bringing about those effects, but they do not *have* phenotypes, that is, they are not the subjects of phenotypic predication; phenotypes do not *belong to* genetic replicators, but to the organismic vehicles in which they are housed. Eye color, running speed, and wing shape are all phenotypes of individual organisms; but so too are the extended phenotypes of web morphology (spiders), shell choice (hermit crabs), and dam size (beavers). If this is correct, then organisms are *presupposed* by the extended phenotype view, in that they are the entities to which these phenotypes are ascribed. This means that organisms are not simply the means by which genes are packaged and propagated through generations; rather, they are central to making sense of the extended phenotype. What we might call the *mere vehicles* view of individual organisms doesn't do justice to the overwhelmingly nonrandom distribution of the bearers of extended phenotypes, bearers who will, of course, be the subject of generalizations about the phenotypes, extended or otherwise, that they instantiate.

There is an ironic even if implicit admission of this point in the final chapter of *The Extended Phenotype* when Dawkins turns to consider the question, "Why organisms?" Given that there was nothing requiring replicators to be packaged into these nice, discrete bundles that we call (paradigmatic) organisms, why are they so packaged? Dawkins's answer is that organisms have a regular life cycle, that is, a sequence of development that "permits a new beginning, a new developmental cycle and a new organism which may be an improvement, in terms of the fundamental organization of complex structure, over its predecessor" (1982, p. 262). Organisms reproduce, rather than simply grow, and the developmental bottleneck that reproduction creates allows for the possibility of the intergenerational modifications that constitute adaptations. Here Dawkins acknowledge his debt to Bonner's *On Development*, and one irony of Dawkins's interesting discussion is that it is Bonner's student, Leo Buss, whose *The Evolution of Individuality* (1987) not only answers Dawkins's question, "Why organisms?" in more detail than any other work but also provides an insightful critique (pp. 171–197) of Dawkins's selfish gene view.

Third, Dawkins contrasts organismically bounded phenotypes with those that reach into the world at large, identifying the extended phenotype with the latter. This creates the worry that extended phenotypic effects, unlike their bodily bounded kin, will be unsuited for systematic study, since the effects of genes on the world at large are infinite in number and various in strength. Call this the *dissipative concern* about the extended phenotype: systematic study of an organism's extended phenotype is precluded, because once we move beyond the boundary of the organism the phenotypic effects such study would require dissipate into the world at large.

For example, the science of extended *genetics* that Dawkins dreams of will remain merely a dream. If the reach of the gene were viewed as extending into the world beyond the organism, then the organism's phenotype would include all sorts of greater and lesser effects that those genes have. Conventional population genetics is largely concerned with phenotypic variance within a population, particularly that portion due to genetic variance, and the organism serves as a clear boundary for individuating (and so measuring) phenotypic characters of study. But in an extended genetics with dissipative genetic effects this presupposition is absent, and so what variation is to range over becomes unclear. Similar problems would arise in other areas of systematic study that seem to presuppose a circumscribed conception of the phenotype, such as evolutionary taxonomy or developmental genetics.

The problem here stems, I think, from Dawkins's own dichotomy between organismically bounded phenotypes and phenotypes that reach into the world at large. This dichotomy is not exhaustive, and so the forced choice it presents is a misleading one. For we can see extended phenotypes as bounded by systems *larger than the individual organism*, and so as *not* dissipating into the world at large. That is, by recognizing systems, even individuals, that include individual organisms as proper parts, as the units at which extended phenotypes end, we can extend the phenotype beyond the boundary of the organism without losing the focus on a bundle of phenotypic effects that could be subject to systematic study. We can make this suggestion clearer, perhaps, by considering a range of Dawkins's own examples.

In *every* example that Dawkins provides—caddis fly house shape, spider web morphology, beaver dams, termite mounds, fluke parasitism in snails (and parasitism in general)—the phenotypic effects are part of some well-defined and bounded system: caddis fly + house, spider + web, beaver + dam, termite(s) + mounds, parasite + host. Thus, although phenotypes are extended in the sense of extending beyond the boundary of the individual organism to which they belong, they are not to be identified, in general, as "all the effects that [a gene] has on the world" (1989, p. 238). Rather, extended phenotypes are circumscribed by individual entities larger than (and that contain) the organisms to which they belong. This addresses the dissipative concern expressed above by identifying an organism-like unit within which one can locate (and so taxonomize and quantify) extended phenotypes.

In making this point, I have restricted myself to considering Dawkins's own examples. In all of these examples, the extra-individualistic or what I have called elsewhere (Wilson 1994, 1995) *wide* systems are exhaustively composed of an individual organism, an organismic artifact (such as a shell, a dam, a mound), and the relations between them. In doing so I do not mean to prejudge the forms that wide systems can take, or to suggest a tidy formulaic account of when they should be invoked in science. In fact, since the wide *cognitive* systems that we and other animals have are not made up of individual organisms plus individual artifacts in those organism's environments (see Wilson 1999 and below), and thus differ from the above examples in significant ways, theorizing formulaically about the full range of wide systems available to one in rethinking the relevant sciences in nonindividualistic terms would seem premature.

I have thus far suggested thinking of these extra-individualistic systems as providing a boundary for the corresponding extended phenotypes as a way of addressing what I am calling the dissipative concern. On this view, the individual organism remains the bearer of the extended phenotype, as I noted it should in making the previous point. But might we go further and posit these *wide* systems themselves as the bearers of extended phenotypes? After all, to return to the language of replicators and vehicles, such wide systems—individual organisms plus their environmental appendages—are no less vehicles for the delivery of replicators than are individual organisms themselves. On this view, phenotypes would extend beyond the body of individual organisms, but they would fall inside the boundary of these wide systems and so *not* be extended with respect to them.

To accept this view would require a more far-reaching revision of our conception of the individual in biology than the relatively modest revisions I have been suggesting so far. In effect, these larger systems of which organisms are a part would *replace* organisms in biological theory. This would imply, given the traditional view of the unit of selection, that *these* systems were the units of selection; alternatively, given genic selectionism, it would be these systems that were the vehicles via which genes were selected. It would be these systems, not organisms per se, that have their various places in the Linnaean hierarchy, and these wide systems that underwent life cycles, formed ecological communities, and had innate behavioral repertoires. Moreover, at least in the case of systems that include multiple organisms—host-parasite systems, predator-prey systems, mutualistic and symbiotic systems—talk of manipulation, deception, and cooperation would seem less appropriate, since we would now be characterizing the relationships that held between two parts of one overall system, not one organism and something that *it* acted on in some way. I pass no judgment on the plausibility of this shift in perspective, but simply point to some of its implications.

Since discussion of these three points has been somewhat lengthy, let me bring them together by way of an interim summary of the chapter so far. What I am suggesting is a version of the extended phenotype that (a) is divorced from its association with the selfish gene, (b) explicitly acknowledges the centrality of individual organisms, and (c) facilitates the prospects for a systematic study of extended phenotypes by recognizing the reality of individual entities larger than organisms.

By no small coincidence, these three suggestions parallel claims that I have defended about computationalism in contemporary cognitive science (Wilson 1994, 1995 [ch. 3], 1999). In reverse order, they are: (c′) a call for the exploration of *wide* computational systems, systems of computational, cognitive states that extend beyond the boundary of the individual; (b′) an acknowledgement of the place of the individual (or parts of that individual) as the subject of those states; and (a′) a general plea for the divorce of the computational theory of mind from the individualistic company that it often keeps. In the next section I turn to some of the metaphysics that lies beneath the surface and the metaphors that bubble up to the surface of the individualistic views of biology and psychology to which I am opposed.

2   Causal Powers and Encoding

There is an illusive cluster of views of the scientific exploration of the mind that involve an appeal to the notions of causal powers and encoding whose discussion sheds some light on corresponding views in the biological sciences.

As I have said, individualism in psychology is the view that psychological states should be individuated or taxonomized so as to supervene on the intrinsic, physical properties of the individuals who have those states, and since it implies that physically identical individuals must have the same psychological states, it is often taken to be a *minimal* materialist constraint on psychology or cognitive science. Individualism is sometimes glossed as the view that psychological kinds are demarcated "by causal powers," meaning that psychological states with the same causal powers must belong to the same kind. So glossed, it has been claimed (e.g., by Fodor 1987, ch. 2) to gain support from a general thesis about scientific kinds, namely, that they are taxonomized by causal powers. This view articulates the intuitions that psychology stops at the skin—really, at the skull—and that environmental variables are relevant to psychology only insofar as they impinge on the internal, physical states of individuals. The name "individualism" is used, in part, to suggest the idea that individuals serve as the uppermost boundaries for the entities that are relevant to do psychology.

This view of the role of causal powers in psychology provides support for a familiar view of mental representation: that it involves *encoding* information about objects, properties, events, or states of affairs. A well-known version of the encoding view is the picture or copy theory of mind, according to which to have a mental representation of *m* is to have a mental picture or image of *m* in your head, where the picture is "of *M*" just because it looks like *m*. A version of the encoding view prevalent in cognitive science is the language of thought hypothesis, according to which to have a mental representation of *m* is to have a token in your language of thought, *M*, that stands for or refers to *m*. Unlike the copy theory of mental representation, on this view there need be no resemblance between the representation and the represented. On either view, because mental representations encode information about the world, cognitive scientists can (and should) explore *these* properties rather than the relationships that exist between organisms and environments.

I have argued at length elsewhere (e.g., Wilson 1995, 1999) that both of these views are false, but that is not my plaint here. At the end of my introduction I characterized causal powers and encoding as *metaphors*, and it is viewing them as such that allows us to make a connection back to the biological sciences. In calling them metaphors, I mean to suggest both their literal falsity and that they create a certain overall conception of what sorts of things mental states are: they are encapsulated in individuals, located in the brain, buried away from direct impingement from the world. Thus they can be investigated as self-contained entities causally insulated from—yet reflective of—the world beyond the organism. This sort of metaphor should be familiar to biologists, since it is the dominant metaphor governing the conception of genes. In genetics, this metaphor has its root in Morgan's school in the 1920s and was developed through the incorporation of the informational language of codes, templates, instructions, and programs in the 1940s and '50s (see Fox Keller 1995). But I think that a variation on the metaphor that places less emphasis on the idea of encoding per se also structures contemporary thought about cells, organs, and even organisms themselves, with influential historical antecedents in Schwann's doctrine of the cell as the unit of living systems in his physiological investigations of the 1830s, and the development of theories of cellular respiration in the 1920s (see Bechtel and Richardson 1993, esp. chs. 3–4).

With this conception comes a certain view of these self-contained entities as *loci of action*: since they are the things in which the relevant causal powers are located, their investigation is central to understanding the corresponding phenomena. In cognitive science, this plays out in the fantasy of being able to read the language of thought off of the brain; in genetics, it is manifest in the rhetoric of DNA as the "code of codes" and the exclusive concentration on the gene as "the" mechanism mediating inheritance, embryogenesis, and more generally, development. In both cases, environments are relevant only insofar as they are encoded by the corresponding entities—neural states or genes—and more complicated systems, such as neural circuits and developmental pathways, are conceived of as spatial aggregates of neutrons or temporal aggregates of genes.

Despite the reluctance of practicing scientists to view themselves as engaging in metaphysics (Hey, that's for *philosophers!*), there is a general metaphysics in the background of this conception that is perhaps worth identifying more explicitly. The general metaphysics here is a twentieth-century update of seventeenth-century corpuscularianism, a view that we might call *smallism*, discrimination in favor of the small and so against the not-so-small. Small things and their properties are seen to be ontologically prior to the larger things that they constitute, and this metaphysics drives both explanatory ideal and methodological perspective; the explanatory ideal is to discover the basic causal powers of particular small things, and the methodological perspective is that of reductionism. In the days of Locke and Boyle, corpuscles were the very small things and the properties they had were referred to as *primary qualities*, these being taken to inhere in the corpuscles themselves, and derivatively in the things they compose.

The problem with smallism as a general metaphysics is that many of the kinds of things that there are in the world—modules, organisms, species, for example—are *relationally* individuated, and thus what they are *cannot* be understood solely in terms of what they are constituted by. Moreover, regardless of how the entities themselves are individuated, many of their most salient properties—their functionality, their fitness, their adaptedness, for example—are relational properties, which, since they don't inhere in the entities that have them, can't be discovered by focusing exclusively on what falls inside the boundaries of those entities.

To illustrate what this objection is getting at, let us return to the extended phenotype. If the phenotype literally extends beyond the body of the organism that has it, then what that phenotype is can't be explored and understood solely by examining an organism's causal powers or its intrinsic, physical properties. Rather, one needs to shift one's focus to the relations between the organism and its environment, to the extra-organismic *system* of which the extended phenotype is a part. This is not to imply that an individual's intrinsic causal powers are not relevant to what extended phenotypes it has, but to point out that the object of study contains individuals as proper parts, not boundaries beyond which one may not venture.

One might look to defend smallism from this objection by moving to a larger individual that, effectively, makes these relations intrinsic properties of this larger individual, a view what I entertained at the end of my discussion of the extended phenotype. The idea of this reply is perhaps best illustrated with another example. Consider Ernst Mayr's biological species concept: "a species is a reproductive community of populations (reproductively isolated from others) that occupies a specific niche in nature" (1982, p. 273). Since the property of being a member of a reproductive community (conspecifics) is not an intrinsic property of an individual organism (I_1)—for it could be lost in a given case simply by changing those conspecifics—being a member of a particular species appears to be neither individualistic nor intelligible in smallist terms. But we can now move to consider the larger "individual," the whole breeding population or even species (I_2) itself, to revive both a sort of individualism and smallism. By shifting up to a larger individual, we convert a relational property of I_1 to an intrinsic property concerning the relations between parts of I_2. And we *can* understand I_2 in terms of its parts and their relations to one another. Smallism is thus defended.

Briefly, there are two problems with this strategy for defending smallism. First, this response simply supposes that one is able to "convert" relational properties into intrinsic properties in this way in general, but there is a range of examples (being highly specialized, being a spandrel, being a face-recognizer) for which this seems a little too reminiscent of reconstructive philosophy of science in the name of saving a general philosophical thesis. It *may* be that one is able to defend smallism in this way, but that is something that one will known only after one has explored the range of individuals that there *are* across the various sciences. (One can't simply make up what counts as an individual.)

Second, this will be smallism defended only if there are no further explanatory important properties of I_2 that are relational, for, if there are, then we need to move to some larger individual still, I_3, that includes I_2 (and hence I_1) as a proper part. If relational properties of *any* individual are significant properties of that individual, then there will always remain something that the smallist view leaves out. My hunch now is that the antecedent of this conditional is true, but at the moment it is not much more than a crude guess at what we will find when we examine the relevant sciences. It is a hypothesis to be confirmed or falsified like any other.

3   Individuality, Complexity, and the Locus of Control

In his short book *The Individual in the Animal Kingdom*, Julian Huxley suggested three minimal conditions of biological individuality: heterogeneous parts whose significance derive from the whole individual to which they belong; self-maintenance and continuity, either of its self or of its progeny; and some level of independence of merely inorganic nature (1912, p. 28). Huxley also suggests that there has been an increase in these qualities, a heightening of individuality, over evolutionary time and through the process of evolution by natural selection. It is this latter idea, and the idea of environmental independence, that I want to discuss here, returning first to the case of psychology.

The idea that cognition affords creatures some measure of autonomy from the immediate worldly envelope in which they find themselves seems uncontroversial enough. As cognitive creatures, we are not bound by the here and now, by mere stimulus and response. Belief and memory allow the past and the distant to influence what we feel and do, and desire and expectation do the same for the future and distant. And although we are sometimes compelled to act and feel as we do by such cognitive states, there is always an internal complexity mediating emotion and action that would seem at least typically to create a space for choice and decision, independent of the particular environmental details impinging on one at that time. I would like to explore the move from this view to a more substantive and interesting thesis about cognition and environmental autonomy that parallels Huxley's evolutionary thesis about individuality, complexity, and environmental independence.

Consider in particular the idea that heightened cognitive complexity brings with it increased environmental autonomy, culminating ultimately in symbolic capacities that can be (and should be) construed individualistically. We might express this claim in terms of a correlation between increasingly sophisticated cognitive abilities and the independence from the environment of what we might think of as one's cognitive *locus of control*, a claim that I shall make more vivid in the rest of this paragraph. Lest what follows be confused with a serious attempt to explore real-life cognitive evolution, the real history of the mind (where the determiner serves to pick out *human* minds), let me put this in terms of an entirely *imaginary* evolutionary move: that from *reactive* through *enactive* to purely *symbolic* cognitive systems. The claim that I want to consider is that in the move from reactive to enactive to symbolic cognition the locus of control shifts from the environment through the body to the mind. Table 17.1 expresses this claim more succinctly and explicitly than I could with more sentences.

With this much (or this little) said about the psychological case, return now to the case of biological complexity. We might see much the same sort of correlation between biological complexity and environmental independence: as we move from biologically simple to biologically complicated creatures, we see organisms that increase their biological autonomy from their environment. Prokaryotes and micro-

Table 17.1

Locus of Control	Type of organism/ representational system	Example in humans
environmental	reactive	reflexes
bodily	enactive	mimetic skills
cranial	symbolic	beliefs, desires

organismic eukaryotes react to the world beyond their boundaries; "higher animals," in François Jacob's words, "literally live within themselves" (1970, p. 188), since they have evolved the internal machinery to enable themselves to delay brute reaction and so gain relative autonomy from their environments. The description of other complicated biological entities—genes come to mind—as "literally living within themselves" would also seem natural on what, in light of the previous section, we might call an *encodingist* conception of them.

Here I believe that the psychological case is instructive for the biological case. For there it is relatively clear that the internal locus of control that characterizes symbolic capacities is compatible with a rejection of individualism. That is, organisms that clearly have an internal, cranial locus of control for the core of their mental life may also possess what we might think of as a cognitive loop extending into the world beyond its own boundaries. In fact, I think the point can be strengthened modally: not only can symbolic representational systems with an internal locus of control be *wide* rather than *narrow* cognitive systems, but in some cases they *must* be wide. These are cases in which organisms have developed strategies of shifting the representational load from inside their heads to their external, symbol-laden environments through the development of what Merlin Donald (1991) calls *external storage systems*, such as writing systems, conventional symbols, and gestures. In short, creatures like us who posses cognitive systems with an internal locus of control can instantiate internal, bodily, and world-involving cognitive capacities. Table 17.2 puts this graphically in terms of what physically *realizes* these various capacities.

Granted that cognitive complexity, as epitomized in the sorts of symbolic capacities that adorn our own cognitive architecture, does provide for an internal locus of control for mentation and behavior, symbolic capacities themselves can be world-involving (and so world-dependent) in that they can require more than a mere brain to be realized. As with the case of the extended phenotype, here there is the suggestion of looking at the larger system of which the individual cognizer is a part; the individual's intrinsic causal powers (and what physically realizes them) are only part of the story to be told.

Table 17.2

Cognitive Capacities in Symbol-Using Creatures	Realization of the Capacity
purely internal	internal cognitive arrangement of the brain
bodily	cerebral + bodily configuration
world-involving	cerebral arrangement + external symbol tokens

Table 17.2 should also indicate what is problematic about the encoding view of mental representation, for in neither the case of enactive, bodily skills nor that of world-involving capacities do parts of the brain encode for the other constituents of the realization of that capacity. Rather, in both cases what is inside the head and what is outside of it are related as parts of an integrated whole, with information flowing between those parts.

I want to suggest that the inference from the presence of internal loci of control to individualism in biology is likewise problematic, and that there are ways in which biological sophistication actually brings with it a *deeper* reliance of the individuals of interest on their environments. Population structures emerge, ecological dependencies are established, and the individual organism can no longer be viewed as a self-contained cluster of causal powers. The relations between individuals, and between individuals and their environments, emerge as significant.

I shall close with two brief (and very different) examples of the sort of shift in perspective that I have in mind in suggesting the abandonment of the metaphors and metaphysics of individualism and its biological equivalent, and what this shift implies in terms of more concrete research programs. Here I consider contemporary work on morphological development (beyond the gene but within the individual) and a snapshot of the history of ecology.

4   Concluding Examples and Remarks

In the study of morphogenesis—clearly a process that happens within the boundaries of an individual—Webster and Goodwin (1996) have recently argued for a return to rational morphology, advocating a shift in focus from genes and the notion of "gene action" to that of morphogenetic fields as complex dynamic systems. They argue that rather than concentrating exclusively on genes and their encoding powers, those interested in the development of biological form (e.g., tetrapod limbs, to take a classic case) should explore the relational principles that govern and constrain the construction of biological form, where these are not properties encoded in the genome.

As in the psychological case, it is not that genes and their powers are ignored or deemed irrelevant; rather, they are not viewed as the exclusive or even necessarily the primary locus for morphogenetic processes. The project here involves going beyond the causal powers of the gene to examine the broader principles governing development. To make this more concrete, consider an example that Goodwin discusses in chapter 5 of his *How the Leopard Changed Its Spots: The Evolution of Complexity*, that of leaf formation, particularly the positioning of leaves on a stem (phyllotaxis).

Although there are diverse shapes that leaves can and do take, there are only three ways in which leaves are arranged on the stem of a plant: in a spiral form, in a decussate form, and in a distichous form. Since one finds plants (such as those in the *Bromeliad* family) with more than one of these three ways instantiated by its various parts, and whose leaves shift from one to another form as development proceeds, it is plausible to think that there is an overall mechanism governing phyllotaxis across different species of plants that can operate in any one of three modes. Goodwin argues, following Green (1987, 1989), that this mechanism takes the form of a morphogenetic field located in the meristem, the tip of the developing plant, a field itself that is not determined solely by the information in the genes of the plant. Goodwin suggest that such a field is governed by physical forces shared by the biological and nonbiological world, forces that lead to the emergence of only some forms and not others.

There are more radical and less radical versions of the research program that emphasizes the role of morphogenetic fields and the principles that govern them over self-contained genes and their role in natural selection. The more radial version sketches these two views as alternatives to one another, such that the former might replace the latter as a general approach to understanding heritability and development. The less radical version—which I take Goodwin to advocate in *How the Leopard Changed Its Spots*—presupposes that the two views can be seen to supplement each other, since it claims more particularly that the morphogenetic field approach can explain phenomena that are simply assumed or ignored by the gene-centered conception of natural selection.

Consider my second example, one from the history of ecology. Although "ecology" was coined by Ernst Haeckel in 1866, then term gained a foothold only in the 1890s to designate a sort of "outdoor physiology" (to use Cittadino's 1980 term) involving the measurement of the responses of individual plants and animals to particular environmental variables. This early conception of ecology was compatible with an individualistic conception of the discipline, as it was still concerned with the causal powers of individual organisms. As ecology turned not only to incorporate a study of units larger than the individual—the population, the community, the predator-prey system—but also to introduce ways of talking about individual organisms that presupposed their location in a broader environment—as competitors, as niche-occupiers, as coevolvers—it became more difficult to conceptualize the science in terms of self-contained individuals that encode aspects of their environments. A closer examination of the concept of a *niche* will perhaps locate some debates within the history of ecology (see Griesemer 1993; Schoener 1989) within the framework of the current discussion.

When Grinnell introduced the concept of an ecological niche in the 1910s, he used it to refer to a place or space preexisting in an environment that actual or possible organisms could be slotted into. Elton's more extensive treatment of the ecological niche in his *Animal Ecology* shares this conception of a niche with Grinnell, although Elton emphasizes both the relation between an animal's niche and "what it is *doing* and not merely what it looks like" (1927, p. 64) as well as the availability of a niche across species, even across higher taxa. On the Grinnell-Elton conception, niches can be empty, characterized as they are independently of the intrinsic properties of particular organisms. By contrast, consider the concept of a niche as it features in the so-called niche theory of MacArthur and Levins in the 1960s, with its roots in Hutchinson's (1957) "formalization of the niche." On this theory, niches are utility distributions, being defined for particular populations or species. Basically, on this conception, a species' niche is the way in which that species uses the resources in its environment, and it defines a species' niche in terms of the role of the species in the overall community of organisms. Two points about the shift in the meaning of "niche" are noteworthy.

First, there is a shift from a conception of the niche as a space or "recess" in a habitat that an organism or species could fill to that of a niche as a (highly complex) property of that organism or species. This shift brought with it a focus on the measurement of things that organisms did and could do with their environments, rather than a concentration on the character of the habitats that organisms lived in. Second, in niche theory there is the potential to reduce what we might think of as population-level concepts (such as the niche itself, or the ecosystem) and phenomena (such as niche-overlap or ecosystem balance) to properties of individual organisms or species and their relations. This is because such concepts are already defined in terms of individual organisms and species, and the phenomena can thus be conceived in terms of relations between these. For example, niche-overlap can be conceived in terms of competition between individuals or species, and ecosystem balance in terms of niche occupation. In community ecology more generally, this fits with the sort of view that Gleason advocated in botany in the 1920s, one whereby entities larger than individual organisms, such as symbiotic pairs, communities, and ecosystems, are viewed as relatively transient entities whose dynamics are to be understood exhaustively in terms of those of the individual organisms that constitute them (Taylor 1993).

My point in ending the chapter briskly with these examples is certainly *not* to try and argue that the views that I am presenting as nonindividualistic in character are a priori preferable over those that are individualistic, or even to pretend to have the relevant, missing empirically driven arguments in support of that preference. Rather, it is to gesture at two distinct areas of biology proper where one can see the contrast between something like individualistic and nonindividualistic perspectives on the subject matter exemplified in alternative research programs. I suspect that the same will be true of many areas of biological inquiry, and that considerations that have been raised for or against individualism in psychology will be relevant to many of these areas. But they remain suspicions to be substantiated elsewhere.

Acknowledgments

Versions of this paper were given at the International Society for the History, Philosophy and Social Studies of Biology in Seattle, July 1997, and to the History and Philosophy of Science group at Northwestern University in February 1998. I thank audiences on both occasions for helpful feedback.

References

Bechtel, W., and Richardson, R. (1993) *Discovering Complexity: Decomposition and Localization as Strategies in Scientific Research*. Princeton, NJ: Princeton University Press.

Bonner, J. (1974) *On Development*. Cambridge, MA: Harvard University Press.

Burge, T. (1979) "Individualism and the Mental," in P. French, T. Uehling Jr., and H. Wettstein (eds.), *Midwest Studies in Philosophy, Vol. 4: Metaphysics*. Minneapolis, MN: University of Minnesota Press.

Buss, L. (1987) *The Evolution of Individuality*. Princeton, NJ: Princeton University Press.

Cittadino, E. (1980) "Ecology and the Professionalization of Botany in America, 1890–95," *Studies in the History of Biology* 4:171–198.

Dawkins, R. (1982) *The Extended Phenotype*. Oxford: Oxford University Press.

Dawkins, R. (1989) *The Selfish Gene*, 2nd edition. Oxford: Oxford University Press.

Donald, M. (1991) *Origins of the Modern Mind: Three Stages in the Evolution of Culture and Cognition*. Cambridge, MA: Harvard University Press.

Elton, C. (1927) *Animal Ecology*. New York: Macmillan.

Fodor, J. A. (1980) "Methodological Solipsism Considered as a Research Strategy in Cognitive Psychology," *Behavioral and Brain Sciences* 3:63–73.

Fodor J. A. (1987) *Psychosemantics*. Cambridge, MA: The MIT Press.

Fox Keller, E. (1995) *Refiguring Life: Metaphors of Twentieth-Century Biology*. New York: Columbia University Press.

Goodwin, B. (1994) *How the Leopard Changed Its Spots: The Evolution of Complexity*. New York: Simon and Schuster.

Green, P. (1987) "Inheritance of Pattern: Analysis from Phenotype to Gene," *American Zoologist* 27:657–673.

Green, P. (1989) "Shoot Morphogenesis, Vegetative through Floral, from a Biophysical Perspective," in E. Lord and G. Gernier (eds.), *Plant Reproduction: From Floral Induction to Pollination*. Rockville, MD: American Society of Plant Physiologists.

Griesemer, J. (1993) "Niche: Historical Perspectives," in E. Fox Keller and E. Lloyd (eds.), *Keywords in Evolutionary Biology*. Cambridge, MA: Harvard University Press.

Hull, D. (1984) "The Units of Evolution," in R. Brandon and R. Burian (eds.), *Genes, Organisms, Populations: Controversies Over the Units of Selection*. Cambridge, MA: The MIT Press.

Hutchinson, G. (1957) "Concluding Remarks," *Cold Spring Harbor Symposium on Quantitative Biology* 22:425–427.

Huxley, J. S. (1912) *The Individual in the Animal Kingdom*. London: Cambridge University Press.

Jacob, F. (1970) *The Logic of Life*. Princeton, NJ: Princeton University Press. 1993 edition.

Kingsland, S. (1985) *Modeling Nature: Episodes in the History of Population Ecology*. Chicago: University of Chicago Press.

Mayr, E. (1982) *The Growth of Biological Thought*. Cambridge, MA: Harvard University Press.

Schoener, T. W. (1989) "The Ecological Niche," in J. M. Cherrett (ed.), *Ecological Concepts*. Oxford: Blackwell.

Sober, E. and Wilson, D. S. (1998) *Unto Others: The Evolution and Psychology of Unselfish Behavior*. Cambridge, MA: Harvard University Press.

Taylor, P. (1993) "Community," in E. Fox Keller and E. Lloyd (eds.), *Keywords in Evolutionary Biology*. Cambridge, MA: Harvard University Press.

Webster, G., and Goodwin, B. (1996) *Form and Transformation: Generative and Relational Principles in Biology*. New York: Cambridge University Press.

Williams, G. C. (1966) *Adaptation and Natural Selection*. Princeton, NJ: Princeton University Press.

Wilson, D. S. and Sober, E. (1994) "Reintroducing Group Selection into the Human Behavioral Sciences," *Behavioral and Brain Sciences* 17:585–608.

Wilson, R. A. (1994) "Wide Computationalism," *Mind* 103:351–372.

Wilson, R. A. (1995) *Cartesian Psychology and Physical Minds: Individualism and the Sciences of the Mind*. New York: Cambridge University Press.

Wilson, R. A. (1999) "The Mind Beyond Itself," in D. Sperber (ed.), *Metarepresentation*. Oxford: Oxford University Press.

Wynne-Edwards, V. C. (1962) *Animal Dispersion in Relation to Social Behavior*. London and Edinburgh: Oliver and Boyd.

Index